TECHNOLOGY, INNOVATION and POLICY 10

Series of the Fraunhofer Institute
for Systems and Innovation Research (ISI)

The following scientists have also participated in the project:

Jan Benedictus, StB TNO, Delft
Johanna Cáceres, UAB, Barcelona
Monique de Leeuw, StB TNO, Delft
Bep Essenstam, StB TNO, Delft
Stefanie Giessler, FhG ISI, Karlsruhe
Dr. Bärbel Hüsing, FhG ISI, Karlsruhe
Christa Knorr, FhG ISI, Karlsruhe
Prof. Dr. Maria Angeles Lizón, UAB, Barcelona
Anna Maina, NSPH, Athens
Panagoula Paschali, NSPH, Athens
Elena Recchia, CERISS, Milan
Elke Strauss, FhG ISI, Karlsruhe
Despina Vagianou-Angelaki, NSPH, Athens

TECHNOLOGY, INNOVATION and POLICY

Series of the Fraunhofer Institute
for Systems and Innovation Research (ISI)

Klaus Menrad · Demosthenes Agrafiotis Christien M. Enzing · Louis Lemkow Fabio Terragni

A Delphi Survey

Springer-Verlag Berlin Heidelberg GmbH

K. Menrad · D. Agrafiotis · C.M. Enzing
L. Lemkow · F. Terragni

Future Impacts of Biotechnology on Agriculture, Food Production and Food Processing

A Delphi Survey

Final report to the Commission of the European Union, DG XII

With 49 Figures
and 139 Tables

Springer-Verlag Berlin Heidelberg GmbH

Dr. Klaus Menrad
Fraunhofer Institute for Systems and
Innovation Research (ISI)
Breslauer Straße 48
D-76139 Karlsruhe, Germany

Prof. Dr. Demosthenes Agrafiotis
National School of Public Health
Department of Sociology
196 Alexandras Avenue
GR-11521 Athens, Greece

Dr. Christien M. Enzing
TNO-STB
Schoemaker Straat 97
NL-2600 JA Delft, The Netherlands

Prof. Dr. Louis Lemkow
Universidad Autònoma de Barcelona
Edifici A
E-08193 Bellaterra, Spain

Prof. Dr. Fabio Terragni
CERISS
Via Andegari 18
I-20121 Milan, Italy

ISBN 978-3-7908-1215-2

Die Deutsche Bibliothek – CIP-Einheitsaufnahme
Future impacts of biotechnology on agriculture, food production and food processing: a Delphi survey; final report to the Commission of the European Union, DG XII; with 139 tables / K. Menrad

(Technology, innovation and policy; 10)
ISBN 978-3-7908-1215-2   ISBN 978-3-642-52474-5 (eBook)
DOI 10.1007/978-3-642-52474-5

The use of general descriptive names, registered names, trademarks, etc. in this publication does not imply, even in the absence of a specific statement, that such names are exempt from the relevant protective laws and regulations and therefore free for general use.

Softcover design: Erich Kirchner, Heidelberg
SPIN 10741226   88/2202-5 4 3 2 1 0 – Printed on acid-free paper

Table of contents   **page**

List of tables

List of figures **page**

XVIII

1.   Introduction

Although the first Agro-Food products based on modern biotechnology (e. g. recombinant chymosin for cheese production; tomato puree based on genetically engineered tomatoes; herbicide-resistant, genetically modified soybean; insect-resistant maize) have been introduced in the EU markets in recent years, the application of this technology is still being intensively discussed in the European Union. Recent opinion polls indicate as well that consumers' acceptance of genetically engineered food and agro-products still is relatively low (e. g. European Commission 1997, Hampel et al. 1997), at least in some member states of the EU. In contrast, representatives from politics and industry underline the necessity to apply modern biotechnology in the Agro-Food sector as well, mainly to ensure the competitiveness of EU agriculture and food industry and for employment reasons.

Against this background there seems to be a need for a scientific analysis of the future impacts of modern biotechnology in the Agro-Food sector of the EU. Recent studies trying to analyse this issue (e. g. OECD 1992, Teuber 1992) usually comprise extrapolations of status-quo analyses. What has not been exploited so far in this context are systematic technology forecasting approaches which do not include only one single country, but get information on an international level. Therefore, the impacts of modern biotechnology on the Agro-Food sector in five member countries of the EU (Germany, Greece, Italy, the Netherlands, and Spain) have been analysed with the help of the Delphi methodology which represents one of the most reliable tools for technology forecasting.

Since there are different wordings and terminologies in the field of Agro-Food biotechnology, the following definitions of "modern biotechnology" and "genetic engineering" are given in order to clarify their meaning within the scope of this project. "Modern biotechnology" is used for the application of new scientific and engineering principles for the processing of materials by biological agents to provide goods and services. In this definition techniques like genetic engineering, bioprocessing, monoclonal antibodies, protein engineering, tissue culture, and protoplast fusion of immobilized enzymes are included. "Genetic engineering" is defined as the characterization, isolation, new combination and multiplication of genetic material. In this sense, "genetic engineering" is regarded as one possible technique in "modern biotechnology" and not as a synonym.

This project has been financially supported by the European Commission (DG XII) as a shared-cost research project in the FAIR programme from 1996 to 1998.

The project was carried out by a consortium of the following five research institutions (the names in brackets refer to the responsible scientists in the different countries):

- Germany: Fraunhofer Institute for Systems and Innovation Research (FhG ISI), Karlsruhe
  (Dr. Klaus Menrad, project co-ordinator)

- Greece: National School of Public Health (NSPH), Athens
  (Prof. Dr. Demosthenes Agrafiotis)

- Italy: Centre for Education, Research and Information on Science and Society (CERISS), Milan
  (Prof. Dr. Fabio Terragni)

- The Netherlands: Center for Technology and Policy Studies (TNO-STB), Apeldoorn
  (Dr. Christien M. Enzing)

- Spain: Universidad Autonoma de Barcelona, Department of Sociology (UAB), Barcelona
  (Prof. Dr. Luis Lemkow)

Structure of the report

The procedure and results of the Delphi survey entitled "Future impacts of biotechnology on agriculture, food production and food processing" are presented in this report. After this introduction, information concerning the objectives of the project, the Delphi methodology, the structure and development of the questionnaire, the selection of the experts who have been asked in the Delphi survey, the realization of the survey, the response behaviour of the experts as well as the statistical analysis is provided in chapter 2.

In chapter 3 an overview is given of the basic results of the Delphi survey in the individual countries. This overview mainly follows the categories included in the questionnaire. Firstly, the self-estimated degree of knowledge of the answering experts is analysed. In the following parts a general overview is given, related to the estimated personal attitude and the expected time of realization of the statements included in the questionnaire. In addition, the most important influential factors relevant for the future application of modern biotechnology in the Agro-Food sector are analysed. Afterwards, the importance of the visions mentioned in the statements of the questionnaire for future knowledge creation in science and technology, competitiveness of the respective national economies as well as the environmental situation is discussed.

Common features and differences between the five involved countries are analysed in chapter 4. The presentation of the results is done according to the structure used in chapter 3.

In chapter 5 the results of an in-depth analysis of selected key issues are presented. These key issues relate to the scientific and technical development of Agro-Food biotechnology as well as its future impacts on society, economy, environment and health. Firstly, the future acceptance of modern biotechnology (5.1) as well as regulatory issues (5.2) are analysed. Afterwards, the impacts of Agro-Food biotechnology on economy (5.3), environment (5.4) and the health of consumers, users and employees (5.5) are discussed. Chapter 5 is finished by a presentation of future scientific and technical trends in four selected areas of Agro-Food biotechnology and the most important prerequisites for these developments (5.6).

The most important findings of the project are summarized in chapter 6. This relates to future scientific and technical developments, their impacts on the health of consumers and the environment, the future acceptance of Agro-Food biotechnology, regulatory aspects as well as economic implications. In addition, the suitability of the applied methodology is discussed. Based on these results recommendations for politics, industry and other social actors are elaborated on how to use the obtained results.

The report concludes with an executive summary. In the annex the questionnaire of the Delphi survey is presented.

This report was jointly written by all partners participating in the project. The division of labour between the involved research institutions during this process is shown in table 1.1. This working step was organized in such a way that each partner who was responsible for a specific step worked out a draft version of the respective chapter or sub-chapter of the report, which were then combined by the project co-ordinator.

Table 1.1:   Co-operation in production of the final report

Chapter	Title	Responsible partner
1.	Introduction	FhG ISI
2.	Concept and realization	FhG ISI
2.1	Objectives	
2.2	Description of the Delphi methodology	
2.3	Structure of the questionnaire	
2.4	Creation/selection of the statements	
2.5	Selection of experts	
2.6	Realization of the survey	
2.7	Returned questionnaires/response rate	
2.8	Statistical analysis	
3.	Results in different countries	
3.1	Germany	FhG ISI
3.2	Greece	NSPH
3.3	Italy	CERISS
3.4	The Netherlands	TNO-STB
3.5	Spain	UAB
4.	Comparative analysis	FhG ISI
4.1	Differences between countries	
4.2	Differences between expert groups	
5.	Future impacts of biotechnology in the Agro-Food sector	
5.1	Acceptance of biotechnology	CERISS
5.2	Legal framework conditions	UAB
5.3	Economic implications	TNO-STB, FhG ISI
5.4	Environmental impacts	NSPH
5.5	Health	NSPH
5.6	Scientific/Technological development	FhG ISI
6.	Conclusions and recommendations	All partners
7.	Executive summary	FhG ISI
8.	Literature	All partners

Due to organizational reasons (mainly the size of the publication), it is not possible to present the entire results of this survey within this publication. Readers who are interested in the detailed results of each country are welcome to contact the authors of this report in order to get the data in a specific report.

2. Concept and realization

2.1 Objectives

In the Agro-Food sector of the European Union the first products based on modern biotechnology or genetic engineering have entered the market (e. g. recombinant chymosin for cheese production, tomato puree, herbicide-resistant plants). In the coming years an increasing influence of biotechnology on agriculture, food production and food processing is expected. However, the application of modern biotechnology in agriculture and the food industry is still being intensively discussed in the EU countries. One of the latest indicators in this context represent the results of the Eurobarometer survey which was conducted in 1996 (European Commission 1997), the activities accompanying the market introduction of the herbicide-resistant soybean by Monsanto in autumn 1996, as well as the intensive public debate after the birth of the cloned sheep "Dolly" in spring 1997 (Wilmut et al. 1997).

Previous studies trying to assess future impacts of biotechnology on the Agro-Food sector usually comprise extrapolations of status-quo-analyses. In the present project the impacts of modern biotechnology on the Agro-Food sector in five different countries of the European Union (Germany, the Netherlands, Italy, Spain, Greece) are analysed, using the Delphi methodology. The project has the following objectives:

- Analysis and forecast of the impacts of biotechnology on the agricultural and food sector in different countries of the European Union.
- Consideration of the scientific and technical development as well as the development of framework conditions for food production and consumption.
- Elaboration of common features and differences between the involved countries.
- Identification of different views of various social groups (e. g. industry, researchers, farmers, consumers, biotechnology critics).
- Elaboration of recommendations for decision-makers in politics, industry and society, allowing anticipation of emerging problems or adaption of business activities.

2.2 Description of the Delphi methodology

The Delphi approach is one of the most effective methods of technology foresight. Especially for the longer-term perspectives of about thirty years, the Delphi approach seems to be the most suitable one (BMFT 1993). Although, originally developed in the United States, Japan has the longest tradition of using this method, where it has been used for a systematic forecast of scientific and technological developments in almost all technology fields for more than 25 years (Cuhls et al. 1995, Cuhls and Kuwahara 1994).

In 1993 the first German Delphi study on the development of science and technology was conducted by the Fraunhofer Institute for Systems and Innovation Research, Karlsruhe, on behalf of the German Ministry of Research and Technology. In this project major future scientific and technical trends in 16 technological fields were analysed, taking into consideration the broad experiences available in Japan with the Delphi methodology (BMFT 1993). Similar studies were carried out in France (Sofres 1994) and the United Kingdom (Loveridge et al. 1995). In 1997 the second German Delphi study on the development of science and technology was carried out, actualizing the trends and data gathered in the first survey in 1993 (Fraunhofer Institute for Systems and Innovation Research 1998). In all these studies a broad range of future developments in major technological fields is considered. In addition, these surveys limit themselves mainly to future scientific and technological trends, without taking the development of the respective economic, social or environmental framework conditions or impacts into consideration.

Experiences with Delphi surveys focused on agriculture and the food industry are available in Western Europe as well. Neubert (1991) analysed major technological trends in the biotech field with relevance to agriculture in Germany, using the Delphi approach. Gotsch and Rieder (1989, 1990) used a Delphi survey to analyse the long and medium-term biotechnological developments in arable farming which will be available for Swiss farmers. In a second step these scientists estimated the economic impacts of these techniques by using a Linear Programming model (Gotsch et al. 1993).

The Delphi approach is based on the assessment of possible future developments by selected specialists, so called "experts", who exchange their opinions with the help of a written questionnaire. The specific feature of this method is that the experts are asked to answer the questionnaire several times, whereby the results of the former rounds are presented in the questionnaire of the next round. The questionnaire contains several statements concerning possible future developments, which have to be assessed by a sophisticated scheme of answering categories.

By proceeding in this way the experts are able to re-examine their views in the light of the other experts' opinions and, if necessary, correct any deviations without losing their face. This procedure supports a consensus mechanism and favours majority opinions. The validity of the results strongly depends on the selection of the specialists to be asked. In this respect it is important to know that specialists who are directly involved in a particular development tend to rather optimistic estimates (BMFT 1993). Therefore, well-informed specialists, who are not actively involved in a particular research area should be encouraged to express an opinion on the future development in this area as well.

It should be taken into account that the future relevance and acceptance of a new technology does not only depend on the scientific and technological development, but also on framework conditions, e. g. their social, ecological and health-related, legal and economic impacts. Conventional Delphi surveys focus on technological developments. This is, however, inadequate for the subject covered in this survey. Therefore, economic questions, ecological impacts, the acceptance of biotechnology and the legal framework conditions have been taken into account. Besides, the combination of technology forecasting and the forecast of framework conditions as well as the selection of different expert groups are the new features in this study. For the first time not only scientific experts of research institutions or industry are engaged in the survey, but also representatives of other relevant groups related to the production and distribution chain (farmers, retailers) and members of different social groups (e. g. consumers, biotechnology critics, journalists, politicians, educational sector) were regarded as "experts" in this context.

When interpreting the results of a Delphi survey it should be remembered that they do not predict the future, but provide a more rational and objective basis for priority-setting within government, industry and science and they should initiate a dialogue on strategic options in the science and technology area. By proceeding in this way, it is possible to assess future developments at a rather early stage, to discuss these developments and their consequences in public and think about additional activities in order to speed up or prevent these developments (Fraunhofer Institute for Systems and Innovation Research 1998).

2.3　　Structure of the questionnaire

The selection of the topics and the design of the questionnaire is extremely important for the success of a Delphi survey because the questionnaire represents the medium through which the experts asked interact with each other. In figure 2.1 a part of the questionnaire (of the second round) of this survey is presented for illustration. 71 statements representing different possibilities of future developments or situations in the Agro-Food sector have been defined for this survey which should be assessed by five answering categories. In addition, the experts had the opportunity to submit written comments (see figure 2.1). At the beginning of the questionnaire information on the targets of the project in general, an explanation of the survey procedure, important definitions and detailed instructions for the answering of the questionnaire have been presented.

Figure 2.1: Questionnaire of the Delphi survey

		Statement	Degree of knowledge: very familiar	Degree of knowledge: average familiar	Degree of knowledge: less familiar	Degree of knowledge: not familiar	Personal attitude: development / situation is positive	Personal attitude: development / situation is indifferent	Personal attitude: development / situation is negative	Personal attitude: not able to express
			Please mark with a cross				Please mark with a cross			
Acceptance	1	A governmental institution which provides information on modern biotechnology in the Agro-Food sector to all interested persons and institutions is established in your country. %	17	46	28	8	74	20	4	1
Acceptance	2	Most consumers in your country have quickly got used to all kinds of food and beverages produced with the help of genetic engineering. %	25	50	23	2	22	25	51	2
Acceptance	3	The widespread diffusion of information on genetic engineering from different sources (e. g. public and private institutions, several interest groups and associations, media) has increased the acceptance of food products made with the help of genetic engineering in your country. %	25	50	22	3	33	28	39	2

Statement	Time of realisation: in the following 5 years	in the following 6 - 10 years	in the following 11 - 15 years	in the following 16 - 20 years	not realised during the following 20 years	not feasible at all / no realisation	not able to judge
	Please mark with a cross						
1	68	20	3	0	1	3	4
2	18	39	20	7	6	6	3
3	28	41	15	4	4	4	3

Statement	Most important influencing factors in your country: R & D infrastructure	personnel (education, skills)	technology transfer	industrial innovativeness	markets	funding	regulation / standards	policy	ecology	social / ethical acceptance	information	international collaboration
	Please mark with a cross (max. 3 crosses)											
1	3	8	3	2	3	27	6	21	2	14	9	3
2	1	0	1	2	9	0	7	13	8	35	24	1
3	2	3	2	2	5	3	5	16	8	32	20	2

Statement	Importance of the statement to: knowledge creation in science & technology	competitiveness of economy	protection of environment / sustainable development	Comment
	Please mark each field with: + = very important, 0 = average or - = not important			
1				
2				
3				

About 40 % of the statements of the questionnaire are dedicated to scientific and technological developments in the biotech area. Around 60 % of the statements refer to the development of the framework conditions in the Agro-Food sector. The statements are thematically structured into the following six domains:

(1) Acceptance   7 statements

(2) Regulation   6 statements

(3) Economy   11 statements

(4) Environment   10 statements

(5) Health   7 statements

(6) Scientific/Technological Development   30 statements

The formulation of the statements contains stereotype wordings which distinguish between different phases of the scientific, technical and economic innovation process. In this context the following phases have been considered:

Elucidation:	scientific and theoretical identification of principles or phenomena
Development:	attainment of a specific technological target or completion of a prototype
Practical use (= practically used):	first market introduction or use under practical conditions of an innovative product or service
Widespread use (= widely used):	significant market penetration to a level that a product or service is commonly used
Matter of routine:	there is reliable experience with a product or service because it has been used for a certain period of time

The experts have been asked to assess each statement with the help of a sophisticated scheme of answering categories so that the design of the questionnaire is comparable to a large table (see figure 2.1). In principle, the assessment of each statement was based on the following six answering categories:

(1) Degree of knowledge

(2) Personal attitude

(3) Time of realization

(4) Influential factors

(5) Importance to:

  – knowledge creation in science and technology

  – competitiveness of economy

  – protection of environment/sustainable development

(6) Comments

It is a common feature of Delphi surveys to ask the experts for their individual degree of knowledge related to a specific statement. This self-estimation of the experts' knowledge has a certain relevance because the statistical analysis of the data is based on the answers of experts who estimate themselves as being "very, average or less familiar" with a certain situation or development (see also chapter 2.8). This self-estimation should not be regarded in the sense of an objective estimation trying to solve the problem of "expert definition" but it contributes at least to an internal standardization. The self-estimation of the individual level of knowledge of the respondents is based on the criteria described in table 2.1. In the second category, the experts have been asked to express their personal attitude on the development/situation described in the statements. For this purpose the definitions mentioned in table 2.2 have been used.

Table 2.1:   Definition of the sub-categories of "degree of knowledge"

Sub-category	Definition
Very familiar:	You actively work in this area or with this issue at present. (This is one of your regular fields of work).
Average familiar:	You are not working in this area, but you are very well informed about the arguments dealing with the issue in the statement.
Less familiar:	You have read articles in the newspaper or in popular magazines about the area/issue covered in the statement.
Not familiar:	You have insufficient knowledge about the area/issue. In this case please continue with the next statement.

Table 2.2:   Definition of the sub-categories of "personal attitude"

Sub-category	Definition
Development/situation is positive:	All in all, I appreciate the development/situation.
Development/situation is indifferent:	Altogether, I have no negative or positive feeling about the development/situation.
Development/situation is negative:	I reject the development/situation.
Not able to express:	I am not able to express my personal opinion on the development/situation.

One of the most important aspects of Delphi surveys is the assessment of the time of realization of a specific development. In the third answering category the respondents have been asked to mark the time period, in which they estimate the described development/situation will be realized. They could choose one of the sub-categories "in the following five years", "in the following six to ten years", "in the following eleven to 15 years", "in the following 16 to 20 years", "not realized during the following 20 years", "not feasible at all/no realization" and "not able to judge". The sub-category "not feasible at all/no realization" has been added because it seemed to be very important for this survey that the respondents had the opportunity to express that there will be no realization at all. If the development or situation described in a specific statement has already been realized in the view of the experts, they should cross the sub-category "in the following five years" and make a corresponding note in the "comment" category.

In the fourth answering category "influential factors", the experts should assess (as objectively as possible) the current conditions for realization of the development/situation mentioned in the statement. In this category up to three influential factors should be marked which are decisive for the future development in the respective country, i. e in this category multiple answers have been possible. In total, twelve possible influential factors have been included in the questionnaire which are defined as shown in table 2.3.

Table 2.3:   Definition of the sub-categories of "influential factors"

Sub-category	Definition
R&D infrastructure:	Quantity and quality of relevant scientific and technological institutions (R&D = research and development).
Personnel (education, skills):	Availability of educated and skilled staff which is required to realize a certain topic.
Technology transfer:	Organization of know-how transfer among universities, research institutions and industry.
Industrial innovativeness:	Ability of the national companies to make a commercial success or new/improved products, processes or services out of scientific/technological findings.
Markets:	Conditions on the relevant markets (e. g. future perspectives, competition), entrance to these markets and chances of new/improved products, processes or services.
Funding:	Availability of private and public research funds and investment capital (e. g. venture, risk capital).
Regulation/standards:	The international and national standards, intellectual property rights and other laws and regulations which influence the topic mentioned in the statement.
Policy:	Activities and measures of the national/federal government and other politicians.
Ecology:	Environmental situation in your country.
Social/ethical acceptance:	Social and cultural opinions, ethical concerns, attitudes arising from general public discussion or pressure groups.
Information:	Access to, and diffusion of, scientific, technological and background information on modern biotechnology (e. g. articles, broadcasts in mass media, information material, databases).
International collaboration:	Number and/or quality of cooperations with research institutions/companies in other countries.

In the fifth answering category the importance of the development/situation described in the statement should be estimated to the following areas:

- Knowledge creation in science and technology

- Competitiveness of economy

- Protection of environment/sustainable development

The experts should assess in each sub-category whether the development/situation described in the statement is very important (+), average (0) or not important (-) according to the definitions provided in table 2.4.

Table 2.4: Definition of the sub-category "importance of the statement"

Sub-category	Definition
Knowledge creation in science and technology:	Research in the direction mentioned in the statement is an essential precondition of additional scientific and technological developments or it is an important contribution to scientific/technological progress by itself.
Competitiveness of economy:	The issue can contribute to the competitiveness of the national economy (e. g. development of marketable products and/or creation of new jobs).
Protection of environment/ sustainable development:	The issue mentioned in the statement can contribute to improving the environmental situation in the respective country.

Additionally, comments of all kinds (e. g. related to a certain statement, a specific category, the methodology) have been appreciated from the experts. The comments give detailed hints to major argumentation lines of the experts and proved to be very useful in the interpretation of the statistical results of the Delphi survey.

2.4   Development of the questionnaire

The questionnaire used in this Delphi survey has been developed in an interactive procedure between the different project teams and national expert committees in the five countries involved. The general procedure of this process is illustrated in figure 2.2.

Figure 2.2:   Procedure of the development of the questionnaire

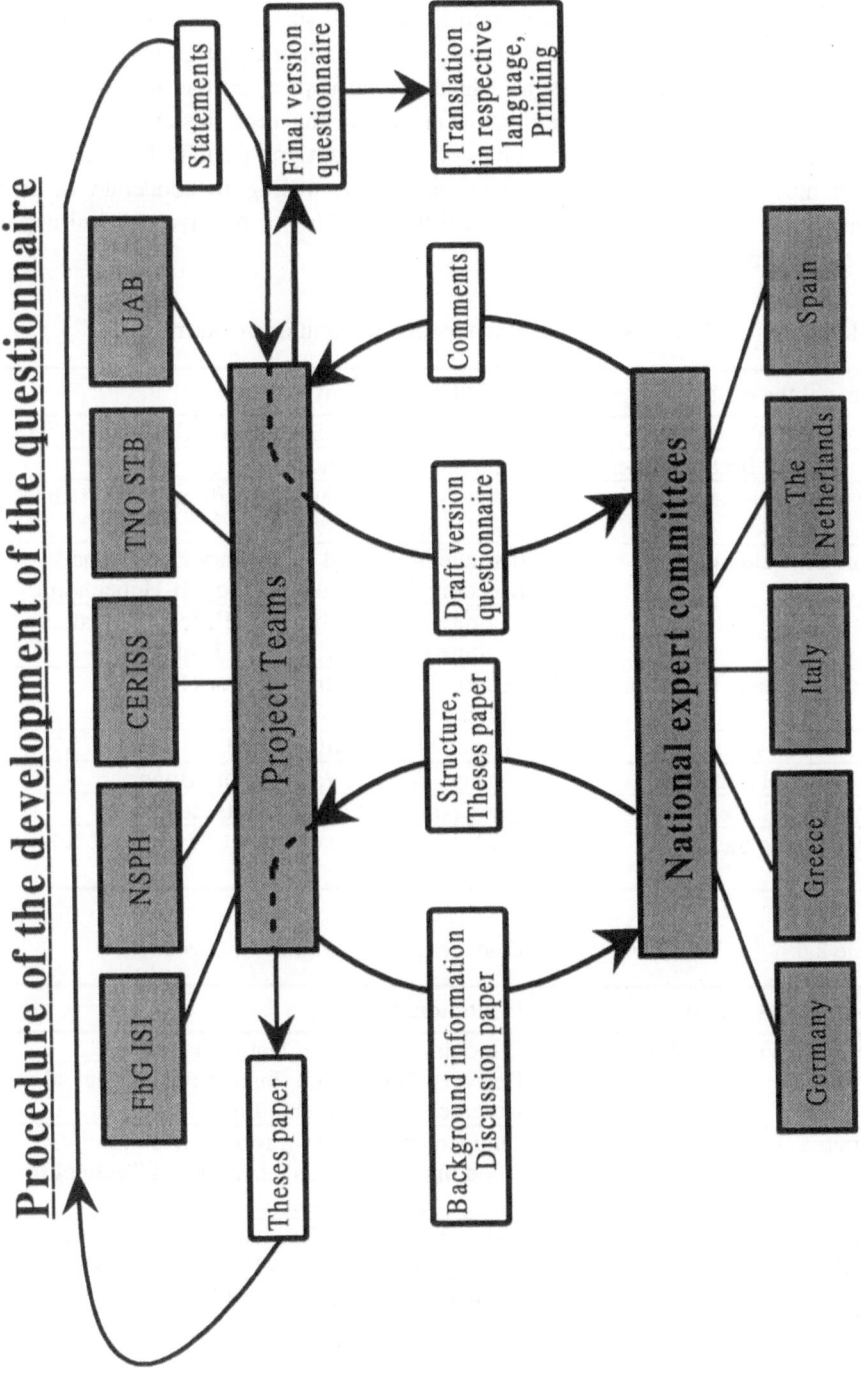

The project teams in the five countries involved started to establish national expert committees in order to assist the project teams in the development of the questionnaire in January/February 1996. The main targets while selecting the members of these committees have been to cover all scientific and technical fields included in the questionnaire, as well as to consider all relevant groups included in the survey. Although the persons asked to participate in the national expert committees often had only limited time available for this task, this approach proved to be successful because representatives of all relevant groups as well as specialists of the most important scientific areas have been willing to cooperate in each country. In the following tables 2.5 to 2.9 the members of the five national expert committees are mentioned.

Table 2.5:   Members of the national expert committee in Germany

Name of the expert	Institution/company
Dr. Sabine Albrecht	Kleinwanzlebener Saatzucht AG, Einbeck
Dr. Günther Bretschneider	Hipp KG, Pfaffenhofen
Mr. Ernst-Michael Epstein	Arbeitsgemeinschaft der Verbraucherverbände (AgV), Bonn
Prof. Dr. Geldermann	Institute of Animal Husbandry and Animal Breeding, University of Stuttgart-Hohenheim
Prof. Dr. K. J. Heller	Institute for Microbiology, Federal Research Institute for Milk Research, Kiel
Prof. Dr. Klaus-Dieter Jany	Federal Research Institute for Nutrition, Karlsruhe
Mr. Dan Leskien	Friends of the Earth Europe, Hamburg
Dr. Michael Lohse	German Farmers Association, Bonn
Dr. Michael Metz	Gewürzmüller GmbH, Stuttgart
Dr. Gabriele Sachse	BRAIN GmbH, Darmstadt (Adviser of Unilever Deutschland)

Table 2.6:   Members of the national expert committee in Greece

Name of the expert	Institution/company
George Sakellaris Ph.D	Biologist. researcher, National Hellenic Research Foundation, Institute of Biological Research and Biotechnology
Dimitris Ikonomou Ph.D.	Engineer/researcher in food biotechnology (food biotechnologist), Institute of Technology of Agricultural Products of NAGREF
Cleanthese Israilides, M.S., Ph.D.	Industrial, food and agriculture microbiologist, National Agr. Research Foundation, Institute of Technology of Agricultural Products
Vasilios Kontos Ph.D	Professor of Veterinary Public Health, National School of Public Health

Name of the expert	Institution/company
Polydefkis Hatzopoulos Ph.D	Assoc. Professor, Dept. of Agricultural Biology and Biotechnology Laboratory of Molecular Biology, Agricultural University of Athens
Roubelaki - Angelakis Kalliopi Ph.D	Professor of Plant Molecular Physiology and Biotechnology, Department of Biology, University of Crete
Koukios Emmanuel Ph.D	Chemical engineer, Professor of National Technical University of Athens, Bioresearch Technology Unit
Sekeris Konstantinos Ph.D	Professor of Biology, Director Institute of Biological Research and Biotechnology, National Hellenic Research Foundation Director, Laboratory of Biological Chemistry, University of Athens, School of Medicine
Balis Konstantinos, Ph.D. (Cambridge)	Professor of Microbiology, Laboratory General and Agricultural Microbiology, Dept. of Agricultural Biotechnology, Athens University of Agriculture
Megapanos Alexandros M.S.	Wine technologist
Kazakopoulos Leonidas Ph.D	Assistant Professor of Sociology of Agriculture, Dept. Agricultural Economy and Sociology, Athens University of Agriculture
Nafsika Tsoulka-Antonopoulou	Professor of Home Economics and Nutrition, Member of the National Council of Consumers, GR., Vice President of the Consumer Union of the city of Volos, member of Nutrition Committee of the National Council

Table 2.7:   Members of the national expert committee in Italy

Name of the expert	Institution/company
Dr. Alessandra Carini	Istituto Sperimentale Lattiero-caseario, Lodi
Prof. Giovanni Cassani	Tecnogen Spa
Dr. Aldo Ceriotti	Consiglio Nazionale delle Ricerche
Dr. Roberto Furlani	WWF Italy
Prof. Gustavo Ghidini	Movimento Consumatori
Prof. Vincenzo Lungagnani	Assobiotec
Prof. Guiseppe Rognoni	Faculty of Veterinary Medicine, University of Milan
Dr. Cesare Sala	Consiglio Nazionale delle Ricerche
Prof. Francesco Sala	Department of Biology, University of Milan
Dr. Luisa Villa	Comitato Difesa Consumatori
Prof. Alessandro Viotti	Consiglio Nazionale delle Ricerche

Table 2.8:   Members of the national expert committee in the Netherlands

Name of the expert	Institution/company
Ir. Arie L. Breure	Managing Director MOGEN International nv, Leiden
Dr. Monica M. de Heide	Vice-director of the Nationale Cooperative Raad voor land- en tuinbouw (NCR), Den Haag
Dr. B. de Vet	Director External Communications Unilever, Rotterdam
Ir. Anneke Hamstra	Stichting voor Wetenschappelijk Onderzoek van Consumentenaangelegenheden (SWOKA), Leiden
Dr. Brigitte Lander	Alternatieve Konsumentenbond (AKB), Amsterdam
Dr. Mei-Lie M.C. Tan	Business development manager MOGEN International nv, Leiden
Prof. Dr. Gerard M.A. van Beynum	Vice President R&D Pharming Health Care Products, Leiden; Professor Industrial Biotechnology University of Groningen; Director of the Dutch Association of Agricultural and Industrial Biotechnology Companies, Leidschendam

Table 2.9:   Members of the national expert committee in Spain

Name of the expert	Institution/company
Pere Arús	Head of the Plant Genetics Department, Institute for Agro-food Research and Technology (IRTA)
Elisa Barahona	Biosecurity National Commission (governmental body in charge of allowing field experiments with GMO)
Covadonga Caballo	General Sub-directorate of Health, Ministry of Health
Juan Antonio García	National Centre of Biotechnology (plant section)
Juan Jordano	Institute for Natural Resources and Agrarbiotechnology (National Council for Research)
Ana Lázaro	Biotechnology Monitoring Office (Spanish Office for Patents)
Mercedes Marín	Department of Animal Biology (Universitat Autònoma de Barcelona)
José Angel Martínez Escribano	Research Centre for Animal Health
Javier Martínez Vassallo	National Institute for Agricultural Research
José Miguel Martínez Zapater	Interministerial Commission on Science and Technology (National Programme on Biotechnology)

Name of the expert	Institution/company
Luis Navarro	Valencian Institute for Agriculture Investigation (IVIA)
Joseph Passolas i Farrerons	Applicationes Biotecnológicas (Research Department of the company)
Lenadro Peña	Valencian Institute for Agriculture Investigation (IVIA)
Daniel Ramón	Institute of Agrochemistry and Food Technology
José Juan Rodríguez	Department of Food Hygiene and Technology (Universitat Autònoma de Barcelona)
Josep Sancho i Valls	Agricultural Engineering Department (Universitat Politècnica de Catalunya)

For the first meetings of the national expert committees of Germany, Greece, Italy and Spain in February/March 1996, the German project team had worked out background material concerning the general target of the project, as well as a description of the Delphi methodology. In addition, discussion papers concerning the general structure of the Delphi questionnaire and the topics which should be addressed in the survey had been elaborated by the project teams. In Greece two meetings of the national expert committees were held, due to the high interest of the members in the project. In the Netherlands the members of the national expert committee were personally interviewed for the fear arose that a fruitful cooperation would be very difficult between the supporters and critics of modern biotechnology.

The national expert committees gave detailed recommendations on these materials as well as pointing out additional relevant issues to be considered in the survey, which were collected and summarized by the project co-ordinator. Based on this information, a first version of a theses paper was elaborated by the German project team which was discussed and modified by the five European project teams during a common project meeting in March 1996. This modified theses paper was the basis for the structure of the Delphi questionnaire (e. g. answering categories, relevance of the different domains in the statements), as well as for the creation of the statements. All project teams made suggestions for statements to be used in the questionnaire. The suggested statements were checked, commented and modified by the project partner who was responsible for a certain domain. Based on these modified statements the German research team elaborated a first version of the questionnaire at the end of April 1996.

A major problem was the adaption of the questionnaire to the specific characteristics of the project. For this purpose new answering categories had to be defined, like e. g. "personal attitude". In general, in conventional Delphi surveys mainly rather "objective" answering categories are considered. Since the acceptance and acceptability of modern biotechnology was regarded as one of the most important aspects in the project, the project teams decided - in accordance with the different national expert committees - to ask the experts directly for their personal opinion and attitude to the described future developments because this approach proved to be successful in a recent Delphi survey in a controversially discussed thematic area (future health care systems) as well (see Jaeckel et al. 1995).

Another difficulty represented the exact formulation and wording of the statements, in order to ensure that all experts belonging to different groups in five different countries are able to accept and understand them. Each statement describes a specific future development related to biotechnology in the Agro-Food sector. Since the considered scientific and technological developments are in different phases of the innovation process (like elucidation, development, practical use, widespread use, matter of routine), the exact wording of the statement had to take into account the present phase of the respective technological development. In addition, the statements had to be formulated in such a way that they were oriented towards future situations and trends.

The elaborated draft version of the questionnaire was discussed by the national expert committees during their respective second (or in Greece, third) meetings in May/June 1996. The recommendations given on these meetings as well as written comments provided by the European Commission were summarized in a synopsis by the project co-ordinator. Based on this, the final version of the questionnaire was defined by the five project teams in the third project meeting in June 1996. The English version of the questionnaire was translated into the respective national language and the questionnaires were printed according to a common layout. Due to this interactive approach the questionnaires in the five countries have a common structure and contain identical statements and answering categories. This standardized structure is essential to get comparable results in the five countries involved.

2.5   Selection of experts

An important aspect for the success of a Delphi survey is the selection of the experts to be asked. It could be shown in Delphi surveys in the USA that by including a high number of experts individual variations in the estimation of the experts can be averaged and the probability of an "accurate prognosis" will rise (Dalkey 1969, Dalkey et al. 1969, Dalkey and Helmer 1963). On the other hand, a certain expertise is required, in order to enable the respondents to assess the statements considered in a Delphi survey. This aspect generally limits the number of persons to be asked in a Delphi survey.

In conventional Delphi surveys experts in the sense of scientists and research specialists from research institutions or industrial research departments are involved. This definition of "expert" is extended in this survey since in addition to scientific and technological developments the impacts of biotechnology applications are analysed in the survey which have a rather high relevance for other groups as well. Therefore, the following main groups of experts have been involved in the survey:

(1) Industrial companies

(2) Research institutions

(3) Farmers

(4) Consumers and users

(5) Biotechnology critics and experts for social aspects

(6) Other experts (e. g. experts from the policy, administration and educational sectors, journalists, biotechnology advisers, patent lawyers etc.)

In general, the project teams in the different countries screened the following information sources for relevant experts:

• Electronic databases and address lists of biotechnological organizations

• Other databases (e. g. research programmes, company guides)

• Handbooks, national and international directories (e. g. of research institutions, biotechnology companies)

• Participants and speakers of national and international conferences and seminars on biotechnology

• Authors in specialized magazines (e. g. of biotechnology critics, organic farming organizations)

In addition, relevant organizations and institutions in the Agro-Food sector as well as specific key persons were contacted (such as industrial associations, trade unions, farmers associations, consumer organizations, environmental groups, animal welfare groups as well as the concerned ministries) and asked to deliver member lists or addresses for the relevant groups. In general these activities proved to be successful because in most groups in the different countries more experts could be selected than needed due to statistical reasons.

In Spain this broad definition of "expert" created some problems because the application of modern biotechnology in the Agro-Food sector is a rather unknown issue even among persons who directly work in this branch. In addition, the industrial companies which work in the biotechnology field in Spain have been very reluctant to release any kind of information about their activities. Therefore, it wa relatively time consuming to identify the relevant experts in Spain.

In addition, the German project team worked out a coding scheme for the expert groups. Based on this coding system the expert panels in all five countries included were coded and checked for a balanced composition by the national project teams.

The databases of the experts were completed in August/September 1996. In table 2.10 the final structure of the expert panels in the involved countries is presented. In total, more than 7,800 experts have been asked in the first round to participate in the Delphi survey. The size of the expert panels range from around 1,200 experts in the Netherlands and Spain to almost 2,400 experts in Germany. Due to the differing situation in the involved countries and the availability of experts in the different fields, the structure of the expert panels varies to a certain extent without endangering the comparability of the results between the five countries.

Table 2.10:   Final structure of the expert panels in the involved countries

Expert group	Germany No. of experts	Germany %	Greece No. of experts	Greece %	Italy No. of experts	Italy %	The Netherlands No. of experts	The Netherlands %	Spain No. of experts	Spain %
Industry	523	21.8	255	17.5	203	12.6	221	18.5	386	32.1
Research institutions	487	20.3	376	25.8	385	23.8	197	16.5	250	20.8
Farmers	354	14.8	244	16.7	196	12.1	171	14.3	144	12.0
Consumers and users	259	10.8	199	13.6	327	20.2	210	17.6	90	7.5
Biotechnology critics and experts for societal aspects	242	10.1	191	13.1	178	11.0	195	16.3	100	8.3
Others (e. g. biotechnology consultants, politics, administration, media, education)	463	19.3	195	13.4	326	20.2	200	16.8	230	19.1
Total	2,398	100.0	1,460	100.0	1,615	100.0	1,193	100.0	1,200	100.0

2.6   Realization of the survey

The Delphi survey was organized and carried out by each project partner for his respective country. This included the selection of the experts to be asked in the survey (see chapter 2.6), the translation of the questionnaire into the respective language, the printing and mailing of the questionnaire to the experts as well as additional measures which have been undertaken in several countries in order to raise the number of answers. The statistical analysis was centrally performed by the project co-ordinator in order to ensure a homogenous handling of the data. An overview of the general organization and timing of the Delphi survey is given in figure 2.3.

Figure 2.3:   Organization and timing of the Delphi survey

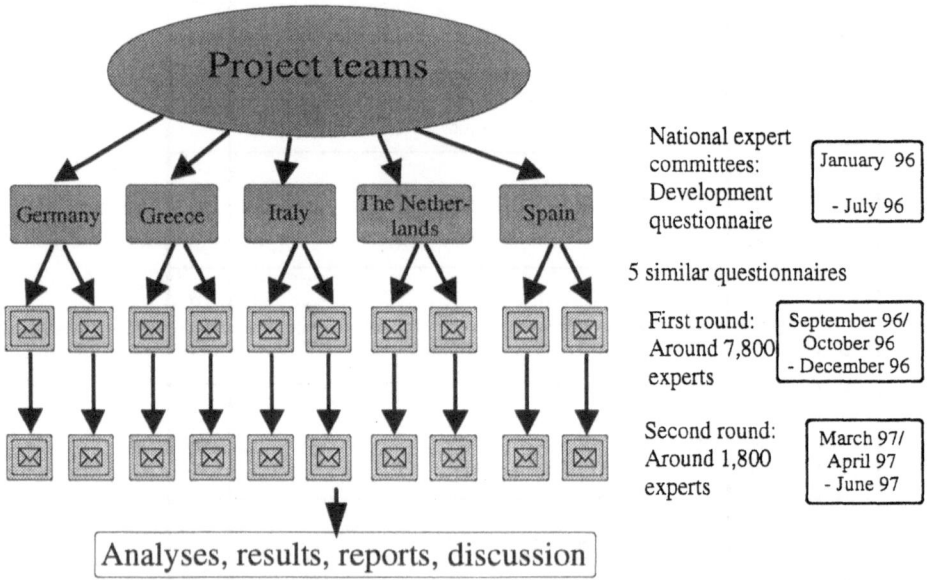

The questionnaires of the first round were mailed to the experts in the involved countries during autumn 1996. Due to differing summer vacations and specific events in some countries (e. g. the Parliamentary elections in Greece in September 1996), it was not possible to have a totally similar timing of the mailing in each country. The questionnaires were mailed on the following dates:

Germany:	September 23rd, 24th, 1996
Greece:	October 7th to 15th, 1996
Italy:	September 16th, 1996
The Netherlands:	October 18th, 1996
Spain:	October 15th, 1996

In each country specific measures have been carried out to raise the number of respondents. In every country a reminding letter (after around four weeks after the mailing of the questionnaires) was sent to the experts, who had not answered by this date. In the Mediterranean countries additional efforts were made to motivate the experts to respond (e. g. call directly to the experts, mobilize the national expert committees, intensify the personal contacts). In addition, the questions and comments of the experts were answered by phone or mail in each country.

Moreover, the project teams in the five countries received a lot of written and verbal comments from the panel. In Germany, for example, there were more than 350 reactions of panelists, in addition to those answering the questionnaire. The reactions of the experts have been very heterogeneous. In general, the following groups of reactions occurred:

- The experts do not want to participate (e. g. not enough time, questionnaire is too difficult and complex, general opposition to genetic engineering).
- The experts have difficulties to answer the statements (e. g. interpretation of the first category, statement is too complex or contains different aspects; statement is difficult to understand or - in the eyes of the experts - wrong; categories do not fit to the statement).
- Some statements do not fit the national situation.
- The experts want more copies or have additional names/addresses of persons/institutions who also will be interested in participating.
- The experts have transferred the questionnaire to another person.
- The address of the expert has changed.
- The experts cannot return the questionnaire in time and want to send it later.

Each project partner was responsible for the registration of the returned questionnaires, of the respondents and for the translation of the comments.

Data input and the basic statistical analysis of the first round for all countries was performed in Germany and was finished in January 1997 (see chapter 2.8). The results of the first round were returned to the project partners, who had to include them when printing the questionnaire for the second round.

The mailing of the questionnaires for the second round was continued in March/April 1997 in each country by the respective project partner. The reminding letter followed around one month after mailing the questionnaires. In Spain and Italy a lot of efforts were undertaken by the project teams to induce the members of the panel to fill in the questionnaire. In this context especially personal contacts and phone calls proved to be very effective as the experts became aware of the importance of their participation in the survey for a second time. In the other three countries specific activities (besides the mentioned reminding letter) have not been necessary in order to get a sufficient number of filled in questionnaires.

2.7   Returned questionnaires/response rate

Due to the specific situation in the involved countries as well as the differing degree of familiarity with the Delphi methodology, the response rates significantly differ between the Mediterranean and the Central European countries (see table 2.11). In the latter a surprisingly high response rate can be registered (especially in Germany) whereas a slower and lower rate of response occurred in the Mediterranean countries.

In total, 2,398 questionnaires have been sent to the selected experts in Germany (table 2.11). 726 of these questionnaires were returned and used for statistical analysis after the first round, which corresponds to a response rate of about 30 %. Compared to other Delphi surveys, in which response rates between 24 % and 40 % were reached, the rate achieved in the first round of this survey represents an average level (BMFT 1993, Cuhls et al. 1995, Jaeckel et al. 1995, Strauss and Jaeckel 1996, Fraunhofer Institute for Systems and Innovation Research 1998). In the second round 522 questionnaires were returned and useful for statistical analysis. This corresponds to a response rate of about 72 %, which is comparable to former Delphi surveys in Germany, with response rates between 70 % and 80 % in the second round (BMFT 1993, Cuhls et al. 1995, Jaeckel et al. 1995, Strauss and Jaeckel 1996, Fraunhofer Institute for Systems and Innovation Research 1998).

Table 2.11:   Response rates in the involved countries

Country	Size of expert panel	1st round		2nd round	
		No. of experts	Response rate (%)	No. of experts	Response rate (%)
Germany	2,398	726	30.3	522	71.9
Greece	1,460	373	25.5	192	51.5
Italy	1,615	189	11.7	149	78.8
The Netherlands	1,193	296	24.8	204	68.9
Spain	1,200	191	15.9	151	79.0

In the Netherlands, response rates almost comparable to Germany were registered. From the 1,193 questionnaires sent out, 296 were returned after the first round, which equals a response rate of around 25 %, which is at the lowest level of the range achieved in Delphi surveys in Germany. This response behaviour corresponds with the low appreciation of the Delphi methodology in the Netherlands (Daniels and Duijzer 1988). In addition, the fact that the application of biotechnology has already had its peak of public interest in the Netherlands may play a role in this context. The response rate of around 69 % in the second round in the Netherlands (see table 2.11) corresponds to the general experiences with this instrument in Germany (BMFT 1993, Cuhls et al. 1995, Jaeckel et al. 1995, Strauss and Jaeckel 1996, Fraunhofer Institute for Systems and Innovation Research 1998).

In Italy 1,615 questionnaires were sent to the experts in September 1996. In this country the rate of response was rather slow. Therefore, the Italian project team started to call the single experts by phone, asking them to fill in the questionnaire. This strategy proved to be successful in order to get the required number of answers although in total, the lowest response rate of around 12 % was registered in Italy during the first round (see table 2.11). Due to an intensive personal contact to the experts who answered during the first round, a relatively high response rate of almost 79 % was achieved in the second round in Italy (see table 2.11).

An almost similar response behaviour can be registered in Spain, where of the 1,200 questionnaires sent out, around 16 % have been filled in during the first round (see table 2.11). The low response rate in this country was mainly due to the fact that many of the experts included in the panel do not have deep knowledge or experience in relation to the application of modern biotechnology in the Agro-Food sector. This result should not be interpreted as a consequence of a week design of the expert panel selected in Spain. There is only a limited number of biotechnology specialists available in Spain. Mainly representatives of consumer organizations, farmers associations, environmental groups and even political institutions do not have much information on biotechnology. On the other hand, biotechnology is a rather specialized field of work, many researchers in that field would not answer the questionnaire because the Agro-Food sector is not their specific area of work. Consequently, only 191 questionnaires were sent out in the second round, in which a rather high response rate of 79 % was achieved (see table 2.11). This can be explained by the intensive field work during the second round (promoted by the limited size of the sample) as well as an increased interest of the experts who already participated in the first round.

In Greece a rather specific response behaviour can be found, resulting in a surprisingly high response rate of more than 25 % in the first round (see table 2.11), which was mainly achieved with the help of intensive personal contacts of the Greek project team as well as involved key persons in the biotech scene in Greece (like the members of the national expert committee). In contrast to the first round, the willingness of the Greek experts to participate again in the second round of the Delphi survey was the lowest among the involved countries (see table 2.11). Although the Greek project team carried out an intensive and systematic telephone action in order to motivate the experts, many of the experts refused to fill in the questionnaire, since they did not understand the sense of answering the questionnaire a second time or they considered it too difficult to answer this questionnaire again. In total, the response rate achieved in Greece can be regarded as a good result, because in general surveys responses rarely exceed 5 % of the total sample in this country.

Although in all countries involved a sufficient number of respondents could be achieved in total, the number of filled in questionnaires per expert group has to be considered as well, because additional statistical analyses should be carried out on this level. Figure 2.4 illustrates the response behaviour of the different expert groups in the five countries. In table 2.12 the distribution of experts in the five countries involved after the second round of the survey is presented.

The response rates differ between the various expert groups in Germany, but no group presents statistical problems. The initial distribution changed, in that the portion of the research and the critics group increased, mostly at the expense of the consumer, industry and farmer group (see figure 2.4). In the Netherlands, an almost similar response behaviour of the different expert groups can be registered as in Germany, whereby the critics and research group gained in relative importance mostly at the expense of the consumers. In this context it has to be taken into account that after finishing the second round, a small part of the original "other experts", namely the professional advisers of economic aspects related to modern biotechnology, have been added to the former critics group. Only 6 representatives of the consumer group participated in the second round of the survey (see table 2.12), so that this group is too small for a separate analysis in the Netherlands. Therefore, the consumer and critics group have been analysed as a combined expert group.

In Italy an overproportional response rate can be registered in the industry and research group (see figure 2.4). Due to an absolute number of experts below 30 in the farmer, consumer and critics group after finishing the second round (see table 2.12), it was decided not to analyse these expert groups individually and to combine the "consumers and critics" in a new group, thereby recoding some respondents who originally belonged to the "other experts". In addition, the farmers were excluded from the statistical analysis on expert group level in Italy.

Researchers on biotechnology have the highest response rates in Spain whereas farmers and representatives of consumers answered very rarely mostly due to lack of knowledge and expertise in the issues mentioned in the questionnaire (see figure 2.4). Because of the low representation of farmers, consumers and biotechnology critics in the Spanish sample after finishing the second round, it was decided not to analyse these expert groups individually. Instead, a combined group of "consumers and critics" was analysed whereby farmers were not included in the statistical analysis separately. In order to get a higher statistical significance of the analysed expert groups, the Spanish project team regrouped the answers of some experts originally belonging to the "other experts", thereby taking into account the self-estimation of the answering experts which was asked at the end of the questionnaire. Twelve experts from this group have been added to the consumer group because they actually work in the Civil Service dealing with consumers or health issues related to food production and processing. In addition, some of the members of the original "other experts" were recoded to the industry and research group as well.

Table 2.12:   Structure of the expert panels in the five countries (after the second round)

Expert group	Germany		The Netherlands		Greece		Italy		Spain	
	No.	%	No.	%	No.	%	No.	%	No.	%
Industry	99	19	34	17	17	9	27	18	43	28
Research Institutions	142	27	51	25	38	20	39	26	58	38
Farmers	66	13	31	15	22	11	19	13	8	5
Consumers and users	37	7	13	6	66	34	22	15	16	11
Critics	75	14	44	22	49	26	21	14	21	14
Others	89	17	31	15	0	0	12	8	5	3
Without code	14	3	0	0	0	0	9	6	0	0
Total	522	100	204	100	192	100	149	100	151	100

A rather specific response behaviour of the different expert groups can be registered in Greece. In this country the consumer and critics group show overproportional response rates (see figure 2.4), whereas especially farmers and representatives of industry rarely filled in the questionnaire due to low knowledge in the covered areas. Because in both expert groups the absolute number of respondents is below 30 after finishing the second round (see table 2.12), these groups were not statistically analysed individually. In addition, thirteen representatives of the original "other experts" have been regrouped by adding them mostly to the consumer and critics groups due to the answers of the experts made on the last page of the questionnaire.

Figure 2.4:   Response behaviour of expert groups in five countries

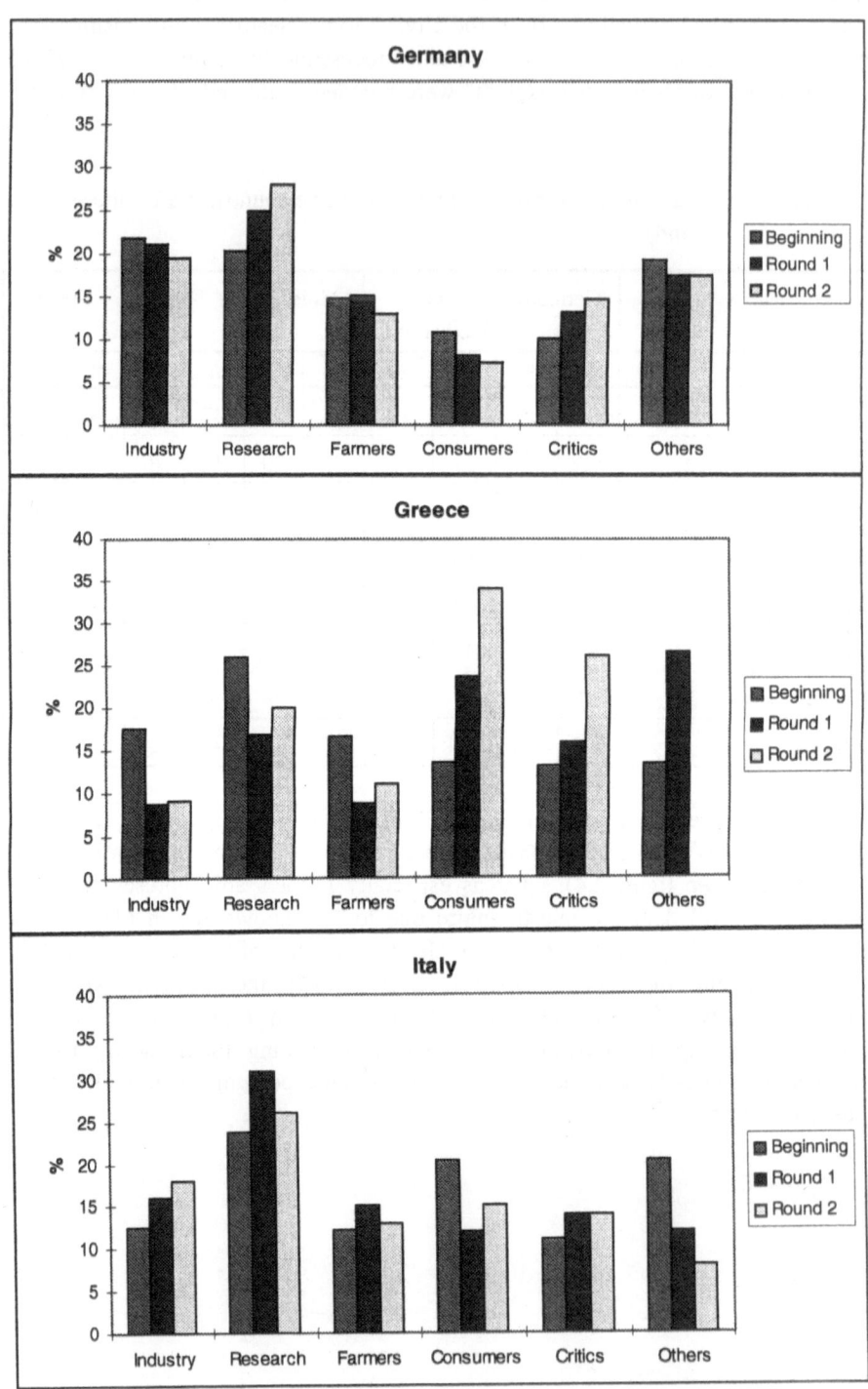

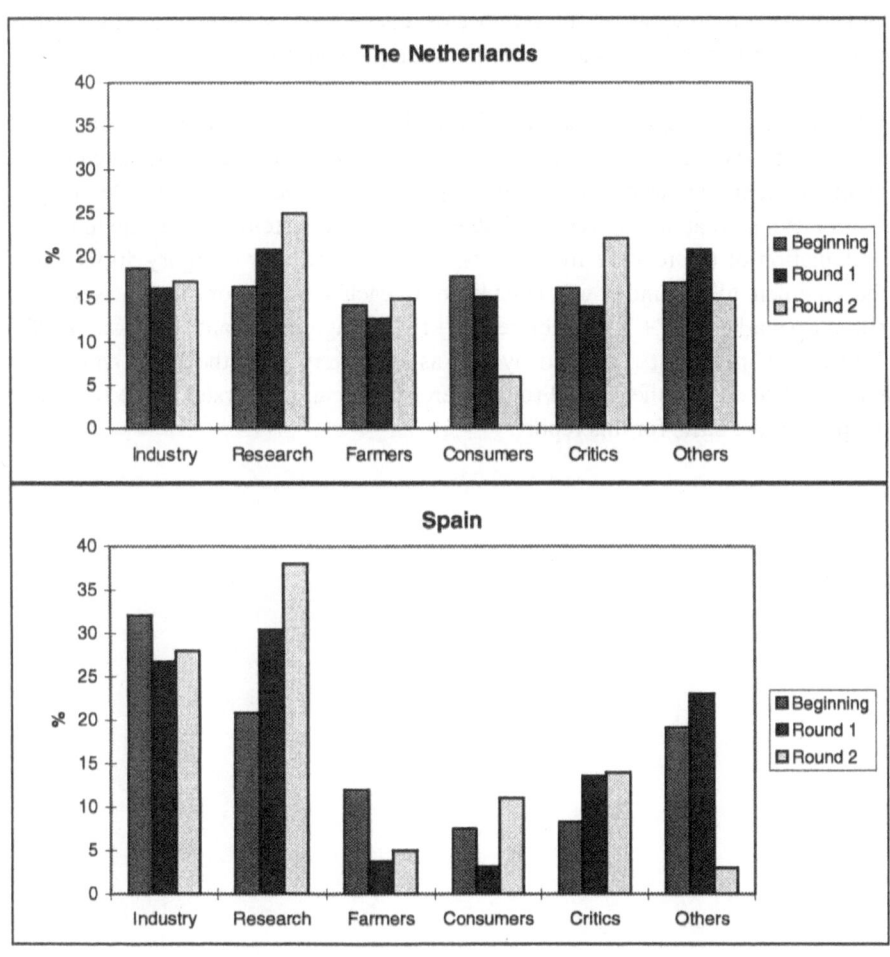

2.8 Statistical analysis

A basic statistical analysis was performed for the data of the first round because the results had to be presented in the questionnaire of the second round. In a first step all data of the available questionnaires in the single countries were used to calculate the relative frequencies of the four sub-categories of the "degree of knowledge" for each statement. Afterwards, all data sets were selected in which the degree of knowledge was assessed by the experts as "very familiar", "average familiar" or "less familiar". This represented the basis for the analysis of the following response categories, whereby answers of experts who assessed themselves as being "not familiar" with a specific statement were excluded from the statistical analysis of this statement. As a consequence, the number of answers per statement always differs from the total number of questionnaires in the single countries.

For the answering categories "degree of knowledge", "personal attitude" and "time of realization", the relative frequencies for the sub-categories were calculated on the basis of the actual answers. In the category "influential factors", multiple answers were allowed, so that all answers available for a single statement were included for the calculation of the relative frequencies. In the last answer category "importance of the statement to", an index was calculated. In each sub-category the answer "very important" (marked by "+") was defined as +1, "average important" (marked by "0") as 0 and "not important" (marked by "-") as -1. Afterwards, the mean value was calculated. The data of the second round were statistically analysed in the same way and represent the basis for this report.

3. Results in different countries

In this chapter, an overview is given of the basic results of the Delphi survey in the participating countries, which are sorted in alphabetical order. This overview mainly follows the categories included in the questionnaire in each country. Firstly, the self-estimated degree of knowledge of the answering experts is analysed. In the following parts a general overview is given related to the estimated personal attitude and the expected time of realization of the statements included in the questionnaire. In addition, the most important influential factors relevant for the future application of modern biotechnology in the Agro-Food sector are analysed. Afterwards the importance of the visions mentioned in the statements of the questionnaire for future knowledge creation in science and technology, and competitiveness of the respective national economies, as well as the environmental situation, is described. Most of the country chapters conclude with a short discussion on how the results of this Delphi survey fit in the general situation of Agro-Food biotechnology in the respective country.

3.1 Germany

Degree of knowledge

In table 3.1 the overall degree of knowledge of all experts in Germany is presented. On average, 12 % of the German experts estimate themselves as being "very familiar", 45 % as being "average familiar", 29 % as being "less familiar" and 14 % as being "not familiar" with the issues mentioned in the questionnaire. There is only a slight variation in the distribution of the degree of knowledge between the different expert groups in Germany. Even though there might be differences in the "realism" of self-estimates, this result is rather surprising, because it is generally assumed that experts from science and industry are dealing with issues related to modern biotechnology more often and intensively than those of the three other groups and therefore, these expert groups should have a higher degree of knowledge in this field.

The answers of experts, who described their degree of knowledge related to a specific statement as "not familiar" are excluded from the statistical analysis. More than 80 % of the statements mainly referring to framework conditions were answered by 400 and more, 10 % of the statements were answered by 350 to 400 and three statements (4 %) were answered by less than 250 interviewees. The latter refer to the application of modern biotechnology in fish breeding and aquaculture, mostly due to a limited number of experts in this field in Germany.

Table 3.1:   Overall degree of knowledge of all expert groups in Germany

Degree of knowledge	All experts in Germany	Industry	Research	Farmers	Consumers	Critics
Very familiar	12 %	12 %	12 %	8 %	11 %	13 %
Average familiar	45 %	44 %	47 %	40 %	41 %	42 %
Less familiar	29 %	29 %	29 %	34 %	32 %	28 %
Not familiar	14 %	15 %	12 %	18 %	16 %	17 %
Number of interviewees	522	99	142	66	37	75

In addition, a low number of answers was obtained for statements referring to very specific application fields in genetic engineering and molecular biology. In total, it can be concluded that most of the statements included in the questionnaire were assessed by a great majority of the German experts and that they tried to answer the whole questionnaire as completely as possible.

Personal attitude

Figure 3.1:   Personal attitude of the experts in Germany

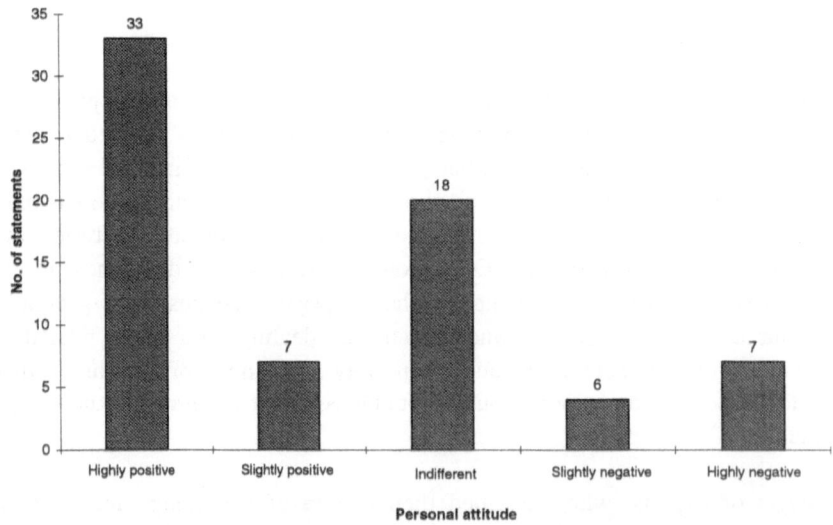

Figure 3.1 gives a general overview of the personal attitude of the German experts. More than half of the statements are appreciated by the German experts, mostly

relating to scientific and technological developments in food technology and plant production, as well as to positive impacts of modern biotechnology on health and environment. The German experts highly appreciate the development of monitoring and control systems based on modern biotechnology (statements 62, 43, 42, 57) (table 3.2). On the other hand, they see a strong necessity in the investigation of

Table 3.2:   List of "top 10" statements for "positive attitude" in Germany

Statement No.	Content	Positive attitude (%)
62	Rapid test systems based on modern biotechnology are widely used for pathogen identification in plant production and animal husbandry.	92.8
58	The molecular basis of virus resistance mechanisms in economically important perennial plants like e. g. vine, olive and fruit trees is elucidated.	92.0
41	The long-term health impacts of the use of Agro-Food products made with the help of genetic engineering on consumers, farmers and employees are investigated (e. g. by epidemiological surveys).	89.4
43	Techniques based on modern biotechnology are practically used for on-line control of quality parameters in food processing (e. g. content of micronutrients or harmful substances).	89.2
42	The hygiene monitoring of food processing is significantly improved due to the widespread use of modern biotechnological analytical methods.	88.7
57	New monitoring and control techniques for genetically engineered plants in open fields have improved the reliability of risk assessment significantly.	88.3
32	Modern biotechnology significantly contributes to the transformation of 40 % or more of the organic agricultural waste into marketable products (e. g. energy, secondary raw materials).	86.9
35	Food companies inform the consumers in detail about the specific health relevant effects of food made with the help of genetic engineering.	86.8
36	Allergy-sufferers are offered special food, of which the allergenic potential has been decreased by genetic engineering.	84.9
1	A governmental institution which provides information on modern biotechnology in the Agro-Food sector to all interested persons and institutions is established in Germany.	83.7

long-term health impacts of modern biotechnology, e. g. by carrying out epidemiological surveys (statement 41). In addition, a high proportion of all German experts agrees on widespread information concerning biotechnology provided by food companies and governmental institutions (statements 1, 35). Moreover, they appreciate selected basic research activities like the elucidation of the molecular basis of virus resistance mechanisms (statement 58), as well as the contribution of biotechnological approaches to the improvement of the environmental situation in animal husbandry (statement 32), or the decrease of food-related allergies (statement 36).

In 18 statements the interviewees have an indifferent personal opinion (see figure 3.1). German experts have no strongly negative or positive feelings in the case that modern biotechnology fails in producing food and beverages satisfying traditional taste preferences of large consumer groups in Germany. The same applies to the production of beer with genetically engineered yeast (statements 47, 45). In addition, the German experts do not express any strong emotions if food products based on modern biotechnology achieve considerable market shares (statements 14, 38) or only few new jobs are created in the food industry due to this technique (statement 17).

Table 3.3: List of "top 5" statements for "negative attitude" in Germany

Statement No.	Content	Negative attitude (%)
40	Unintended impacts on the health of consumers occur in some products or manufacturing processes due to the use of genetic engineering during food processing.	90.0
37	The increased use of modern biotechnology in food production and processing results in additional allergies in people actively involved in these processes.	85.5
26	The widespread deliberate release of genetically engineered organisms results in a significant transfer and recombination of the introduced genes with unintended negative impacts (e. g. development of resistances to herbicides and pathogens in weeds).	85.2
24	Due to the hesitant application of genetic engineering (compared to other EU member states) the Agro-Food industry cuts 10 % or more of the jobs in this sector in Germany.	82.1
20	The widespread use of modern biotechnology in animal and plant production leads to an approximate 30 % decrease in traditional jobs in agriculture.	80.3

The German interviewees strongly oppose around 10 % of the statements (see figure 3.1). These statements mainly deal with negative impacts on health (e. g. additional allergies) (statements 37, 40), environment (statement 26) and economy

Figure 3.2:   Statements with the most significant differences between the expert groups in Germany

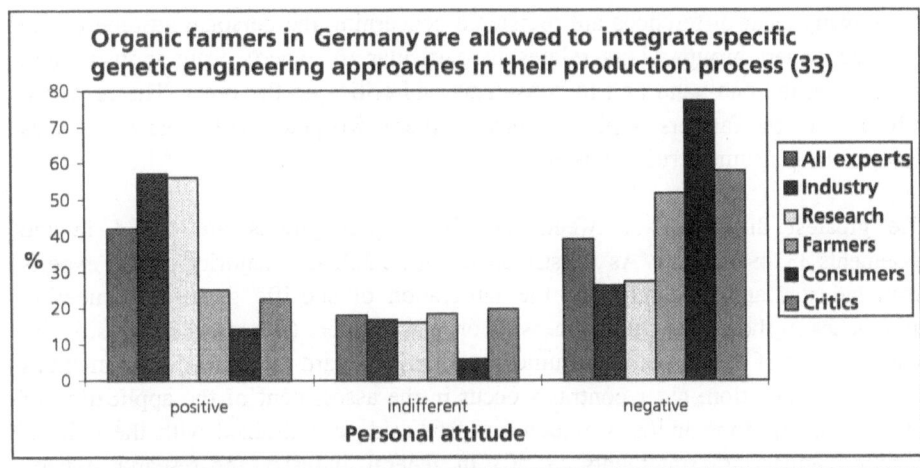

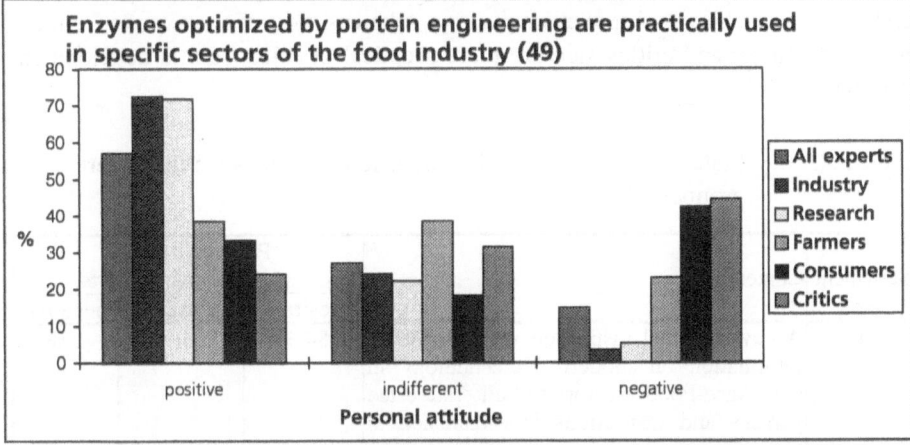

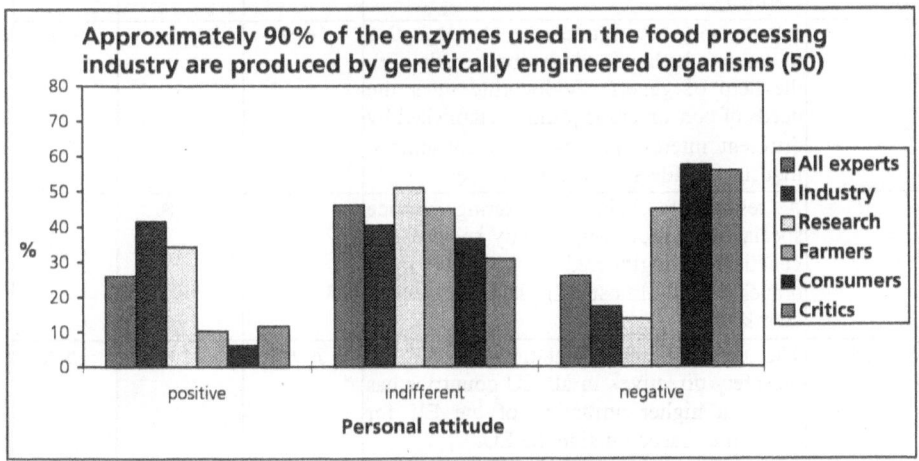

(e. g. reduction of employment) due to the use of modern biotechnology (statements 20, 24) (see table 3.3).

In Germany clear differences are registered concerning the personal attitude of the analysed expert groups. This relates in particular to experts from industry and research on the one hand and to consumers and critics on the other. The response behaviour of the farmers is placed between these two poles with clear tendencies towards the consumer/critics cluster.

The greatest differences between the five expert groups are found in the statements 33, 49 and 50. As illustrated in figure 3.2, the majority of the experts from industry and research sees the integration of specific genetic engineering approaches in the production process of organic farmers (statement 33) positively, whereas most of the farmers, consumers and critics regard the same development as negative. In addition, clear contrasts occur in the assessment of the application of enzymes in the food industry which are produced or optimized with the help of genetic engineering (statements 49, 50). In general, industry and research experts tend to assess such a development as positive, whereas most of the experts from the farmer, consumer and critics side are more sceptical or reject such a development (see figure 3.2).

Table 3.4:   Statements with no significant differences between the expert groups in Germany

Statement No.	Content	Personal attitude of all German experts		
		positive	indifferent	negative
1	A governmental institution which provides information on modern biotechnology in the Agro-Food sector to all interested persons and institutions is established in Germany.	84 %	15 %	1 %
5	An advisory committee with the objective to assess whether food products made with the help of genetic engineering meet the needs of consumers is jointly established by different interest groups (e. g. consumers, industry, retailers) in Germany.	76 %	14 %	9 %
7	A restaurant chain or catering service specialized in offering trendy meals with genetically engineered ingredients open branches in almost all large cities in Germany.	8 %	40 %	51 %
13	The uniform implementation of the EU biosafety directives in all EU countries has led to a higher attraction of the EU for companies based outside the EU.	50 %	31 %	15 %

Statement No.	Content	Personal attitude of all German experts		
		positive	indifferent	negative
20	The widespread use of modern biotechnology in animal and plant production leads to an approximate 30 % decrease in traditional jobs in agriculture.	3 %	16 %	80 %
30	Despite the widespread use of methods and techniques related to modern biotechnology, all in all, no negative effect on the maintenance of biodiversity in agriculturally influenced ecosystems can be discovered.	76 %	13 %	8 %
32	Modern biotechnology significantly contributes to the transformation of 40 % or more of the organic agricultural waste into marketable products (e. g. energy, secondary raw materials).	87 %	8 %	3 %
35	Food companies inform the consumers in detail about the specific health relevant effects of food made with the help of genetic engineering.	87 %	9 %	3 %
37	The increased use of modern biotechnology in food production and processing results in additional allergies in people actively involved in these processes.	2 %	9 %	86 %
40	Unintended impacts on the health of consumers occur in some products or manufacturing processes due to the use of genetic engineering during food processing.	1 %	7 %	90 %
41	The long-term health impacts of the use of Agro-Food products made with the help of genetic engineering on consumers, farmers and employees are investigated (e. g. by epidemiological surveys).	89 %	8 %	1 %
61	After years of experimentation with genetic engineering, the majority of plant breeders strongly prefers the combination of marker-assisted breeding with traditional breeding methods, compared to the use of genetically modified plants.	53 %	36 %	8 %

In spite of the differences, there are areas with almost similar personal attitudes among the experts (see table 3.4). All expert groups highly appreciate information of consumers concerning modern biotechnology in general (statement 1), or the health effects of this technology in particular (statement 35). The same applies to the participation of different social groups in the assessment of genetically modified food (statement 5). A great majority of all German experts agrees as well if modern biotechnology contributes to the improvement of the environmental situation (statements 30, 32), or the investigation of possible long-term health effects of genetically engineered products (statement 41). All expert groups in Germany highly reject a potential job decrease in agriculture (statement 20) as well as negative health impacts (statements 37, 40) due to modern biotechnology.

Time of realization

Figure 3.3 presents the overall distribution of the median class of the time of realization of the 71 statements estimated by the German experts. Around two thirds of the statements are indicated by the German experts as realizable in the next six to ten years. 20 % of the statements are classed as realizable in the next eleven to 15 years. 11 % of the statements are expected to be realized in the next five years. One statement is seen as not being realistic at all.

The short-term developments are dominated by information activities concerning modern biotechnology in general and health effects of gene food in particular (statements 1, 35). In addition, initiatives to involve the public in the debate concerning the use of modern biotechnology in the Agro-Food sector to a higher extent (statements 5, 12) as well as specific labelling activities are seen as realizable in the next five years, i. e. the creation of specific labels for food produced without genetically modified organisms and, conversely, the compulsory labelling of food made with genetically modified organisms (statements 6, 8). From the scientific and technically oriented statements only the use of enzymes optimized by protein engineering in the food industry is expected as realizable in the short term (statement 49). Since the first protein engineered enzymes (amylases) for food purposes have just been introduced in the market (Heldt-Hansen 1998), this estimation of the expert panel has just been realized.

Statements which are assessed as realizable in eleven to 15 years mostly belong to the domains "economy", "environment" and "scientific/technological development". The German panel estimates that the use of modern biotechnology enables small and medium-sized companies of the Agro-Food sector to introduce a large variety of innovative processes and products due to the use of modern biotechnology by the end of the next decade. The same applies to an approximate 30 % decrease in traditional jobs in agriculture due to the use of modern biotechnology, as well as a 30 % market share of food made with the help of genetic engineering in Germany (statements 16, 20, 14). In addition, a high proportion of the German interviewees assumes that the imports of fish from outside the EU will decrease within the next eleven to 15 years, as well as that during the same time period, renewable resources will be produced on 20 % or more of the arable land in Germany (statements 21, 22).

Figure 3.3:   Time of realization of the statements in Germany (according to the median class)

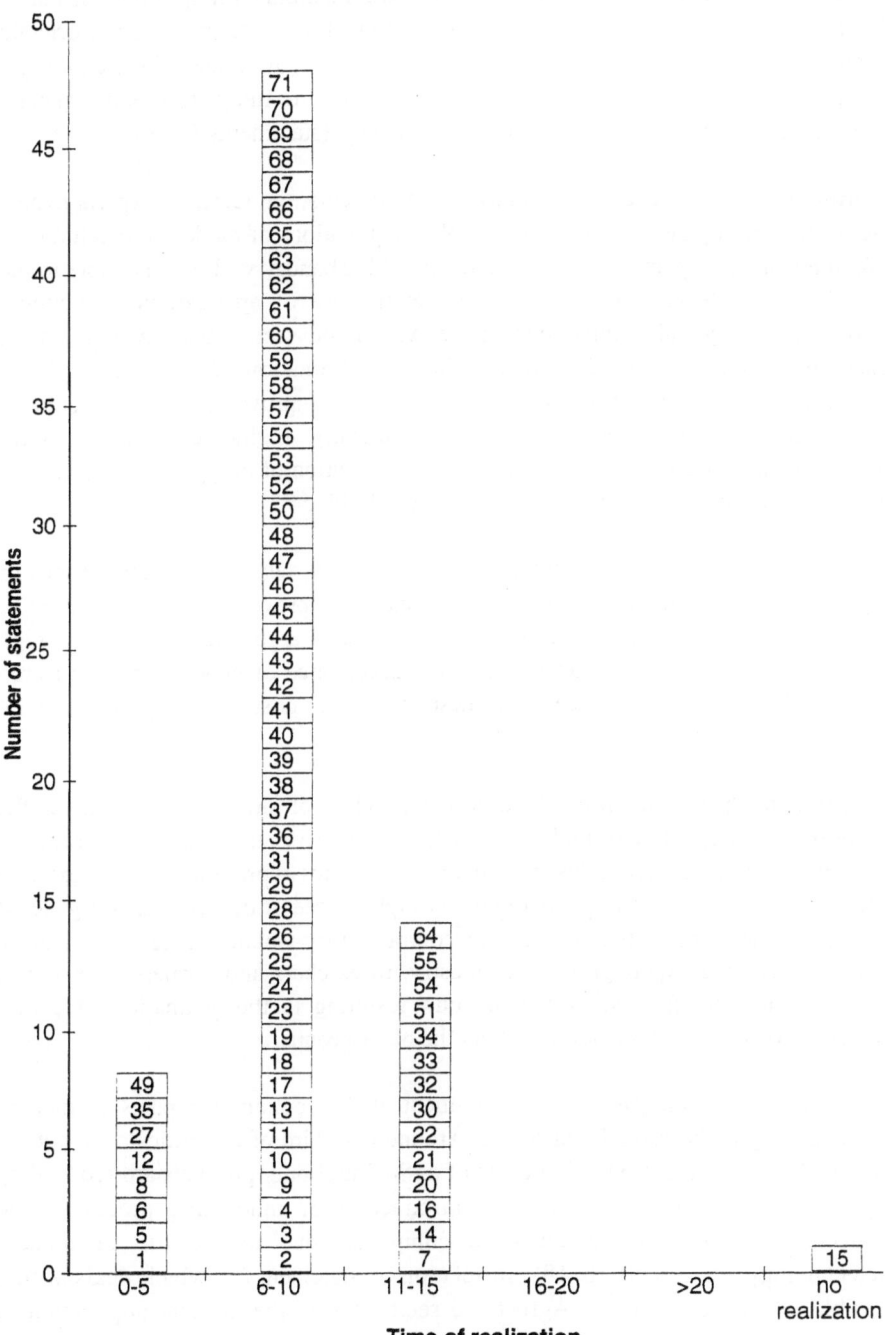

With respect to environmental statements the German respondents expect that the use of genetically modified organisms in food production aggravates the loss of traditionally used organisms (statement 34) in the medium to long term (eleven to 15 years). The same applies to the transformation of 40 % or more of the organic agricultural waste into marketable products contributed by modern biotechnology, as well as that organic farmers in Germany are allowed to integrate specific genetic engineering approaches in their production processes (statements 32, 33).

Referring to scientific and technological developments, the German experts expect that mostly developments relating to complex applications of modern biotechnology in food technology, plant production and animal husbandry will be realizable in the next eleven to 15 years as well. These are: the use of artificial polyfunctional enzymes for analytical applications in the Agro-Food sector, the development of genetic engineering approaches which allow the alternation of polygenic traits in most economically important plant species, the improvement of agricultural productivity in arid evironments, using genetically engineered plant varieties resistant to salinity and drought, as well as the alteration of polygenic traits in farm animals by genetic engineering (statements 51, 54, 55, 64).

Statement 15, which is seen as unrealistic in Germany, deals with higher prices of food products, of which the quality is significantly improved by the application of modern biotechnology. Most of the German experts argue that due to the high competition on the food market and the low acceptance of genetically engineered food in Germany higher prices for these products most probably will not be realized.

Compared to "personal attitude", a lower number of differences occur in the answering category "time of realization" between the five analysed expert groups in Germany. In this respect industry/research and critics represent the two extreme poles among the expert groups. A relatively high degree of differences is registered between farmers and critics as well. In contrast, only isolated differences can be found between the expert groups of industry, researchers and farmers. Consumers show a rather specific response behaviour, resulting in the estimation of earlier realization of specific developments than the other groups.

There are clear differences in the assessment of the economic impacts of modern biotechnology in the Agro-Food sector. Around one third of the critics does not see the possibility that the production costs of agricultural bulk products are reduced by approximately 30 % as well as the introduction of innovative processes and products in small and medium-sized companies due to the use of modern biotechnology (statements 16, 18). In relation to statement 24, which deals with a possible job reduction in the Agro-Food sector due to the hesitant application of genetic engineering, more than one third of the critics see such a development as unrealistic compared to 14 % or 8 % of industry and research experts (see table 3.5).

Table 3.5: Statements assessed (to a high proportion) as being "not realized" in Germany

Statement No.	Content	Percentage of sub-category "no realization"		
		Critics	Industry	Research
4	Research on genetic engineering of farm animals is not funded by public institutions in Germany due to ethical reasons (e. g. animal rights, animal welfare, preservation of Creation).	36 %	21 %	15 %
7	A restaurant chain or catering service specialized in offering trendy meals with genetically engineered ingredients open branches in almost all large cities in Germany.	40 %	22 %	25 %
8	The labelling of all food products made with the help of genetic engineering is compulsory and fully implemented in Germany.	36 %	9 %	16 %
9	All food and beverages which are produced with the help of genetic engineering are subject to an official case-by-case procedure for approval in Germany.	49 %	26 %	30 %
10	For the marketing approval of Agro-Food products made with the help of genetic engineering, concerns on their social impacts and potential ethical conflicts are generally taken into account.	40 %	24 %	28 %
11	The responsible authorities in Germany practically use specific technical equipment and standardized methods to identify food and beverages which were produced with the help of genetic engineering.	24 %	20 %	21 %
15	The prices of food products in specific market segments, of which the quality is significantly improved by the application of modern biotechnology are, at least, 30 % higher than that of corresponding conventional products.	59 %	52 %	52 %
16	The widespread use of modern biotechnology enables small and medium-sized companies of the Agro-Food sector in Germany to introduce a large variety of innovative processes and products.	30 %	11 %	5 %
18	The widespread use of modern biotechnology in agriculture reduces the production costs for agricultural bulk products (e. g. cereals, milk) by approximately 30 %.	32 %	7 %	8 %
24	Due to the hesitant application of genetic engineering (compared to other EU member states) the Agro-Food industry cuts 10 % or more of the jobs in this sector in Germany.	34 %	14 %	8 %

Statement No.	Content	Percentage of sub-category "no realization"		
		Critics	Industry	Research
29	The widespread cultivation of genetically engineered field crops resistant to herbicides results in an approximate 50 % reduction of environmental pollution in plant production.	51 %	9 %	11 %
30	Despite the widespread use of methods and techniques related to modern biotechnology, all in all, no negative effect on the maintenance of biodiversity in agriculturally influenced ecosystems can be discovered.	61 %	12 %	14 %
33	Organic farmers in Germany are allowed to integrate specific genetic engineering approaches (e. g. use of genetically engineered biocides or animal vaccines) in their production process.	43 %	14 %	19 %
47	Modern biotechnology has failed in producing food and beverages satisfying traditional taste preferences of large consumer groups in Germany.	19 %	24 %	23 %

In addition, almost half of the critics sees it as unrealistic that genetically engineered food and beverages are approved after an official case-by-case procedure (statement 9) and that during this process concerns on the social impacts and potential ethical conflicts are taken into account (statement 10). In these two statements a rather high proportion of the industry and research experts expects as well that these developments will not be realized (see table 3.5).

Table 3.6:   Statements with significantly different time estimations of consumers and critics in Germany

Statement No.	Content	Median class of the time of realization	
		Consumer	Critics
4	Research on genetic engineering of farm animals is not funded by public institutions in Germany due to ethical reasons (e. g. animal rights, animal welfare, preservation of Creation).	5-10	>20
7	A restaurant chain or catering service specialized in offering trendy meals with genetically engineered ingredients open branches in almost all large cities in Germany.	5-10	15-20
14	Food made with the help of genetic engineering achieve a turnover share of 30 % or more of all food consumed in Germany.	5-10	10-15

Statement No.	Content	Median class of the time of realization	
		Consumer	Critics
22	Due to the application of modern biotechnology, farmers produce renewable resources (e. g. biofuel, starch, fatty acids) which are used outside the food sector on 20 % or more of the arable land in Germany.	5-10	10-15
24	Due to the hesitant application of genetic engineering (compared to other EU member states) the Agro-Food industry cuts 10 % or more of the jobs in this sector in Germany.	5-10	10-15
29	The widespread cultivation of genetically engineered field crops resistant to herbicides results in an approximate 50 % reduction of environmental pollution in plant production.	5-10	- [1]
31	Modern biotechnology is widely used to reduce emissions and waste from animal production (e. g. less manure due to enzymes as feed additives, improved anaerobic waste treatment, biofilter).	5-10	10-15
35	Food companies inform the consumers in detail about the specific health relevant effects of food made with the help of genetic engineering.	0-5	5-10
44	The genomes of most economically important bacteria used in food production are completely sequenced.	5-10	10-15
45	In Germany most of the beer is produced with genetically engineered yeast.	5-10	10-15
48	A large variety of genetically engineered microorganisms, which can be precisely controlled and regulated in their metabolic activities during food production processes has been developed.	5-10	10-15
50	Approximately 90 % of the enzymes used in the food-processing industry are produced by genetically engineered organisms.	5-10	10-15
56	To prevent resistance against biocides in pathogens, new genetic engineering approaches for plant defence mechanisms have been developed (e. g. combination of different resistance genes, increase in pathogen tolerance).	5-10	10-15

[1] More than 50 % of the experts of the group estimate that the situation/development mentioned in the statement never will be realized. Therefore, no median class was calculated.

In the statements summarized in table 3.6, consumers tend to estimate an earlier time of realization than critics. This relates in particular to specific scientific and technological developments. The majority of the consumers expect that within the next five to ten years, e. g. the genomes of important bacteria used in food production are completely sequenced (statement 44), most of the beer in Germany is produced with genetically engineered yeast (statement 45), and around 90 % of the enzymes used in the food industry are produced with genetically engineered organisms (statement 50). The critics expect the same developments around five years later (see table 3.6).

Similar differences can be registered in the assessment of the economic impacts of modern biotechnology between consumers and critics. This relates, e. g. to the market share of genetically engineered food (statement 14), the extended production of renewable resources (statement 22) and possible job reductions in the Agro-Food industry of Germany due to the hesitant application of modern biotechnology (statement 22). Extreme differences in the expected time of realization between consumers and critics are found in the assessment of public funding of genetic engineering of farm animals (statement 4), as well as the environmental impacts of herbicide-resistant plants (statement 29) (see table 3.6) - both areas are heavily disputed in the public debate in Germany.

Influential factors

In figure 3.4 the general importance of the different influential factors related to all 71 statements is illustrated. On average, 21 % of all answers given in this category represent "social and ethical acceptance", 16 % the conditions on the respective markets, 11 % "policy" and 10 % "funding". The environmental situation, availability of educated and skilled staff, as well as international collaboration, are considered less important. In contrast to other answering categories, the analysed expert groups do not show clear differences in the assessment of the influential factors.

Considering the different domains, clear differences occur in the relevance of the included influential factors. An overview of this aspect is given in table 3.7. Not surprisingly, the German experts assess "R&D infrastructure" as more relevant in the statements related to the scientific and technological development in food technology, plant production and animal husbandry compared to the other domains. Especially for basic research activities like the elucidation of the molecular basis of virus resistance mechanisms in perennial plants, the development of genetic engineering approaches to alter polygenic traits in economically important plants, as well as the complete sequencing of the genomes of the most important bacteria in food production (statements 58, 54, 44), the availability and quality of respective scientific and technological institutions is regarded as an essential precondition for future development.

Figure 3.4:   General importance of the influential factors in Germany

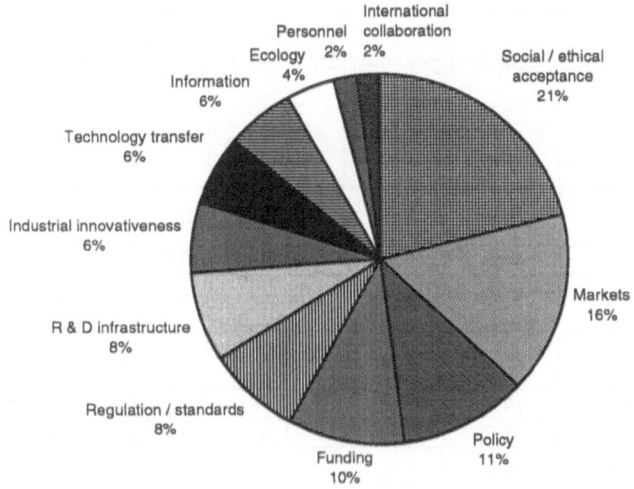

The organization of an efficient know-how transfer between universities, research institutions and industry is seen as a crucial factor for the future application of modern biotechnology in the food industry and plant production (see table 3.7). This relates in particular to the use of specific monitoring and control techniques in this field (statements 62, 51, 43). In addition, technology transfer is seen as an important prerequisite for innovation activities of small and medium-sized companies in general and in plant breeding companies in particular (statements 16, 61).

Industrial innovativeness is regarded as a crucial factor in specific statements of different domains, mostly linked to the introduction of new products or new techniques based on modern biotechnology. This relates in particular to the development of specific enzyme systems in order to improve the environmental performance of conventional food-processing procedures, or the use of plant cell cultures for the production of high-value components (statements 25, 59). Moreover, new market possibilities based on modern biotechnology (e. g. transformation of organic agricultural waste into marketable products or production of pharmaceutical substances with the help of genetically engineered animals and plants), as well as the creation of new jobs in the food industry (statement 17), are mainly influenced by the ability of industrial companies to make a commercial success out of new scientific and technological findings. The same applies to the use of improved starter cultures and enzyme systems in the food industry (statements 48, 49).

Table 3.7:   Relevance of the different influential factors in relation to the domains considered in the questionnaire in Germany

Influential factors for the realization for a specific statement	Domains							
	Acceptance	Regulation	Economy	Environment	Health	Food	Plants	Animals
R&D infrastructure	1 %	2 %	3 %	5 %	6 %	13 %	21 %	14 %
Personnel	1 %	4 %	1 %	0 %	2 %	2 %	1 %	1 %
Technology transfer	1 %	1 %	7 %	6 %	2 %	10 %	12 %	7 %
Industrial innovativeness	2 %	2 %	6 %	8 %	5 %	11 %	8 %	5 %
Markets	15 %	14 %	28 %	10 %	18 %	17 %	9 %	13 %
Funding	9 %	12 %	8 %	4 %	7 %	12 %	20 %	11 %
Regulation/ standards	6 %	13 %	5 %	9 %	11 %	8 %	3 %	8 %
Policy	19 %	30 %	12 %	11 %	8 %	3 %	3 %	7 %
Environment	3 %	1 %	3 %	14 %	2 %	2 %	4 %	3 %
Social/ethical acceptance	26 %	9 %	21 %	25 %	22 %	19 %	15 %	28 %
Information	14 %	5 %	3 %	7 %	15 %	2 %	2 %	2 %
International collaboration	2 %	8 %	2 %	1 %	1 %	2 %	2 %	1 %

The German panel regards the conditions on the markets as highly relevant for the pricing and import policy related to gene food products (statements 15, 21), the introduction or diffusion of new or improved products based on modern biotechnology (statements 14, 22, 38, 39), as well as the possible job creation in the food industry due to these techniques (statement 17). According to the assessment of the German panel the market conditions are highly influenced by information and labelling activities of the food industry (statements 6, 35). Moreover, the production of beer with genetically engineered yeast (statement 45) as well as the success of a restaurant chain or catering service specialized in meals with genetically engineered ingredients (statement 7) depends to a high extent on the current market situation and its future development.

The availability of private and public funds is regarded as a key factor for the future scientific and technological development, especially in the plant production field, as well as the realization of certain regulation activities (see table 3.7). Examples in this direction are the complete sequencing of the genomes of important bacteria strains, the elucidation of the molecular basis of virus resistance in perennial plants, the development of plant varieties resistant to salinity or drought, as well as rapid test systems used for pathogen identification in plant production and animal husbandry (statement 44, 58, 55, 62). In addition, funding is required for the use of specific monitoring or control techniques based on modern biotechnology either by public institutions (statement 11) or the food industry (statements 42, 43). The same applies to the long-term monitoring of health impacts of gene food products (statement 41), as well as providing of information on modern biotechnology by a governmental institution (statement 1).

In the view of the German panel, national and international standards and the regulatory framework have the highest relevance in the health domain (see table 3.7). This relates in particular to the occurrence of unintended impacts on the health of consumers due to the use of genetic engineering as well as additional allergies of people actively involved in food production and processing due to the increased use of modern biotechnology (statements 40, 37). In addition, labelling activities (statements 6, 8), the market approval of gene food products (statement 9), as well as the implementation of EU biosafety directives in the different member countries (statement 13), highly depend on this factor.

Activities and measures of the national or federal government and other politicians are regarded as highly relevant for issues of the regulation and acceptance domain (see table 3.7). Especially the market approval procedures of gene food products (statements 9, 10), the compulsory labelling of these products (statement 8), as well as an increased public participation in the debate and decision-making on the application of modern biotechnology (statement 12), are highly influenced by political activities. The same applies to information activities on modern biotechnology as well as public funding of research activities on genetic engineering of farm animals (statements 1, 4).

As already mentioned, social and ethical acceptance of modern biotechnology is regarded as the major influential factor for its future application in the Agro-Food sector in Germany. This relates in particular to its use in animal breeding and animal husbandry (see table 3.7), like the genetic engineering of early stages of embryo cells, the genetic alteration of polygenic traits in farm animals or the use of cloned embryos (statements 66, 64, 65). In addition, social and ethical acceptance influence the market success of genetically engineered food products (statements 2, 15) and the penetration of a restaurant chain specialized in this field (statement 7) to a high extent. In the environmental field social and ethical acceptance of modern biotechnology is a major issue in the integration of these approaches in the production process of organic farmers (statement 33) and the deliberate release and marketing of genetically engineered microorganisms (statement 27).

Most of the German experts regard information of consumers and the public related to modern biotechnology in general and its environmental and health effects in particular as an important prerequisite to influence the acceptance of this technology. Accordingly, the highest percentual value for the influential factor "information" occurs in the statement describing the increased acceptance of gene food products in Germany due to the widespread diffusion of information on genetic engineering (statement 3). Moreover, information is seen as crucial for the habituation of consumers to this type of products (statement 2). On the other hand, the German experts emphasize the necessity of information concerning the environmental risks of modern biotechnology (statements 27, 30) as well as possible health impacts of these technologies (statements 35, 37, 38, 40).

Importance

In table 3.8 the most important statements for knowledge creation in science and technology are presented. The majority of these statements are developments in plant breeding and production like the creation of basic knowledge in this field, e. g. elucidation of the molecular basis of virus resistance mechanisms in perennial plants or the development of new monitoring and analytical tools based on modern biotechnology (statements 58, 57), as well as the genetic modification of complex regulated characteristics in plants (statements 54, 55, 56).

Additionally, a significant extension of scientific knowledge is expected from the finishing of basic research activities with relevance for food processing like the complete sequencing of genomes of important bacteria or the use of artificial polyfunctional enzymes for analytical applications (statements 44, 51). A high contribution to knowledge creation is expected as well if the long-term health impacts of genetic engineering on consumers, farmers and employees are investigated (statement 41), not least due to the lack of available studies in this field and the needed rather difficult investigation design.

In table 3.9 the statements are presented which are classified as being "very important" for the competitiveness of the German economy. The strongest impacts on the economy are expected if new markets for agriculture and food industry can be made accessible with the help of modern biotechnology. This relates e. g. to the production of pharmaceutical substances with the help of genetically engineered animals and plants ("gene pharming"), optimized renewable resources or genetically engineered vaccines for farm animals (statements 23, 22, 71). In addition, a clear increase in productivity in food processing due to genetically engineered enzymes (statements 25, 49, 50), or the use of plant cell cultures in large-scale bioreactors (statement 59), are regarded as essential for the future competitiveness of the German food industry. Moreover, a strong impact on the German economy is expected if modern biotechnology contributes to the reduction of production costs for agricultural bulk products (statement 18), or the extension of the industrial innovativeness of small and medium-sized companies of the German Agro-Food sector (statement 16).

Table 3.8:   Most important statements for "knowledge creation in science and technology" in Germany

Statement No.	Content	Index
58	The molecular basis of virus resistance mechanisms in economically important perennial plants, like e. g. vine, olive and fruit trees, is elucidated.	0.93
57	New monitoring and control techniques for genetically engineered plants in open fields have improved the reliability of risk assessment significantly.	0.87
62	Rapid test systems based on modern biotechnology are widely used for pathogen identification in plant production and animal husbandry.	0.87
44	The genomes of most economically important bacteria used in food production are completely sequenced.	0.86
56	To prevent resistance against biocides in pathogens, new genetic engineering approaches for plant defence mechanisms have been developed (e. g. combination of different resistance genes, increase in pathogen tolerance).	0.86
41	The long-term health impacts of the use of Agro-Food products made with the help of genetic engineering on consumers, farmers and employees are investigated (e. g. by epidemiological surveys).	0.86
54	Genetic engineering approaches are developed which allow the alteration of polygenic traits in most economically important plant species.	0.84
51	Artificial polyfunctional enzymes are practically used for analytical applications in the Agro-Food sector (e. g. rapid quantitative analyses of metabolic activities or the composition of raw materials).	0.82
55	Genetically engineered plant varieties resistant to salinity and/or drought have improved agricultural productivity in arid environments significantly.	0.81
48	A large variety of genetically engineered microorganisms, which can be precisely controlled and regulated in their metabolic activities during food production processes has been developed.	0.80

Table 3.9:   Most important statements for "competitiveness of economy" in
                Germany

Statement No.	Content	Index
16	The widespread use of modern biotechnology enables small and medium-sized companies of the Agro-Food sector in Germany to introduce a large variety of innovative processes and products.	0.86
59	Plant cell cultures in large-scale bioreactors are widely used for the production of high value components (e. g. pharmaceuticals, fine-chemicals, proteins).	0.86
23	Pharmaceutical substances produced by genetically engineered animals and plants (e. g. proteins, enzymes, hormones, antibodies) achieve a turnover share of at least 5 % of the pharmaceutical market in Germany.	0.85
22	Due to the application of modern biotechnology, farmers produce renewable resources (e. g. biofuel, starch, fatty acids) which are used outside the food sector on 20 % or more of the arable land in Germany.	0.82
49	Enzymes optimized by protein engineering are practically used in specific sectors of the food industry (e. g. starch processing, bakeries, breweries, cheese/dairy production).	0.82
18	The widespread use of modern biotechnology in agriculture reduces the production costs for agricultural bulk products (e. g. cereals, milk) by approximately 30 %.	0.82
71	A large variety of genetically engineered vaccines for farm animals is developed which can reduce the use of other pharmaceuticals in animal husbandry (e. g. antibiotics) significantly.	0.82
25	Enzyme systems are specifically developed to improve the environmental performance of conventional food-processing procedures.	0.81
50	Approximately 90 % of the enzymes used in the food-processing industry are produced by genetically engineered organisms.	0.80
46	Genetically engineered microorganisms and enzyme systems are widely used to improve the processing quality of food (e. g. prolonged shelf life of biological products, synchronization of the ripening process).	0.79

In table 3.10 the most important statements for the protection of the environment and sustainable development in Germany are presented. Significant positive impacts on the environment are expected if modern biotechnology contributes to the transformation or reduction of emissions and waste from animal production (statements 31, 32). The same applies if genetically engineered microorganisms and plants producing biocides are used for biological pest control or if salt-or drought-resistant plant varieties are cultivated in arid areas (statement 55). In the food industry the environmental performance can be improved by the application of specifically developed enzyme systems (statement 25).

Table 3.10:   Most important statements for "protection of environment/ sustainable development" in Germany

Statement No.	Content	Index
32	Modern biotechnology significantly contributes to the transformation of 40 % or more of the organic agricultural waste into marketable products (e. g. energy, secondary raw materials).	0.89
31	Modern biotechnology is widely used to reduce emissions and waste from animal production (e. g. less manure due to enzymes as feed additives, improved anaerobic waste treatment, biofilter).	0.89
29	The widespread cultivation of genetically engineered field crops resistant to herbicides results in an approximate 50 % reduction of environmental pollution in plant production.	0.85
30	Despite the widespread use of methods and techniques related to modern biotechnology, all in all, no negative effect on the maintenance of biodiversity in agriculturally influenced ecosystems can be discovered.	0.83
28	Genetically engineered microorganisms and plants producing biocides are widely used for biological pest control.	0.80
57	New monitoring and control techniques for genetically engineered plants in open fields have improved the reliability of risk assessment significantly.	0.78
26	The widespread deliberate release of genetically engineered organisms results in a significant transfer and recombination of the introduced genes with unintended negative impacts (e. g. development of resistances to herbicides and pathogens in weeds).	0.75
25	Enzyme systems are specifically developed to improve the environmental performance of conventional food-processing procedures.	0.73
55	Genetically engineered plant varieties resistant to salinity and/or drought have improved agricultural productivity in arid environments significantly.	0.63
56	To prevent resistance against biocides in pathogens, new genetic engineering approaches for plant defence mechanisms have been developed (e. g. combination of different resistance genes, increase in pathogen tolerance).	0.62

At least in three of the ten statements presented in table 3.10, the question arises whether the majority of the German experts has assessed the sub-category "importance of the statement to protection of environment/sustainable development" according to the instructions given in the questionnaire (see chapter 2). This relates to statement 29, describing a 50 % reduction of environmental pollution in plant production due to the cultivation of herbicide-resistant field crops, as well as to statement 30, in which no negative effects of the maintenance of biodiversity in agriculturally influenced ecosystems, despite the use of modern biotechnology, is mentioned. In addition, statement 26 (describing the deliberate release of genetically engineered organisms results in the transfer and recombination of the introduced genes with unintended negative impacts) has to be regarded rather critically in this context. In the latter case a negative impact of genetic engineering on the environment is described, which has obviously a certain impact on the environmental situation, but not in a positive way. Related to the other two statements, more than 20 % of the German experts expect that the described developments will not be realized at all. In addition, the comments of the German experts document that at least some of them doubt that there will be a positive environmental impact of herbicide-resistant plants and that modern biotechnology will not damage biodiversity in agriculturally influenced ecosystems. With these aspects in mind, it seems to be likely that most of the German respondents see high relevance of these statements for the environmental situation, but some at least expect an impact opposite to the description in the questionnaire.

Particuliarities of the results in Germany

In general, the public debate about the use of modern biotechnology in the Agro-Food sector in Germany is still characterized by rather polarized positions between the supporters and opponents of this technology. On the one hand, there are rather euphoric estimations and expectations of the German government and some representatives of science and industry, on the other hand there are rather sceptical and critical viewpoints especially from consumer organizations, churches and environmental groups. Some examples might highlight the general positions of the two different groups.

The German government has announced the target that Germany shall become the number one in modern biotechnology and genetic engineering in Europe until 2000. Even if in this context major activities are focused on the medical field, the German government has rather high expectations in the use of these technologies in the Agro-Food sector as well. In the year 2000 a worldwide turnover of 170 bill. DM with genetically engineered products is expected, of which around 40 % apply to agriculture and food processing (BML 1997b). According to the estimation of the German government, the Agro-Food sector will become the most important application area of modern biotechnology and genetic engineering during this time frame in the European Union. In this context it is expected that new employment possibilities will emerge in Germany as well (BML 1997b, BMBF 1997).

Table 3.11:   Attitude of the German population to the use of genetic engineering
                in different application areas

Application area of genetic engineering	Personal attitude					
	very good	rather good	neutral	rather bad	very bad	no answer
Production of vaccines	32 %	31 %	21 %	6 %	5 %	5 %
Clinical diagnosis	45 %	29 %	16 %	4 %	4 %	3 %
Treatment of cell diseases	38 %	32 %	18 %	5 %	4 %	3 %
Prenatal diagnostics	29 %	25 %	21 %	10 %	12 %	3 %
Food processing	2 %	6 %	16 %	22 %	54 %	1 %
Improved yields in plants	6 %	14 %	22 %	22 %	35 %	1 %
Resistance in plants	13 %	23 %	29 %	16 %	17 %	1 %
Gene transfer among animals	2 %	4 %	10 %	17 %	66 %	1 %

Source: Hampel et al. 1997

According to population surveys, the application of modern biotechnology and
especially of genetic engineering in the Agro-Food sector is still regarded rather
critically in Germany. This emerges from a representative population poll among
1,500 citizens in Germany during spring 1997. The results indicating the attitude of
the use of genetic engineering in different application areas are shown in table 3.11.
While the use of genetic engineering is mostly accepted in the medical areas, there
is still a strong opposition in the German population to the use of this technology in
the Agro-Food sector. This relates in particular to the gene transfer among animals
and - to a smaller extent - to the application of genetic engineering in food
processing. In contrast, the use of this technology in plant production meets less
opposition, especially, if resistance mechanisms are concerned (see table 3.11).

According to the Eurobarometer-survey carried out on behalf of the Commission in
the EU, no significant change can be registered in the attitude of the German
population regarding the use of modern biotechnology since the beginning of the
90s. In all three surveys carried out in this time period, only around 40 % of the
Germans expect clear benefits from biotechnology, compared to an EU average of
about 50 % to 55 % (European Commission 1997). In contrast, the personal attitude
concerning the use of genetic engineering takes the form of an U-relationship since
1990. After a noticeable slump between 1991 and 1993, the percentage of optimists
rose again between 1993 and 1996 but still reached only 32 % agreement in
Germany compared to an EU average of 43 % (European Commission 1997). In
total, it can be concluded that there is still a lot of scepticism and hesitation among

the German population, especially concerning the use of genetic engineering approaches in agriculture and food processing. At least for this application area the "positive development" of biotechnology/genetic engineering which has been announced by the German government has not yet reached the average citizen in Germany. Qualitative indicators in this context are the on-going resistance of different organizations against field trials with genetically modified plants (Mühlenberg 1997a, b) as well as the intensive public debate which followed the cloning of "Dolly" in the year 1997.

In contrast to the described public debate, the results obtained in the Delphi survey convey a more differentiated picture of the estimation in Germany. Even if it is considered that not average citizens but experts have been participating in the Delphi survey, the results give empirical evidence to the thesis that there are many more similarities and common estimations between the supporters and critics of the so-called "Green biotechnology" in Germany than are apparent in the public debate. This is not only valid for areas with no significant differences in the personal attitude of the different expert groups (like extended information activities concerning modern biotechnology, strong demand for monitoring activities, rejection of negative health impacts of modern biotechnology or high appreciation if modern biotechnology contributes to improve the environmental situation) (see table 3.4), but also for the assessment of the economic impacts of biotechnology. This relates in particular to a possible job decrease in the Agro-Food sector due to the application of modern biotechnology, which is rejected by a great majority of the German experts. In addition, only little chance is seen for additional employment opportunities based on this technology in the food industry. Furthermore, the German experts have rather high expectations in the use of modern biotechnology for non-food purposes. This relates both to renewable resources for energy and chemical applications as well as the production of pharmaceuticals with the help of transgenic animals and plants. In contrast, considerable market shares of gene food products in Germany, as well as reduced EU imports of fish products due to the application of modern biotechnology are differently assessed by the German experts.

The health impacts of modern biotechnology are assessed rather differentiatedly by the German experts as well. The German panel strongly rejects health-related risks of modern biotechnology in food production and processing, resulting in additional allergies in persons actively involved in these processes, as well as the occurrence of unintended impacts on the health of consumers. Despite their personal attitude they expect that negative impacts most probably will occur in around six to ten years. On the other hand, future developments, in which modern biotechnology contributes to positive health effects (like e. g. offering of food, of which the allergenic potential has been decreased) are differently assessed by the German experts who expect a short to medium-term realization, independent of the expressed personal attitude.

Specific information of consumers about health relevant effects of gene food by food companies as well as the investigation of long-term health impacts of these products on consumers, farmers and employees are positively assessed by a broad majority of the German experts from all the included different groups. Most of them expect a short-term realization of information activities by food companies, whereas epidemiological surveys are expected to be realized later. In both cases a considerable proportion of the German experts sees no chance of realization, mostly due to lack of interest of the food industry or political institutions.

The regulation activities related to gene food products are rather controversially assessed by the German experts. Compulsory labelling of all gene food products, an official case-by-case procedure for market approval of these products, the consideration of their social impacts and potential ethical conflicts during this process, as well as the use of efficient control techniques for these products, are positively assessed by the majority of the German experts. While the majority expects a short to medium-term realization of these developments, up to one third of them (mainly consumers and critics) regard these developments as unrealistic. Specific initiatives to involve the public in the debate and decision-making process on modern biotechnology to a higher extent are highly appreciated in Germany and expected to be realized in short term. In contrast, a higher attractiveness of the EU for companies located in non-member countries due to the uniform implementation of EU biosafety directives is regarded as a rather unrealistic scenario by most of the German experts.

3.2   Greece

In Greece agriculture still represents a main sector of economy with 18.7 % of the active population of Greece employed here, and its contribution to the gross national product amounts to about 13 %. In recent years Greek agriculture has faced cyclical fluctuations and reached its best point in the decade of the 70s. Since many of its structural and institutional problems have not been solved yet, Greek agriculture has reached its limits in terms of productivity and competitiveness. This is due to the following major constraints/challenges which it has to face:

• Low pace at the accumulation of capital

• Technological insufficiency in agriculture's infrastructure

• No satisfactory diffusion of the agricultural products (at vertical and horizontal level)

• Missing restructuring of the agricultural sector in order to meet the demand of the food industry in Greece

• No tradition in research projects (especially in the field of biotechnology)

The food industry is also an important sector of the national economy in Greece for the following reasons:

- Food industrial companies achieve relatively high turnover share and profits
- Other sectors of the economy depend on them
- There is a remarkable appearance of food companies among the ten strongest companies in Greece

Despite their successful image they have to face some pressures such as antagonists within the context of globalized economy. They react to this challenge with cooperations and mergers. In addition, they intend to adapt their marketing strategy in direction of final products (name origin, "high aesthetic value") as well as new, innovative products.

In biotechnology, there is a considerable amount of scientific collaboration between research laboratories and other scientific institutions of the Hellenic and international arena. The researchers in this field are competitive also on the international level, mainly due to Greek scientists who worked in the USA and returned to Greece. However, this knowledge is not articulated to the Greek economy which is due among other reasons to problems of the Greek industry in adopting and using this knowledge. At present, the public debate about modern biotechnology is not of considerable significance (not a "heated debate") since many achievements in this field are neither produced nor applied in Greece. Recently there was an important coverage of biotechnology issues by mass media. Considering this debate we can assume the following:

- On the one side, there are those experts (mainly specialists from the various biological disciplines) who have generally a positive attitude towards biotechnological progress.
- On the other side, the majority of other experts express a general hesitation towards the application of modern biotechnology. Their concerns are not sharply focused to a specific issue. Sometimes they refer to something very general (like the "over-industrialization" of life itself), or they imply very specific topics, like hesitations concerning the safety of a particular product or process.

Degree of knowledge

Concerning the experts' self-assessment of knowledge, the prevailing category is the "less familiar" (mean value of answers: > 50 %). The sub-category "not familiar" is the second in class, percentages of answers per statement between 7 % to 77 %. Very few experts assess themselves as being "very familiar" with the statements of the questionnaire (percentages of answers per statement between 3 % and 5 %). The domain "environment" followed by "acceptance" includes statements, which the experts feel slightly more familiar with, whereas the "science and technology" domain include statements with which the experts, in general, feel less familiar.

Personal attitude

Concerning the "personal attitude", 46 out of 71 statements of the questionnaire (equalling 65 % of the total) are assumed "positively" by the Greek experts. Among those there are 37 statements, of which at least 70 % of the experts appreciate the situation/development mentioned in the statement. The "top 10" statements, for which the experts express the highest level of "positive attitude" (percent values between 93 % and 98 %) belong to the "regulation" domain. The statement which got most answers of "positive attitude" (98 %) comes from the "acceptance" domain and has a strong "informative" character (statement 1: a governmental institution which provides information on modern biotechnology in the Agro-Food sector to all interested persons and institutions is established in Greece).

The "top 10" statements, for which experts express the strongest "negative attitude" belong to the "health", "economy" and "science and technology" domains. They mainly refer to unintended, uncontrolled and/or unidentifiable negative health effects (e. g. allergies) and environmental impacts of modern biotechnology. The other group of statements, for which experts express very "negative attitudes" belong to the "economy" domain. More precisely, those statements concern either the reduction of jobs or the low capabilities specific groups have to handle the new situation. The loss of biodiversity of traditional Greek organisms used in agriculture is another statement which experts reject to a high extent. Finally, experts express remarkably strong negative attitudes towards the statements concerning genetic engineering of farm animals.

The sub-category of "indifferent attitude" is the weakest (percent values between 3 % and 15 %). It is worth noting that no statement of the "environment" or the "health" domain is included among them. The "not able to express" answers are extremely limited in the category "personal attitude".

All in all, it could be said that the biotechnological issues included in the questionnaire strongly motivate the experts and make them react with strong attitudes. Moreover, Greek experts show no "indifferent attitude" to statements concerning "environmental" and "health" issues, are generally highly sceptic about statements of ecological and economic content, and are in favour of statements referring to legislation/regulation, public acceptance, information and health-improving as well as hygiene monitoring themes.

Two other facts deriving from the Delphi survey underline that "application of modern biotechnology in the Agro-Food sector" (still) creates and faces strong attitudes:

- The high "rejection" factor (or, equally, the low response rate) that the Delphi questionnaire had from experts of strong theological or religious concerns, pure ecological or strong environmental activities.

- The stability of answers given by the experts during the two rounds of the survey. The modifications concerning either the general picture of "attitudes" or the ranking series of the statements between the two rounds are incremental.

Time of realization

The most remarkable point of the answers concerning "time of realization" is that no statement is expected to "happen" in the long-term horizon, or never in Greece. 69 out of 71 statements included in the questionnaire are expected to be realized within the following ten years. Moreover, 17 statements (24 % of the total) are expected to be realized within the following five years, if not realized already.

The statements with short-time realization "in the following five years" have the following conceptual characteristics:

- Function within and towards the enforcement of the triangle: information about modern biotechnology (especially adverse health effects), protection of health (methods and measurements), and regulation
- Reference to the practical or the widespread use of modern biotechnology on field crops and farm animals
- Reference to dairy products

Yet, it is also worth noting that, even though the time-horizon was already short during the first round, it became even shorter during the second round of the survey. The general "mobility" of answers during the two rounds was very limited as far as the other answering categories are concerned, except the "time of realization".

Taking into account that strong public attention of "biotechnology application in the Agro-Food sector" occurred in-between the two rounds of the survey (articles in newspapers, magazines, references in TV related to "Dolly" or Monsanto's herbicide-resistant soybean etc.), it can be assumed conclusively that this "shortening" in time of realization is mainly due to the aforesaid public interest in "Green" biotechnology.

Another specific Greek response pattern arose concerning statements dealing with information aspects. Accordingly, Greek experts react in a rather reserved way concerning the realization of detailed information of the consumers about specific health relevant effects of food made with the help of genetic engineering by food companies (statement 35).

Influential factors

The prevailing influential factors, on the basis of mean percentual values of answers given to each statement, per domain of interest and totally, are "R & D infrastructure" followed by "funding" and "personnel". In contrast, "industrial innovativeness" and "international collaboration" are seen of minor importance. These two categories are given the lowest percentual mean values of answers both totally and per domain of interest, and are never included among the "top 3" influential factors for any statement (see table 3.12). Thus, they may be assessed as the "weakest" influential factors among those examined in the Delphi survey.

According to table 13.12 it is worth noting that:

- The three prevailing parameters do not necessarily form analogous patterns of combinable factors for the various statements or domains of interest, even though most of them appear as strong couples in many statements and as couples of prevailing factors in many "domains of interest".

- "R&D infrastructure" may be considered as the dominant influential factor, as it is included among the three prevailing factors in 54 (out of 71) statements.

Table 3.12:   Three prevailing influential factors per statement in Greece

Domain	State-ment No.	R&D infrastructure	Person-nel	Tech-nology transfer	Regu-lation	Mar-ket	Ecol-ogy	Fun-ding	Poli-cy	Accep-tance	Infor-mation
	1	X						X	X		
	2					X				X	X
	3					X				X	X
Accep-tance	4	X						X		X	
	5		X					X			X
	6				X	X					X
	7					X				X	X
	8		X		X	X					
	9	X			X	X					
Regu-lation	10				X	X					X
	11	X	X					X			
	12								X	X	X
	13				X	X			X		
	14	X				X				X	
	15					X				X	X
	16	X	X					X			
	17	X	X					X			
	18	X		X				X			
Econo-my	19					X		X	X		
	20		X						X	X	
	21					X		X		X	
	22	X	X					X			
	23	X	X		X						
	24					X		X	X		

Domain	State-ment No.	R&D infrastru cture	Person-nel	Tech-nology transfer	Regu-lation	Mar-ket	Ecol-ogy	Fun-ding	Poli-cy	Accep-tance	Infor-mation
Environ-ment	25	X			X			X			
	26	X					X			X	
	27						X			X	X
	28	X	X					X			
	29	X		X				X			
	30	X					X				X
	31	X		X				X			
	32	X		X				X			
	33	X	X					X			
	34						X			X	X
Health	35					X				X	X
	36	X	X								X
	37	X	X		X						X
	38	X		X		X					
	39	X		X				X			
	40	X			X						X
	41	X	X					X			
Science & Technol-ogy: Food pro-cessing	42	X	X					X			
	43	X	X					X			
	44	X	X					X			
	45	X				X				X	
	46	X		X				X			
	47	X				X				X	
	48	X	X	X							
	49	X	X	X				X			
	50	X		X				X			
	51	X	X	X							
Science & Technol-ogy: Plant pro-duction	52	X		X				X			
	53	X	X					X			
	54	X	X					X			
	55	X		X				X			
	56	X		X				X			
	57	X	X					X			
	58	X	X					X			
	59	X		X				X			
	60	X		X				X			
	61	X	X					X			
	62	X	X	X							
Science & technol-ogy: Animal pro-duction	63	X						X		X	
	64	X	X				X			X	
	65	X						X		X	
	66	X		X				X			
	67	X					X			X	
	68	X	X					X			
	69	X	X					X			
	70	X						X		X	
	71	X	X					X			

Importance of the statements

First of all, it has to be noted that this answering category is not easily perceived by the experts: in their comments some of them ask whether "importance" means "crucial for further development" or "important, but with serious negative side effects". Thus, some confusion accompanies the answers given by the experts: a lot of experts regard "important" as "critical for the development", while others see it as "strongly effective". Nevertheless, it can be assumed that most experts answered this category according to the instructions given in the questionnaire.

Generally, according to the mean total value, the biotechnological issues (statements of the questionnaires) are not of very high importance according to the views of the Greek experts. Especially, for the "improvement of the environment/sustainable development", the aforesaid conclusion is pushed towards the extremes. In this case, the mean total value is nearly zero (0.04) (see table 3.13). Moreover, Greek experts rank the three fields of reference in the following series of "importance": "knowledge creation in science and technology " > "competitiveness of economy" > "environmental protection and sustainability" (see table 3.13).

Table 3.13:   Overall importance of the statements in Greece

	Knowledge creation in science and technology	Competitiveness of national economy	Environmental protection/ Sustainable development
Mean value	0.44	0.33	0.04
Maximum value	0.93	0.81	0.80
Minimum value	-0.39	-0.61	-0.49

This ranking is confirmed by the values of the "importance-indexes" given to the prevailing statements for each field. More analytically, the "importance-index" value for the ten prevailing statements concerning the "knowledge creation" in the "science and technology" field, fluctuates from 0.93 to 0.81, the analogous values for the "top 10" statements concerning the "competitiveness of economy", fluctuate from 0.81 to 0.64 and the relevant values for the prevailing statements referring to the "environmental protection and sustainability" range from 0.80 to 0.52.

Differences between expert groups

Strong differences in assessing and reacting towards the biotechnological issues covered in the Delphi questionnaire do not appear among the Greek expert groups. Yet, despite the lack of strong differences, three "levels" of attitude have been located. Generally, the "researchers" feel more comfortable, the "biocritics" more sceptical, and the "consumers" are located in-between. Yet, the fact that most answers given originate from a relatively shallow depth of knowledge, brings the following delicate issue to the surface: it is possible that the lack of strong differences is due to the remarkable distance that both the various groups and the whole corpus of experts feel they have in relation to the issues included in the questionnaire.

Moreover, the fact that the participation of the various expert groups in the total sample has been strongly unbalanced, because from two expert groups (farmers, industry) not enough answers could be collected in order to produce statistically reliable results (and therefore are excluded from a separate and specific analysis), support the appearance of higher "homogeneity" and "uniformity" between the overall results and the results devising from the prevailing groups, diminish the range of possible differentiating results and decrease the relevant range of possible comparisons (that, if realized, might have led to distinctive characteristics of the expert groups).

A possible third reason why strong differences do not appear, is the character and the structure of the Delphi questionnaire: it has been already mentioned that experts of either strong theological or ecological concerns refused to answer. Some others, although they answered, express their hesitations about the scope of the survey (notes included in the "comments" which assess the role of the survey as a preliminary stage of biotechnology-promotion in the Agro-Food sector, are characteristic), and their difficulties in expressing these hesitations within the questionnaire. Thus, some kind of process that smooths out the oppositions/objections seems to have penetrated the questionnaire.

Domains of interest

The various "domains of interest" (acceptances, regulation, economy, environment, health, science and technology) are unequally answered. The domains "environment" and "acceptance" include, on average, the highest number of answered statements - whereas "science and technology" is the domain which is least answered. It is worth noting that the experts express, generally, "positive attitudes" towards the referred statements included in the domain "regulation". Strong negative attitudes are mostly met in statements included in the domains "economy", "environment" and "health". Statements for which the experts seem to have "contradictory" attitudes are mostly found in the domain "science and technology". On the other hand, no such statements are located in the "regulation" domain and only one statement (with this characteristic) in the "economy" domain.

Even though time of realization does not appear to be strictly linked to the domain "origination of the statements", it can be observed that the "acceptance" and the "science and technology" domain include the relatively highest number of statements, for which the realization horizon is relatively short. In contrast, a relatively high number of statements of the "economy" domain are expected to be realized in the long term.

Particularities of the results in Greece

This survey reveals the following Greek specificities which are considered from a sociological viewpoint as well:

- Greek experts react towards the current survey with a pronounced attitude or otherwise the expert groups "participate" in the survey with strong and concrete attitudes. In this sense, the total absence of certain social groups may be easily understood.

- This strong attitude has often a lack of strong goals of positive importance - strong goals which might have been either the overall sum of minor goals or a dominant goal itself. Thus, strong attitudes seem either to remain "unfounded" or represent comprehensive protest. In other words, the challenge of biotechnology application is not recognized as a factor capable of offering integration and orientation of social action.

- Yet, various rationalities of minor level/range of reference/application have been developed. The inter-science and inter-technological rationality mostly acknowledged and supported by the "researchers" seems to be the most clear one. Generally, as the field of examination moves towards less autonomous or interlimited areas of social activity, rationalities become more vague and unstable.

- Finally, three elements seem to us very significant for the status of modern biotechnology as a social issue in Greece:
  - The features of the issue examined, where possible conflicts or competitions among social actors might have been raised, are not clearly defined by this study remaining, thus, hidden or even not formulated.
  - The lack of information concerning the two "productive" social groups (industry, farmers) may have led possibly to an analogous lack of locating such features. Nevertheless, if these features are essential for the expert groups' self-determination such a "lack" would not have appeared.
  - On the other hand, we can assume that the lack itself may be well assessed as the non-existence of such features of common interest and interaction, meaning that the application of modern biotechnology in the Agro-Food sector is not located yet within the framework of social interactivity or it is not a strong pole for social interactivity.

Biotechnology in the Agro-Food sector is still a social issue addressed by each "individual" social group in a quite "lonely" direction of "self-orientation" towards its socially non-determined potential. That is to say, the positions are not taken in relation, in consideration of the positions of the others, the positions do not yet have a strategic character. This lack of comprehensiveness of biotechnology in the Agro-Food sector as a social object is strongly linked to its lack of sufficient inter-societal origination "rationale" - in this consists the Greek specificity.

In conclusion, we can say that the application of modern biotechnology in the Agro-Food sector is not yet formed as a "social object" - has not yet gained the structural capacity to be recognized as a social object - among the Greek social actors.

3.3   Italy

Degree of knowledge

The number of answers per statement is useful to understand the familiarity of the respondents with the concerned issues. The lowest rates of answers are obviously found for statements with a very specific technical development, such as statements 51, 60 and for statements related to aquaculture and fish production (statements 21, 67, 68, 69), (see table 3.14). The Italian respondents appear very unfamiliar with all the issues related to fish. This seems to indicate a delay in this specific research area in Italy, but also a general lack of knowledge about fish production. In addition, low rates of answers are observed for statements involving quantitative forecasts of future economic and market developments (statements 14, 15, 20, 22, 23, 24, 38). A low rate of answers may also indicate that the issue addressed is a particularly sensitive one. This is likely to be true for statements 37 (allergies in workers in biotechnological plants) and 40 (unintended negative impacts on consumers' health). The rates of answers to these statements are very low, even among the researchers.

Table 3.14:   Statements with low numbers of answers in Italy

Statement No.	Content	No. of answers
67	New cell-biological methods are developed for the production of polyploid fish.	52
68	Genetic maps and molecular test systems are practically used for the identification of economically important traits and marker-assisted breeding of fish.	56
60	The content of erucic acid in rapeseed reaches 65 % or more due to breeding systems based on genetic engineering.	58

Statement No.	Content	No. of answers
37	The increased use of modern biotechnology in food production and processing results in additional allergies in people actively involved in these processes.	60
21	Due to the widespread use of modern biotechnology in EU fish breeding and aquaculture the imports of fish and fish products from outside the EU are significantly reduced.	65
51	Artificial polyfunctional enzymes are practically used for analytical applications in the Agro-Food sector (e.g. rapid quantitative analyses of metabolic activities or the composition of raw materials).	66
7	A restaurant chain or catering service specialized in offering trendy meals with genetically engineered ingredients open branches in almost all large cities in Italy.	69
69	Monogenic traits are specifically and stably altered in fish important for aquaculture by genetic engineering as a matter of routine (e. g. anti-freeze gene, growth hormone gene).	69
45	In Italy most of the beer is produced with genetically engineered yeast.	72
47	Modern biotechnology has failed in producing food and beverages satisfying traditional taste preferences of large consumer groups in Italy.	75
66	The elucidation of the principles underlying the developmental processes of oocytes leads to the genetic engineering of early stages of embryo cells from farm animals.	79
61	After years of experimentation with genetic engineering, the majority of plant breeders strongly prefers the combination of marker-assisted breeding with traditional breeding methods, compared to the use of genetically modified plants.	79

Personal attitude

As indicated in figure 3.5, the Italian respondents show mostly a positive attitude towards the majority of the statements. Over 63 % of the developments are highly appreciated, while only 14 % of the statements are negatively assessed. This basic attitude pattern is observed for all the expert groups without major differences (see figure 3.6).

All expert groups seem aware that the level of public information and debate concerning modern biotechnology in Italy is not sufficient, and generally appreciate the statements concerning initiatives aimed at improving this situation (statements 1, 5, 8, 10, 12). In this context, the majority of Italian experts expresses a positive attitude toward the creation of ad hoc institutions and bodies (statements 1, 3, 5). A majority (though not so clearly: 59 % of the experts appreciate this statement) seems persuaded that the provision of more and better public information would improve the acceptance of Agro-Food biotechnology in Italy (statement 3).

Figure 3.5:   Personal attitude of the Italian respondents

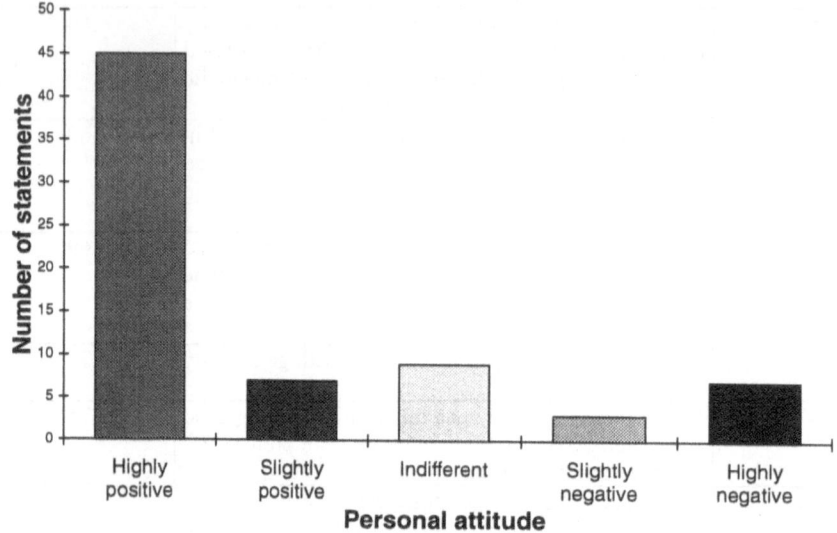

A rather unexpected finding concerns the broad consensus (75 % "positive attitude") expressed towards statement 10 (consideration of social and ethical implications in marketing approval procedures of gene food products). This indicates that the awareness that Agro-Food biotechnology has implications which go beyond its technical aspects is relatively widespread among all the concerned groups.

This favourable attitude toward public information seems also to prevail as regards labelling, even among the members of the group "industry" (statement 8). All in all, a widespread need of openness and transparency in relation to modern biotechnology is expressed by the Italian respondents. In this context, one of the main worries expressed in their comments concerns the neutrality of the institutional bodies created to promote public information and the unbiasedness of the information provided. Public confidence in the trustworthiness of public institutions seems therefore not very high and this could affect the efficacy of any new policy action undertaken in this direction.

An important and not anticipated finding concerns the positive attitude expressed by the group "industry" toward regulatory developments. Such a result, associated by the high importance assigned by this group to the influential factor "regulation" compared to the other groups, suggests that a clear and certain regulatory framework is seen as a positive development (a sort of guarantee), rather than as a constraint by the Italian industry. In fact, until the implementation of directives 219/90 and 220/90 in 1993, industrial biotech R&D activities in Italy were seriously affected by regulatory uncertainty, which caused a virtual moratory of field trials with genetically modified plants.

Significant differences are observed between the Italian expert groups concerning the assessment of many statements related to the degree of penetration and diffusion of genetic engineering in food production (statements 2, 3, 14, 19, 33, 47 regarding general developments, and statements 46, 48, 50, 54, 61, 63, 64, 67, 69, 70 regarding specific applications to microorganisms, plants and animals). The group "consumers and critics" appears clearly less favourable to the use of genetic engineering in food production and processing by comparison with the group "industry".

Figure 3.6:   Personal attitude of the respondents by expert groups in Italy[1]

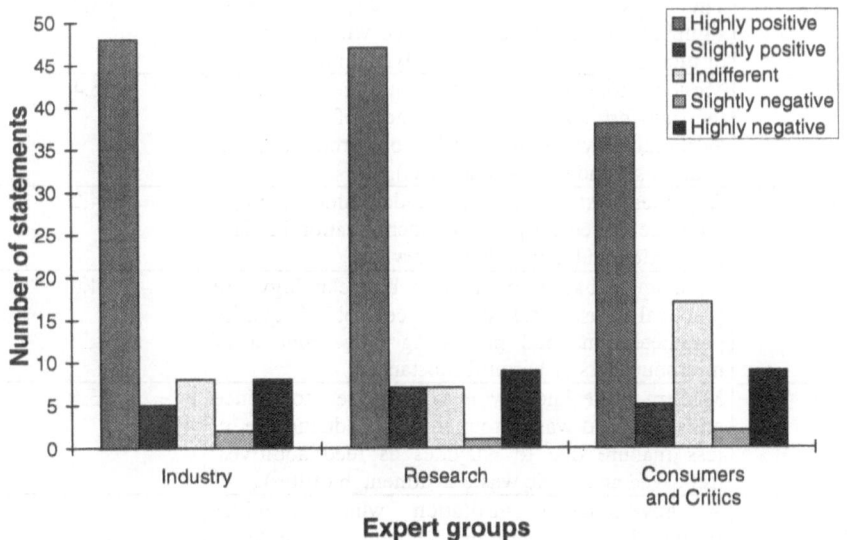

---

[1] Consumers and critics originally formed two separate expert groups. However, due to the insufficient number of respondents in these groups in Italy, it was necessary to aggregate the two groups to create a larger group including those respondents who show a more critical attitude toward food biotechnology than the general public.

Among the developments most appreciated in Italy (see table 3.15) are two statements concerning the improvement of public information and of scientific knowledge on the risks of food biotechnology (statements 1, 41). This request for adequate research and information seems to suggest that the generally positive attitude of Italian experts towards Agro-Food biotechnology is accompanied by widespread perception of possible risks.

Table 3.15:   "Top 10" statements for "positive attitude" in Italy

Statement No.	Content	Proportion of "positive attitude" (%)
58	The molecular basis of virus resistance mechanisms in economically important perennial plants like e. g. vine, olive and fruit trees is elucidated.	99.0
41	The long-term health impacts of the use of Agro-Food products made with the help of genetic engineering on consumers, farmers and employees are investigated (e. g. by epidemiological surveys).	98.0
25	Enzyme systems are specifically developed to improve the environmental performance of conventional food-processing procedures.	97.3
42	The hygiene monitoring of food processing is significantly improved due to the widespread use of modern biotechnological analytical methods.	95.7
32	Modern biotechnology significantly contributes to the transformation of 40 % or more of the organic agricultural waste into marketable products (e. g. energy, secondary raw materials).	95.4
62	Rapid test systems based on modern biotechnology are widely used for pathogen identification in plant production and animal husbandry.	94.2
43	Techniques based on modern biotechnology are practically used for on-line control of quality parameters in food processing (e. g. content of micronutrients or harmful substances).	94.0
31	Modern biotechnology is widely used to reduce emissions and waste from animal production (e. g. less manure due to enzymes as feed additives, improved anaerobic waste treatment, biofilter).	93.5
1	A governmental institution which provides information on modern biotechnology in the Agro-Food sector to all interested persons and institutions is established in Italy.	90.7
59	Plant cell cultures in large-scale bioreactors are widely used for production of high value components (e. g. pharmaceuticals, fine chemicals, proteins).	89.5

The other highly appreciated statements concern

- An important result in basic ("pure") scientific research (statement 58),
- Applications of modern biotechnology implying some environmental benefit in food processing and animal husbandry (statements 25, 31, 32),
- Analytical applications (statements 42, 43),
- Non-food applications of modern biotechnology (statement 59).

As concerns highly rejected statements, of course the most negative attitudes (see table 3.16) are expressed with regard to negative environmental effects (statements 26, 34), negative health impacts (statements 37, 40) and negative effects on employment and small farmers (statements 20, 24, 19). Also a possible increase in the price of food products due to improvements related to application of biotechnology is negatively evaluated by the majority of respondents (statement 15).

It is interesting to observe that the Italian respondents also reject that food products of biotechnology may be requested by consumers for futile reasons (statement 7), while food applications involving some real benefit for consumers' health are positively evaluated (statement 38). This seems to indicate an implicit need that the applications of biotechnology and genetic engineering in food production are justified on the basis of their usefulness. In fact, pharmaceutical/veterinary applications (which are likely to be considered as more justified than food applications) have a very high level of acceptance (76 % of "positive attitude" for statement 23, 86 % of "positive attitude" for statement 71).

The developments concerning the "habituation" of consumers to gene food products in general (statements 2, 7, 14, 45, 47) are precisely those with the highest score for "indifferent attitude" in Italy (see table 3.17). This suggests that on the one hand the introduction of novel foods in Italy does not arouse any enthusiasm, on the other hand it is not likely to meet any strong opposition. It is rather seen as something inevitable. This sense of "fatalism" and inevitability emerges most clearly for statement 65 (cloning of farm animals): though negative attitudes are prevailing over positive attitudes (35 % vs. 34 %), only 5 % of respondents believe that such a development will never be realized, while 42 % believe that it will be realized in five years, and 38 % think that it will be realized in six to ten years.

Table 3.16:   "Top 10" statements for "negative attitude" in Italy

Statement No.	Content	Proportion of "negative attitude" (%)
26	The widespread deliberate release of genetically engineered organisms results in a significant transfer and recombination of the introduced genes with unintended negative impacts (e. g. development of resistance to herbicides and pathogens in weeds)	89.6
37	The increased use of modern biotechnology in food production and processing results in additional allergies in people actively involved in these processes.	86.4
40	Unintended impacts on the health of consumers occur in some food products or production processes due to the use of genetic engineering during the food production process.	84.8
34	The widespread use of genetically engineered organisms in food production aggravates the loss of traditionally used organisms in Italy significantly.	83.8
20	The widespread use of modern biotechnology in animal and plant production leads to an approximate 30 % decrease in traditional jobs in agriculture.	75.6
24	Due to the hesitant application of genetic engineering (compared to other EU member states) the Agro-Food industry cuts 10 % or more of the jobs in this sector in Italy.	74.7
19	Small farmers in Italy are not able to afford new genetically engineered plants and animals.	72.3
15	The prices of food products in specific market segments, of which the quality is significantly improved by the application of modern biotechnology are, at least, 30 % higher than that of corresponding conventional products.	55.7
4	Research on genetic engineering of farm animals is not funded by public institutions in Italy due to ethical reasons (e. g. animal rights, animal welfare, preservation of the Creation).	55.0
7	A restaurant chain or catering service specialized in offering trendy meals with genetically engineered ingredients open branches in almost all large cities in Italy.	52.2

Table 3.17:   "Top 5" statements for "indifferent attitude" in Italy

Statement No.	Content	Proportion of "indifferent attitude" (%)
45	In Italy most of the beer is produced with genetically engineered yeast.	72.2
47	Modern biotechnology has failed to produce food and beverages satisfying traditional taste preferences of large consumer groups in Italy.	49.3
14	Food made with the help of genetic engineering achieve a turnover share of 30 % or more of all food consumed in Italy.	45.8
2	Most consumers in Italy have quickly got used to all kinds of food and beverages produced with the help of genetic engineering.	37.5
7	A restaurant chain or catering service specialized in offering trendy meals with genetically engineered ingredients open branches in almost all large cities in Italy.	34.8

Taken all together, the acceptance of the applications of Agro-Food biotechnology, as expressed by the personal attitudes of the respondents in Italy, is decreasing according to the following classification:

• Diagnostic/analytical applications

• Genetic modification of microorganisms/enzymes

• Genetic modification of plants

• Genetic modification of animals

Time of realization

According to the estimation of the Italian experts, the majority of the developments will take place in the medium term (six to ten years) (see figure 3.7). In particular, virtually all the applications related to genetic modification of plants and animals fall in this time category.

The developments foreseen in the short term (within five years) are mostly in the following areas:

• Institutional and regulatory initiatives (statements 6, 8, 9, 12 and 35). On the whole, legal and institutional activities (in particular, the establishment of public and private standards concerning labelling) seem to be considered as imminent, both in Italy and within the EU.

• Use of genetically engineered enzymes/microorganisms in food industry (statements 38, 49). In addition, other similar developments (statements 45, 46), though included in the second class, are expected to take place in five years by a significant share of respondents (35 %).

• Analytical applications of modern biotechnology (statements 42, 43, 62).

Figure 3.7:   Time of realization of the statements in Italy (according to the median class)

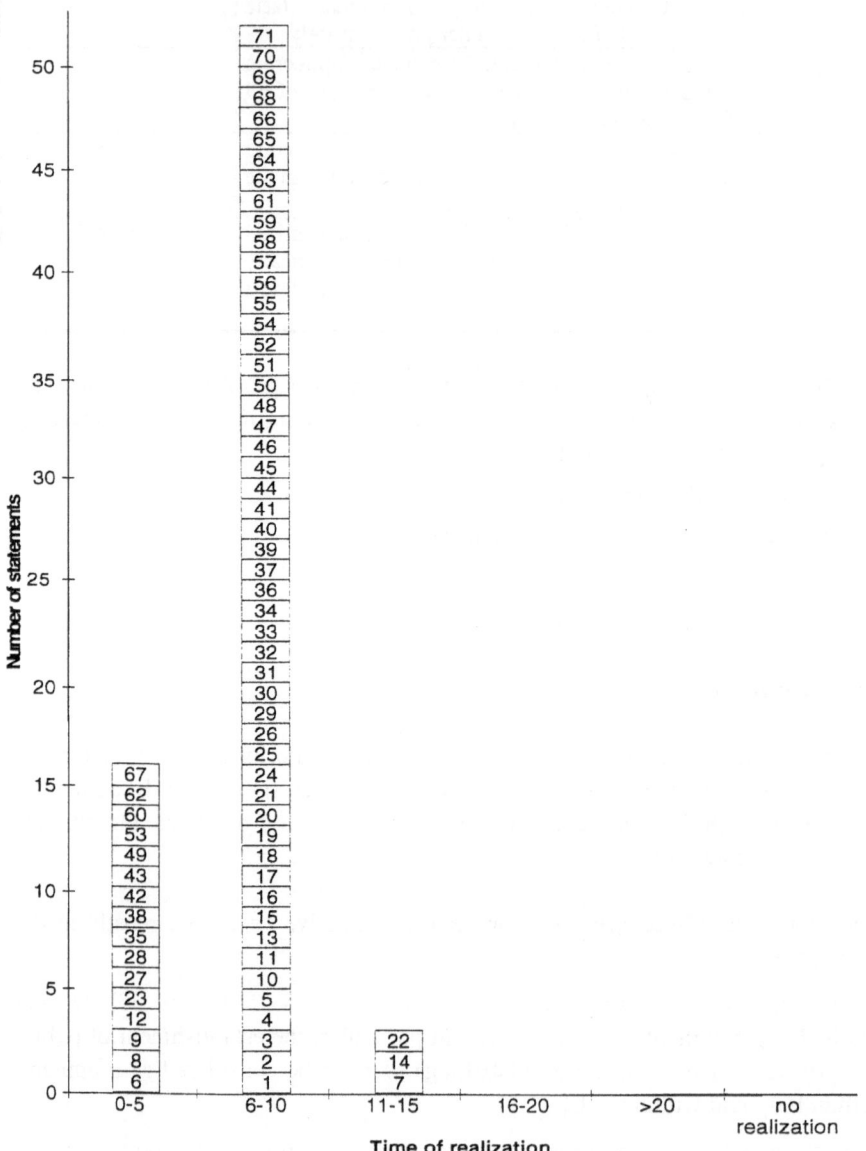

The developments foreseen in the medium term (eleven to 15 years) are very few and concern two statements related to the development of a mass market for food biotechnology (statements 7, 14). Finally, this time category also includes statement 22 (enhanced production of renewable resources used outside the food sector). Other time horizons have a negligible relevance as well as "no realization" in Italy.

70 % of the Italian experts believe that enzymes optimized by protein engineering in specific sectors of the food industry will be realized within five years. Analytical applications of modern biotechnology like quality control and monitoring in food processing, detection of pathogens in plants and animals (statements 42, 43, 62) are estimated by the majority of the Italian experts as short-term developments. The practical use of genetic engineering in plant breeding (for the development of hybrids) is also seen as being realized within five years. However, the possibility to alter polygenic traits in plants is regarded as a rather long-term development (only 15 % see it possible within five years).

The practical use of genetic engineering in animal breeding is expected in the medium term (73 % of the Italian experts expect it will be realized within ten years). Only a small minority sees this development realized within five years (19 %) and the precentage of experts, who believe in the possibility to alter polygenic traits in farm animals within five years is even smaller (8 %). However, such a development is seen as feasible within ten years by the large majority of respondents (71 %). Animal cloning is expected by 42 % of the Italian experts to be widely used in animal production within five years (however, the actual scientific meaning of "cloning" here should be clarified, as it might have been confused with mere embryo splitting).

Other applications expected to be realized in the short term by a significant share of experts in Italy are:

- Increase of erucic acid in rapeseed (52 %)

- Creation of polyploid fish (46 %)

- Production of feed additives with the help of genetic engineering (41 %)

The assessment of the overall economic impact of Agro-Food biotechnology (statement 14) involves a high level of uncertainty among the Italian respondents (only 83 answers). However, a significant market impact of gene food products is not expected in the short term, rather in the medium term. 76 % of the respondents believe that gene food will achieve 30 % market share in Italy within the coming 15 years. Which factors are expected to slow down this development? The most influential factors indicated in this context are "market" and "acceptance", suggesting that the major constraints to the introduction of gene food will be market demand and widespread consumer rejection of such products (in fact only 30 % of respondents have a "positive attitude" toward this development, while 22 % have a "negative attitude"). A significant economic impact in the coming five years is only expected for the production of pharmaceutical substances by genetically engineered animals and plants (statement 2). Concerning the impact of Agro-Food biotechnology on the prices of food (statement 15), a significant uncertainty emerges (only 80 answers). The realization of this statement is expected in the medium term (only 46 % of respondents believes it will occur before ten years). In contrast, the impact of modern biotechnology on agricultural costs of bulk commodities (statement 18) is seen as less uncertain (112 answers) and occurring earlier than the effect on gene food prices (67 % of the experts expect a significant cost reduction within ten years).

Possible adverse effects of modern biotechnology on the environment (creation of resistant weeds and pathogens, loss of agricultural biodiversity) are seen as occurring relatively late (time of realization within ten years: for statements 26 and 34 respectively 58 % and 47 %), with respect to possible environmental benefits of this technology (reduction of pollution from agricultural activities) which are seen as more immediate (time of realization within ten years: for statements 25, 28, 31, 32 respectively 90 %, 95 %, 85 % and 68 %). Similarly, the possible unintended effects of food biotechnology on human health are seen as more remote (time of realization within ten years: for statements 37 (workers) and 40 (consumers) 43 % and 53 % respectively) compared to the possible benefits (creation of food products beneficial to human health) which are seen as more immediate (time of realization within ten years: for statements 36, 38, 39 respectively 82 %, 82 %, and 83 %). On the whole, optimistic expectations seem to prevail over health and environmental concerns.

The most "optimistic" group in Italy seems to be experts from research institutions, who have the highest percentage of the sub-category "realized within five years" for the highest number of statements (statements 25, 36, 39, 43, 46, 48, 52, 58, 59, 60, 63, 64, 65, 66, 67, 69, 70, 71). The group which forecasts immediate realization for the lowest number of developments/applications is "consumers and critics", with the highest percentage of "realized within five years" for a very low number of statements (statements 50, 51, 54, 55, 56).

The experts from industrial companies regard single applications of modern biotechnology as being realized more immediately than the other groups. These applications mainly related to:

• Some applications considered as "environmentally friendly" such as biological control of pests, reduction of pollution from animal farming, energy from biomass, risk assessment methods, tests to detect pathogens in plant production and animal breeding (statement 28, 31, 32, 57, 62)

- Economic and technological developments in fish culture (statement 21, 68)

- Probiotic foods (statement 38)

- Improvement of quality control in food processing (statement 42)

- Genetic engineering in plant breeding (statement 53)

- Use of recombinant enzymes in specific sectors of food industry (statement 49)

- Genetically engineered yeast in beer brewing (statement 45)

For statements with a "negative" content such as reduction of employment, negative health impacts etc., the "industry" group tends to express long-term forecasts with a high share of "no realization" (statements 20, 26, 34, 37, 40). In several cases the industry experts express rather more pessimistic expectations compared to the group "research" as concerns the realization of specific developments in the short term:

- Job creation due to modern biotechnology (statement 17)

- Foods with reduced allergenic potential (statement 36)

- Production of enzymes and high-value substances with the help of genetically engineered plants and plant cells (statements 52, 59)

- Use of genetic engineering in animal breeding (statement 63)

- Modification of polygenic traits in animals (statement 64)

- Cloning of farm animals (statement 65)

- Use of genetically engineered animal vaccines (statement 71)

Influential factors

The results indicate that the factors seen as most influencing the future development of Agro-Food biotechnology in Italy have a scientific/technical character (see figure 3.8): "R&D infrastructure" and "technology transfer". "Funding" is also seen as important, while from the other listed "technology push" factors "personnel" (education, skills) and "industrial innovativeness" are not seen as crucial.

In addition, "regulation/standards" are regarded as quite important influential factors, mostly in the statements included in the domains "acceptance", "regulation", "environment" and "health", and in the statements 63 to 66 concerning the genetic engineering of animals.

"Market" and "social/ethical acceptance" - both related to reactions of the public towards modern biotechnology, though in different ways - play a less significant role: acceptance is seen as crucial factor in genetic engineering of animals, while it is of low relevance in the genetic modification of microorganisms and plants (statements 42 to 62). The role of "market" is even more limited, except in the "economy" domain. Information also plays a minor role, mostly in the domains "acceptance" and "health".

Figure 3.8:   Importance of influential factors in Italy

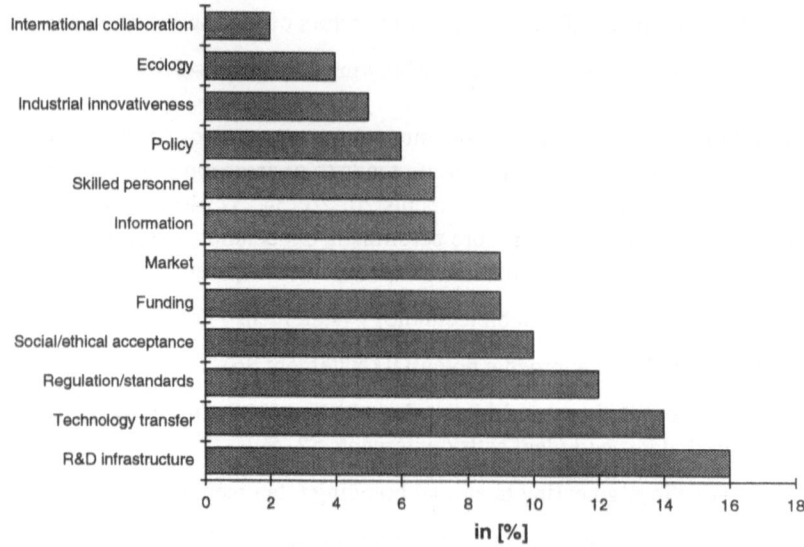

Figure 3.9:   Differences between expert groups related to most important influential factors

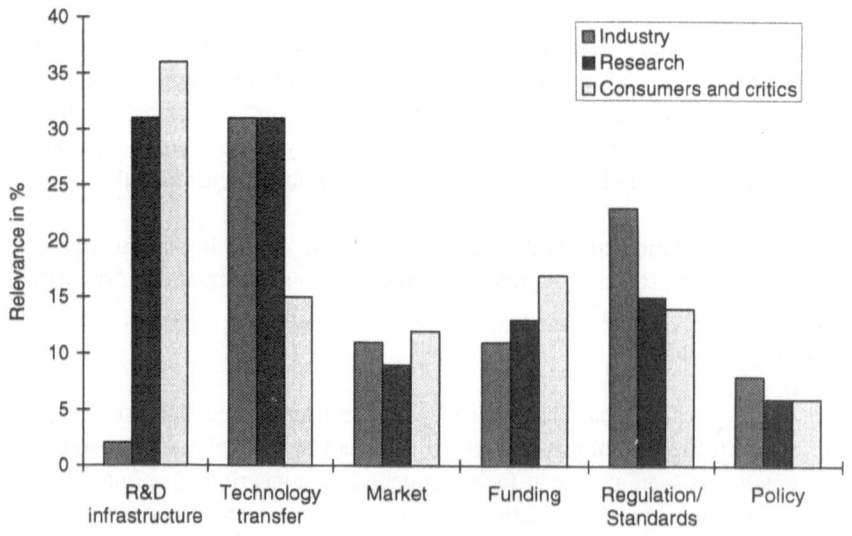

Surprisingly, "ecology" is only regarded as a key factor in the statements of the domain "environment" where it is directly concerned. "International collaboration" is given a negligible importance.

The analysis of the differences in the influential factors between the expert groups (see figure 3.9) reveals that the "industry" experts attach almost no relevance to the factor "R&D infrastructure", while they are more concerned about "regulation/standards" and "policy" than the other groups. Other differences between the expert groups are not statistically significant.

Importance

Which are the most important developments for knowledge creation in science and technology? According to the Italian sample, there seem to be three main "hot" areas (see table 3.18): the first regards the elucidation of basic genomic functions and particularly gene regulation (statements 44, 48, 54, 58). The second field is related to biotechnological monitoring and control techniques applied for different purposes, like epidemiological surveillance of long-term health impacts, identification of pathogens or control of genetically engineered plants in open fields (statements 41, 57, 62). Another area deals with the rDNA support to traditional breeding techniques (statements 61, 68).

The highest consensus among the different expert groups goes to statement 58, regarding the elucidation of the molecular basis of virus resistance mechanism in economically important perennial plants. It is judged as very important for knowledge creation in science and technology by 100 % of the respondents in the expert groups 2 (researchers) and 4 and 5 (consumers and critics), while only 85 % of the respondents in the expert group 1 (industry) recognizes such relevance. A high consensus is also expressed towards statement 44, about the sequencing of the genomes of the most important bacteria used in food production, in particular by the industrialists. Of course researchers, compared to the other two groups, generally express a stronger emphasis on science and technology and particularly on aspects regarding basic knowledge such as the research on polygenic traits (statement 54, 96 %) and new techniques applied to research and production (statement 53: 97 %; statement 56: 93 %; statement 55: 91 %; statement 59: 88 %). On the other side, industrialists focus mainly on technological aspects, mainly with an environmental and health nuance (statement 32: 94 %; statement 41: 94 %; statement 25: 85 %; statement 57: 81 %).

Table 3.18: Statements classified as "very important" for "knowledge creation in science and technology" in Italy

Statement No.	Content	Index
58	The molecular basis of virus resistance mechanisms in economically important perennial plants like e. g. vine, olive and fruit trees is elucidated.	0.96
44	The genomes of the most economically important bacteria used in food production are completely sequenced.	0.93
41	The long-term health impacts of the use of Agro-Food products made with the help of genetic engineering on consumers, farmers and employees are investigated (e. g. by epidemiological surveys).	0.93
48	A large variety of genetically engineered microorganisms which can be precisely controlled and regulated in their metabolic activities during food production processes has been developed.	0.85
62	Rapid test systems based on modern biotechnology are widely used for pathogen identification in plant production and animal husbandry.	0.79
53	For the development of hybrids, genetic engineering is practically used in breeding of most field crops.	0.78
68	Genetic maps and molecular test systems are practically used for the identification of economically important traits and marker-assisted breeding of fish.	0.77
25	Enzyme systems are specifically devoted to improve the environmental performance of conventional food-processing procedures.	0.77
57	New monitoring and control techniques for genetically engineered plants in open fields have improved the reliability of risk assessment significantly.	0.77
54	Genetic engineering approaches are developed which allow the alteration of polygenic traits in most economically important plant species.	0.77

The highest economic importance is attached to two statements related to non-food, high value-added applications of modern biotechnology (statements 59, 52) (see table 3.19). Another key area concerns food applications intended to improve the quality or the processing characteristics of food products by genetically engineered enzymes or microorganisms (statements 46, 48, 49) (see table 3.19).

Table 3.19:   Statements classified as "very important" for "competitiveness of economy" in Italy

Statement No.	Content	Index
59	Plant cell cultures in large-scale bioreactors are widely used for the production of high value components (e. g. pharmaceuticals, fine chemicals, proteins)	0.88
52	The production of industrial enzymes in genetically engineered field crops is developed for large-scale application.	0.87
39	A large variety of food and beverages of superior nutritional value (e. g. vitamin, protein and fibre content; fatty acid ratios) supporting dietary health requirements is produced with the help of genetic engineering.	0.87
48	A large variety of genetically engineered microorganisms which can be precisely controlled and regulated in their metabolic activities during food production processes have been developed.	0.86
53	For the development of hybrids, genetic engineering is practically used in breeding of most field crops.	0.85
49	Enzymes optimized by protein engineering are practically used in specific sectors of the food industry (e. g. starch processing, bakeries, breweries, cheese/dairy production).	0.85
69	Monogenic traits are specifically and stably altered in fish important for aquaculture by genetic engineering as a matter of routine (e. g. anti-freeze gene, growth hormone gene).	0.84
46	Genetically engineered microorganisms and enzymes are widely used to improve the processing quality of food (e. g. prolonged shelf life of biological products, synchronization of the ripening process).	0.84
18	The widespread use of modern biotechnology in agriculture reduces the production costs for agricultural bulk products (e. g. cereals, milk) by approximately 30 %.	0.84
16	The widespread use of modern biotechnology enables small and medium-sized companies of the Agro-Food sector in Italy to introduce a large variety of innovative processes and products.	0.83

There are few significant differences in the assessment of the sub-category "importance to economy" by the various expert groups. The most significant differences concern:

- Labelling of gene food (statement 8): the "industry" experts attach a very low relevance of labelling to economy compared to "consumers and critics".

- Impact on employment due to hesitant introduction of genetic engineering in food production (statement 24): the group "industry" attaches a very low importance to this aspect. All in all, experts from "industry" seem the least inclined to believe that the introduction of Agro-Food biotechnology will have any significant impact (positive as well as negative) on employment in this sector (statements 17, 20, 24).

- Unintended impacts of modern biotechnology on consumer health (statement 40): the group "industry", unlike "research", does not believe that this could have an economic impact.

- Modern biotechnology fails to satisfy traditional taste preferences of consumers (statement 47): "consumers and critics" believe that preserving traditional foods and processes is relevant to the economic development, while the group "industry" seems to neglect the impacts of such activities on the economy.

It is interesting to note that the "industry" group seems also to attach less economic importance to the genetic engineering of farm animals than the other groups (mean index for statements 63, 64, 65, 66: "industry" 0.57, "research" 0.75, "consumers and critics" 0.64). In addition, the very low "importance to the economy" estimated by the "industry" experts to statement 4 (ethical rejection of genetic engineering of animals) seems to confirm this issue.

The fact that more importance for the environment is attached to statements related to applications with a positive impact than to statements concerning unintended negative effects on the environment might indicate that positive expectations about the environmental impact of biotechnology prevail over negative expectations. On the whole, "ecology" is generally given minor importance as an influential factor, suggesting that Italian respondents are not very concerned about possible negative impacts of Agro-Food biotechnology on the environment. The "importance to the environment" given to applications potentially beneficial to the environment (statements 25, 28, 31, 32 with respectively 0.77, 0.76, 0.91 and 0.91) is higher than that given to potentially harmful applications (statements 26, 34 with respectively 0.62 and 0.37). When interpreting these results, it has to be taken into account that the Italian experts did only partly follow the instructions given in the questionnaire exactly like their German colleagues (see chapter 3.1), while filling in this answering category.

On the whole, the "industry" experts assess the factor "ecology" as less influential with respect to the other groups. The group which on average sees "ecology" as more influential with respect to the other expert groups is, not surprisingly, "consumers and critics", while "research" has an intermediate position. Such differences are more significant for some specific statements (statements 9 and 10 concerning authorization procedures, statements 22 and 32 concerning non-food applications, statement 67 concerning polyploid fish). Apparently, the groups "industry" and "research" exclude that non-food applications of modern biotechnology may have any significant impact on the environment.

Table 3.20:   Statements classified as "very important" for "protection of environment/sustainable development" in Italy

Statement No.	Content	Index
31	Modern biotechnology is widely used to reduce emissions and waste from animal production (e. g. less manure due to enzymes as feed additives, improved anaerobic waste treatment, biofilter).	0.91
32	Modern biotechnology significantly contributes to the transformation of 40 % or more of the organic agricultural waste into marketable products (e. g. energy, secondary raw materials).	0.91
57	New monitoring and control techniques for genetically engineered plants in open fields have improved the reliability of risk assessment significantly.	0.84
30	Despite the widespread use of methods and techniques related to modern biotechnology, all in all, no negative effect on the maintenance of biodiversity in agriculturally influenced ecosystems can be discovered.	0.78
25	Enzyme systems are specifically developed to improve the environmental performance of conventional food-processing procedures.	0.77
28	Genetically engineered microorganisms and plants producing biocides are widely used for biological pest control.	0.76
29	The widespread cultivation of genetically engineered field crops resistant to herbicides results in an approximate 50 % reduction of environmental pollution in plant production.	0.72
55	Genetically engineered plant varieties resistant to salinity and/or drought have improved agricultural productivity in arid environments significantly.	0.66
27	The public debate about environmental risks of the deliberate release and marketing of genetically engineered microorganisms (e. g. uncontrollable spread, disturbance of natural balances) has lead to a broad rejection of such activities in Italy.	0.63
26	The widespread deliberate release of genetically engineered organisms results in a significant transfer and recombination of the introduced genes with unintended negative impacts (e. g. development of resistances to herbicides and pathogens in weeds).	0.62

Peculiarities of the Italian results

The level of the public debate on applications of modern biotechnology in Italy, as well as in other Southern European countries, has been more recent and much less intense compared to the average of the other EU member countries. In particular, the Italian public is not familiar with - and not very interested in - the major issues raised by the applications of biotechnology and genetic engineering in the Agro-Food sector, while it is relatively better informed and more involved concerning the medical applications of biotechnology. Therefore, the situation in Italy was not very favourable for a Delphi survey which requires large samples of experts with a high degree of involvement in the concerned issues.

Given such unfavourable conditions, in Italy the Delphi survey on Agro-Food biotechnology has met with some difficulties. The initial panel included 1,615 experts but the rate of responses in the first round (September 1996) was very slow. In order to obtain the needed number of responses (minimum 180 in the first round to ensure statistical significance), many telephone calls to single experts were needed. The total number of filled-in questionnaires at the end of the first round was 189. A second mailing to the respondents to the first questionnaire was made in March 1997. This time the rate of answers was higher and by June 1997 149 filled-in questionnaires were collected, representing the lowest number of responses among the countries covered by the survey.

In particular, such low interest for the subject is due to the limited development of industrial activities in this area. The Italian context has several peculiarities which are not very favourable to the development of Agro-Food biotechnology.

The major constraints seem not to be represented by a limited availability of scientific personnel or by a lack of technical expertise. This is also the prevailing opinion of the respondents to the Delphi survey who have assigned a very low relevance to the influential factor "skilled personnel". Several Italian companies have established expertise and skills in biotechnology, particularly in fermentation techniques for antibiotics production. Also the level of educational and training activities seems quite adequate. Courses in biotechnology have been established at many Italian universities since the late 1980s. As concerns R&D activities, several public research programmes have been launched since 1987: two national research programmes "Advanced Biotechnologies" and several specific programmes of the National Research Council (CNR), with substantial allocation of funds. However, the focus of biotechnological research programmes is mostly on the biomedical area, while the Agro-Food area has secondary importance. Altogether, the quality and quantity of available R&D resources are assessed as insufficient by the Italian expert sample as indicated by the high importance given to the influential factor "R&D infrastructure".

At the industrial level, the development of Agro-Food biotechnology appears rather slow. At the end of 1994, 210 biotechnological companies were active in Italy, but 35 % operated exclusively at the marketing level (the average percentage for the EU being 24 %) and another significant share in engineering and instrumentation. Moreover, medical applications play an overwhelming role with more than 70 % of total sales of the Italian biotech industry (see figure 3.10).

The development of biotechnological R&D activities in the food industry has been affected by the process of "passive internationalization" with the takeover of the major national food companies by multinational corporations. The high importance given in the Delphi survey to the influential factor "technology transfer" by the Italian respondents stresses the difficulties in the industrial application of the new technology.

Table 3.21:   The biotech industry in Italy, 1994

No. of companies [2]	210
Personnel	4,000
Sales (bill. Lit.)	1,130
Applied R&D expenditure (bill. Lit.) [3]	220
Basic R&D expenditure (bill. Lit.)	250

Source: Celestino Spalla, *Le biotecnologie in Italia e nel mondo*, ASSOBIOTEC (1996)

Figure 3.10:   Sales of biotech products in different sectors in 1994

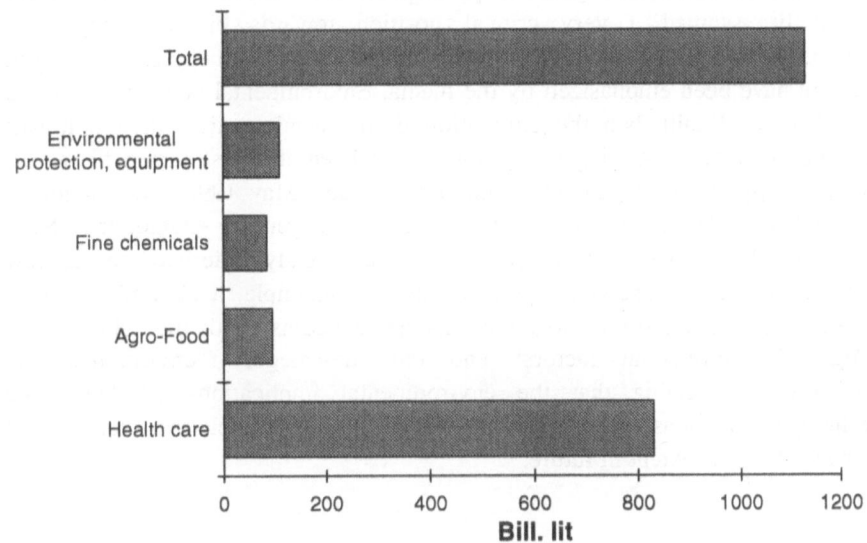

Source: Celestino Spalla, *Le biotecnologie in Italia e nel mondo*, ASSOBIOTEC (1996)

---

2   Including instrumentation and engineering companies (58) and trading companies (75).

3   Including 40 bill. Lit. funded by the Italian government through national research programmes.

In the agricultural sector, the majority of Italian farms consists of a numerous group of small and often marginal farms, while large operations managed according to professional criteria and potentially open to biotechnological innovation represent a minority. A peculiar feature of Italian agriculture is the relevant weight of quality-oriented, traditional productions, many of them recently registered as DOP (Denomination d'Origine Protegée) or IGP (Indication Géographique Protegée). Such productions, which are gaining an increasing importance for Italian food exports, are unlikely to be interested by biotechnological innovation in the short term. Another feature of Italian agriculture is a rich domestic biodiversity, resulting in a high number of different crops and cultivars which are also unlikely to be covered by biotechnological innovation.

Social acceptance has not represented a major constraint to the development of biotechnology so far. This point is confirmed by the findings of the Delphi survey as the influential factor "social-ethical acceptance" is given very limited importance by Italian respondents. In fact, the absence of social opposition has favoured a high number of field trials with genetically modified organisms (GMOs). In 1996 Italy ranked second among EU countries according to the number of authorized releases of GMOs into the environment. However, it is possible that this situation is going to change suddenly. The public debate has become more intense following the first commercial approvals of transgenic crops within the EU and the Italian Parliament has recently assumed a very critical position towards biotechnology. The environmental and social risks related to the introduction of genetic engineering in agriculture have been emphasized by the media. Environmental concerns have led the Ministry of Health ban the cultivation of transgenic maize, though it was withdrawn soon after. A disquieting signal has been the first terroristic action inspired by opposition to genetic technology in Italy (May 1998), in reaction to Nestlé Italy's declared will to use the products of genetic engineering. Such developments began, while the Delphi survey was already underway, so that few repercussions can be observed in the results. For example, a negligible role is assigned in the Italian survey to environmental concerns (both the sub-category "ecology" in "influential factors" and the sub-category "environment" in "importance"), indicating that the environmental implications of Agro-Food biotechnology are not assessed as relevant. However, it is possible that such opinion will change to some extent in future.

3.4   The Netherlands

Degree of knowledge

In table 3.22 the degree of knowledge of the Dutch Delphi panel is presented. The table shows, for each statement and grouped for each of the six domains, the percentage distribution of the degree of knowledge based on the volume of the total respondents group, which is 204, and gives the absolute figure of the sum of the respondents which are "very familiar", "average familiar" and "less familiar" with the subject mentioned in the statement. Answers of panellists who describe their degree of knowledge for specific statements as "not familiar" are excluded from the statistical analysis. For these statements the limited size of the group that has been included should be taken into account when interpreting the results.

Overlooking table 3.22, it can be concluded that the Dutch Delphi panel is most familiar with the subjects mentioned in the statements in the domains acceptance, regulation and environment. For each of these domains, more than 170 of the 204 respondents indicate having a certain level of knowledge about the issues presented in the statements. There was only a small number of subjects in these fields which are not familiar to the Dutch panel. Within the "acceptance" domain, the panel is not familiar with the subject of restaurant chains and catering services and their strategies, such as offering trendy meals with genetically engineered ingredients (statement 7). In the "environment" domain, the Dutch panel appears to have a low expertise in the field of organic agricultural waste and organic farming and the contribution biotechnology can have in this field (statements 32, 33).

The panel is average to less familiar with subjects in the fields of "health, economy and scientific and technological development". The latter finding is very surprising, taking into consideration the fact that the panel consisted of a relatively high number of respondents with an academic degree in natural and technical science.

However, interpreting the results, it should be taken into consideration that each group in the panel has its own specific socio-cultural reasons for calling themselves an expert.

For instance, scientists only consider themselves as an expert in a specific field if they have spent a considerable amount of effort in terms of research and/or publication in that field.

Table 3.22:   Distribution of degree of knowledge and number of answers in the Netherlands

State-ment No.	Very fami-liar (%)	Ave-rage fami-liar (%)	Less fami-liar (%)	Not fami-liar (%)	No. of ana-lysed answers	State-ment No.	Very fami-liar (%)	Ave-rage fami-liar (%)	Less fami-liar (%)	Not fami-liar (%)	No. of ana-lysed answers
Acceptance						Health					
1	12	59	23	5	193	35	6	55	30	9	185
2	10	67	19	5	194	36	6	30	33	31	141
3	12	62	22	5	194	37	3	22	32	43	116
4	10	53	25	12	180	38	7	34	40	19	166
5	9	53	27	10	183	39	3	42	41	14	175
6	6	57	25	12	180	40	2	25	32	41	120
7	3	15	25	57	88	41	1	25	50	24	155
Regulation						42	5	35	38	22	159
8	16	66	13	5	193	43	4	30	37	29	145
9	24	62	12	2	200	Scientific / Technological Development					
10	14	64	17	5	194	44	4	48	28	20	164
11	8	39	29	24	155	45	3	30	36	30	143
12	11	64	19	6	191	46	8	49	31	12	179
13	4	24	36	36	130	47	2	22	31	45	112
14	6	50	18	26	151	48	3	32	33	31	140
Economy						49	11	52	28	9	186
15	2	24	32	41	120	50	6	34	30	30	143
16	3	50	39	8	187	51	0	14	31	54	94
17	1	26	60	13	177	52	4	26	34	35	132
18	4	44	37	15	173	53	21	50	22	8	188
19	8	38	33	21	162	54	11	23	28	38	127
20	5	21	43	31	141	55	7	30	37	25	152
21	0	4	34	61	79	56	12	25	36	26	150
22	7	42	35	16	171	57	5	22	39	34	135
23	6	48	30	16	171	58	5	17	23	55	91
24	4	21	43	32	138	59	3	36	30	30	143
Environment						60	6	16	21	57	87
25	10	61	24	6	192	61	11	17	21	50	101
26	11	53	27	9	186	62	11	34	33	22	159
27	11	63	17	10	184	63	7	35	36	22	160
28	4	60	25	11	181	64	5	23	40	33	137
29	11	43	29	16	171	65	7	60	24	9	185
30	5	52	29	14	176	66	3	24	28	45	113
31	12	58	25	5	194	67	2	10	21	68	66
32	3	33	47	18	168	68	1	7	24	68	66
33	5	19	26	50	103	69	1	8	26	65	72
34	4	50	30	16	172	70	5	34	43	18	167
						71	3	25	45	27	149

In the domains "acceptance, environment and regulation" the relative number of statements answered with "not familiar" is relatively low (15.2 %, 16.3 % and 14.8 %). In the domains "economy" and "health", the responses "not familiar" are 25.4 % and 27.4 % respectively. On "scientific and technological development" these responses are the highest, 34.5 %.

Personal attitude

Table 3.23 shows the twelve statements towards which the Dutch Delphi panel has a very positive attitude (higher than 90 %). Most of these statements belong to the "scientific and technological development" domain and deal with practical devices for monitoring, control and tests based on biotechnology. Also the environmental impact of biotechnology is appreciated by the Dutch Delphi panel: especially the use of biotechnology to help solve the Dutch waste problem and for process integrated clean technologies is scoring high.

Table 3.23:   List of "top 12" statements with "positive attitude" in the
Netherlands

Statement No.	Content	Proportion of "positive attitude" (%)
62	Rapid test systems based on modern biotechnology are widely used for pathogen identification in plant production and animal husbandry.	96
9	All food and beverages which are produced with the help of genetic engineering are subject to an official case-by-case procedure for approval in the Netherlands.	94
42	The hygiene monitoring of food processing is significantly improved due to the widespread use of modern biotechnological analytical methods.	94
57	New monitoring and control techniques for genetically engineered plants in open fields have improved the reliability of risk assessment significantly.	94
1	A governmental institution which provides information on modern biotechnology in the Agro-Food sector to all interested persons and institutions is established in the Netherlands.	94
25	Enzyme systems are specifically developed to improve the environmental performance of conventional food-processing procedures.	94
12	European and national authorities start initiatives which involve the public in the debate and decision-making on the application of modern biotechnology in the Agro-Food sector.	93

Statement No.	Content	Proportion of "positive attitude" (%)
41	The long-term health impacts of the use of Agro-Food products made with the help of genetic engineering on consumers, farmers and employees are investigated (e. g. by epidemiological surveys).	92
43	Techniques based on modern biotechnology are practically used for on-line control of quality parameters in food processing (e. g. content of micro nutrients or harmful substances).	92
32	Modern biotechnology significantly contributes to the transformation of 40 % or more of the organic agricultural waste into marketable products (e. g. energy, secondary raw materials).	92
31	Modern biotechnology is widely used to reduce emissions and waste from animal production (e. g. less manure due to enzymes as feed additives, improved anaerobic waste treatment, biofilter).	91
58	The molecular basis of virus resistance mechanisms in economically important perennial plants like e. g. vine, olive and fruit trees is elucidated.	90

Table 3.24:   List of "top 6" statements for "negative attitude" in the Netherlands

Statement No.	Content	Proportion of "negative attitude" (%)
37	The increased use of modern biotechnology in food production and processing results in additional allergies in people actively involved in these processes.	95
26	The widespread deliberate release of genetically engineered organisms results in a significant transfer and recombination of the introduced genes with unintended negative impacts (e. g. development of resistances to herbicides and pathogens in weeds).	92
40	Unintended impacts on the health of consumers occur in some products or manufacturing processes due to the use of genetic engineering during food processing.	92
19	Small farmers in the Netherlands are not able to afford new genetically engineered plants and animals.	88
20	The widespread use of modern biotechnology in animal and plant production leads to an approximate 30 % decrease in traditional jobs in agriculture.	84
24	Due to the hesitant application of genetic engineering (compared to other EU member states) the Agro-Food industry cuts 10 % or more of the jobs in this sector in the Netherlands.	84

The six statements towards which the Dutch panel has a rather negative attitude are listed in table 3.24. This negative attitude is less pronounced than the positive attitude. The 90 % score in the positive list is reached after twelve statements; in the negative list this is already the case after three statements.

Although in the "top 12" positive list no "economy" statements appear, they are responsible for half of the number of statements in the "top 6" negative list. They deal with negative assessments of cuts in job due to biotechnology (statements 20 and 24) and to excluding small farmers from buying genetically engineered plants and animals (statement 19). The other three statements are in the "health" (2) and in the "environment" (1) domain. All "negative statements" deal with framework conditions; none with scientific and technological development, as in the "positive" list.

In table 3.25 the "top 10" list of statements towards which the Dutch panel has an indifferent attitude is presented. In fact, this is a very interesting list because it contains a number of biotech applications which have been in the centre of public debate during the last deceiders in the Netherlands. For instance, the fact that the Dutch panel is very indifferent towards the use of GMOs in beer production is surprising. In the early 1980s the biggest Dutch beer producer was using these ingredients. This lead to negative reactions of a number of critics from social organizations about the use of these ingredients and received considerable press coverage. If the Delphi panel also reflects the opinion of the Dutch public - which of course has to be assumed with very great caution - and if the company uses products of new biotechnology, this means that a respective message of the beer producer might be expected in the near future.

The Dutch panel is also very indifferent towards food products produced with GMOs whether they have a positive health effect or not. Also remarkable is the fact that the Dutch panel is indifferent towards all fish statements (statements 21, 67, 68, 69), although fish consumption is rising and the use of biotechnology in animals is a hot topic in the Netherlands.

The debate in the Netherlands about biotechnology and its socio-economic, its health and safety impacts and the ethical aspects has been very controversial; nowadays more or less being materialized in a number of organizations and informal consent activities between industry, consumers and environmental organizations where relevant issues such as regulation, labelling etc. are discussed. Against this background it is interesting to find out whether there are still considerable differences between the considered groups.

Table 3.25:   List of "top 10" statements for "indifferent attitude" in the
               Netherlands

Statement No.	Content	Proportion of "indifferent attitude" (%)
45	In the Netherlands most of the beer is produced with genetically engineered yeast.	79
50	Approximately 90 % of the enzymes used in the food-processing industry are produced by genetically engineered organisms.	61
7	A restaurant chain or catering service specialized in offering trendy meals with genetically engineered ingredients open branches in almost all large cities in the Netherlands.	60
67	New cell-biological methods are developed for the production of polyploid fish.	59
21	Due to the widespread use of modern biotechnology in EU fish breeding and aquaculture, the imports of fish and fish products from outside the EU are significantly reduced.	57
14	Food made with the help of genetic engineering achieve a turnover share of 30 % or more of all food consumed in the Netherlands.	56
61	After years of experimentation with genetic engineering, the majority of plant breeders strongly prefers the combination of marker-assisted breeding with traditional breeding methods, compared to the use of genetically modified plants.	54
68	Genetic maps and molecular test systems are practically used for the identification of economically important traits and marker-assisted breeding of fish.	52
38	Food enriched with specific microorganisms having positive health effects achieve approximately 25 % turnover share in their product group (e. g. yoghurt).	50
69	Monogenic traits are specifically and stably altered in fish important for aquaculture by genetic engineering as a matter of routine (e. g. anti-freeze gene, growth hormone gene).	49

Table 3.26 gives a general overview of the distribution of attitudes among expert groups. A slow decrease in positive and an increase in negative attitude towards all subjects as described in the statements is found, going from the industry group to the biotechnology critics. The table shows a general picture of the industry expert group with the highest positive attitude, followed by public researchers. The

biotechnology critics have a flatter attitude profile: less positive, more indifference and some negative.

Table 3.26: Distribution of personal attitudes among expert groups in the Netherlands

Expert groups	Positive (%)	Indifferent (%)	Negative (%)	No opinion (%)
1. Industry	60	23	16	1
2. Public research	56	20	23	0
3. Farmers	52	25	22	1
4. Biotechnology critics	47	24	29	1

Table 3.27: Personal attitude per group and for selected statements in the Netherlands

State-ment No.	Content	Positive %	Indifferent %	Negative %	Expert group
6	**Acceptance** Food companies and retailers proactively create specific labels for food and beverages produced without the help of genetic engineering.	25 56 79 67	59 38 21 24	16 7 0 10	Industry Research Farmers Biotech-nology critics
13	**Regulation** The uniform implementation of the EU biosafety directives in all EU countries has led to a higher attraction of the EU for companies based outside the EU.	35 42 15 48	39 42 77 35	22 17 8 10	Industry Research Farmers Biotech-nology critics
38	**Health** Food enriched with specific microorganisms having positive health effects achieve approximately 25 % turnover share in their product group (e. g. yoghurt).	67 39 57 30	33 59 33 58	0 2 5 13	Industry Research Farmers Biotech-nology critics

For a set of three statements (see table 3.27) considerable, differences between the groups can be observed which are not in accordance with the general picture presented above. For statement 6 the industry group is mainly indifferent towards a situation in which food companies and retailers pro-actively create specific labels for food and beverages produced without the help of genetic engineering, while the three other groups, especially the farmers group, is rather positive about this. For statement 13 the "farmers" are mainly indifferent towards the implementation of the EU Biosafety Directive and the influence this can have on the attractiveness of the EU for companies from outside the EU. The other groups are more or less positive about this (biotechnology critics the most). The last "exceptional" statement in this respect is statement 38 which deals with positive health effects of food enriched with specific microorganisms: "industry" and "farmers" - the most important involved actors in food production - are positive about the situation described in the statement. "Public researchers" and "biotechnology critics" are indifferent.

Time of realization

Figure 3.11 presents the time of realization according to the median for each statement. About half of the statements are expected to be realized between six and ten years. The biggest part of the "acceptance" and "regulation" statements will, according to the Dutch panel, be realized in the next five years. A number of statements have already been realized in the Netherlands, according to the comments. This accounts for statement 1 (the governmental institution which provides information on modern biotechnology in the Agro-Food sector to all interested persons and institutions), statement 8 (the labelling of all food products made with the help of genetic engineering is compulsory and fully implemented), statement 9 (all food and beverages which are produced with the help of genetic engineering are subject to an official case-by-case procedure for approval), statement 12 (national authorities which start initiatives which involve the public in the debate and decision-making on the application of modern biotechnology in the Agro-Food sector), statement 42 (hygiene monitoring of food processing is significantly improved due to the widespread use of modern biotechnological analytical methods), and statement 49 (enzymes optimized by protein engineering are practically used in specific sectors of the food industry). In the case of a number of statements, the situation as described is the subject of research and waits for implementation, such as the rapid test systems of pathogen identification (statement 62), cloned embryos of husbandry animals (statement 65), or is slightly different from the situation described in the questionnaire (like statement 70: feed additives are the case, but for environmental reasons).

Most statements in the "economy", "environment" and "health" domain are expected to be realized in the next six to ten years. Almost one third of the "scientific and technological development" statements will be realized in the next five years, most of the other two-thirds in the following five years. There are two statements which are expected to be realized on the long term (eleven to 15 years): statement 14 dealing with the market share of food made with genetic engineering and statement 64 dealing with the principles regulating the expression of multigenic traits in farm animals.

Figure 3.11:　　　Time of realization of the statements in the Netherlands (according to the median class).

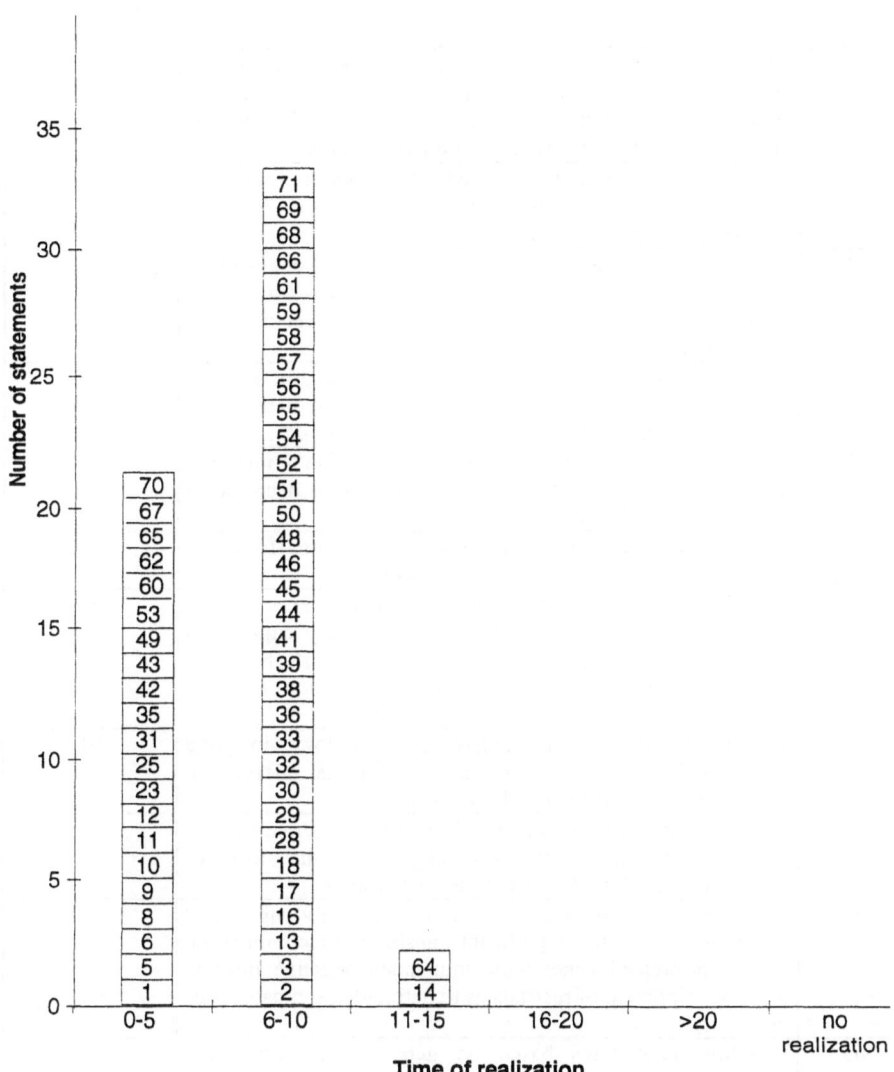

Table 3.28:   Statements with high proportion of "no realization" in the
                Netherlands

Statement No.	Content	Proportion of "no realization" (%)
15	The prices of food products in specific market segments, of which the quality is significantly improved by the application of modern biotechnology are at least 30 % higher than that of corresponding conventional products.	84
47	Modern biotechnology has failed to produce food and beverages satisfying traditional taste preferences of large consumer groups in the Netherlands.	75
24	Due to the hesitant application of genetic engineering (compared to other EU member states), the Agro-Food industry cuts 10 % or more of the jobs in this sector in the Netherlands.	74
7	A restaurant chain or catering service specialized in offering trendy meals with genetically engineered ingredients open branches in almost all large cities in the Netherlands.	69
20	The widespread use of modern biotechnology in animal and plant production leads to an approximate 30 % decrease in traditional jobs in agriculture.	65
19	Small farmers in the Netherlands are not able to afford new genetically engineered plants and animals.	65
27	The public debate about environmental risks of the deliberate release and marketing of genetically engineered microorganisms (e. g. uncontrollable spread, disturbance of natural balances) has led to a broad rejection of such activities in the Netherlands.	64
37	The increased use of modern biotechnology in food production and processing results in additional allergies in people actively involved in these processes.	61
34	The widespread use of genetically engineered organisms in food production significantly aggravates the loss of organisms traditionally used in the Netherlands.	49
26	The widespread deliberate release of genetically engineered organisms results in a significant transfer and recombination of the introduced genes with unintended negative impacts (e. g. development of resistances to herbicides and pathogens in weeds).	43
63	Breeding procedures based on genetic engineering are practically used for those farm animals (cattle, pigs, poultry, sheep, goats) which are of major economic importance for food production in the EU.	42
4	Research on genetic engineering of farm animals is not funded by public institutions in the Netherlands due to ethical reasons (e. g. animal rights, animal welfare, preservation of the Creation).	41
40	Unintended impacts on the health of consumers occur in some products or manufacturing processes due to the use of genetic engineering during food processing.	38

There are 13 statements with a percentage of "will not be realized/no realization", which is higher than one of the other properties of the time-intervals mentioned in the questionnaire. As a conclusion from table 3.28, situations described in four of the eleven "economy" statements will not be realized. The Dutch panel does not expect higher prices for products of modern biotechnology and also is of the opinion that a considerable job decrease (30 %) due to the widespread use of biotechnology will not take place. They also do not find a situation realistic, in which a hesitant application of genetic engineering (compared to other EU member states) leads to a loss of jobs of 10 % in the Dutch Agro-Food sector. We can conclude that the employment impact of modern biotechnology is really an issue for the Dutch panel: there is a relatively high negative attitude towards this issue and the Dutch panel also has a strong opinion about this with respect to the realization. A possible job decrease is not welcome, so it will not happen, seems to be the general attitude of the panel.

The Dutch panel has faith in what biotechnology can do for food products satisfying national taste preferences of large consumer groups. A considerable part (75 %) of the Dutch panel has the opinion that modern biotechnology will not fail to produce these foods.

Statement 7 - on the trendy restaurants selling products of modern biotechnology - is a statement, which the panel (in this case a rather small part of the panel and most of them consider themselves as less familiar with the subject) rejects. As also can be observed from table 3.25, the panel is very indifferent to this statement. It can be concluded that the subject raised in this statement is a typical non-issue for the Dutch Delphi panel.

Table 3.29 shows the average percentages of the time of realization for all 71 statements for the four Dutch expert groups. The table gives information about the expectation of the groups concerning the impact of biotechnology in the future society. The differences are not very remarkable but, nevertheless, they are there.

Table 3.29: Time of realization for the four Dutch expert groups (average percentage over all statements)

Expert group	0-5	6-10	11-15	16-20	>20	No realization	No opinion	Total (%)
1. Industry	39	29	9	3	1	16	4	100
2. Public research	34	33	10	1	0	18	2	100
3. Farmers	38	33	7	3	1	14	5	100
5. Biotechnology critics	34	31	9	2	1	20	5	100

There are eight statements that show a diversity in assessment amongst the groups. The most significant differences are found in those statements, in which the "will not be realized" category plays a significant role (table 3.28). In the domain "acceptance" the whole panel appears to have a polarized opinion, but the farmers and their organizations are less negative about the fact that research on genetic engineering of farm animals will not be financed by public institutions (statement 4) than the other three expert groups. In the domain "regulation" in statement 13 dealing with the implementation of EU biosafety directives, the farmers expect a later realization and also in the domain "economy" the farmers' group takes a different position. This group expects a realization of food made with the help of genetic engineering in the time frame up to ten years. The other groups expect a later realization. With respect to the reduction of production costs of agricultural products (statement 18), the farmers' group expects this as a not realistic situation, while from the other groups - although also with a relatively high percentage in the "no realization" category - the majority expects this to be a realistic option in the time frame of six to ten years. In the domain "environment" the biotechnology critics, for a considerable part, do not expect the situation described in statement 30 to be realized, which means that they do expect negative effects on maintenance of biodiversity. Other groups also have these "no realization" voters, but these figures are much lower and these groups expect this situation to be realized in the next five years. The farmers' group expects that organic farmers are allowed to integrate specific genetic engineering approaches in their production processes in the next five years (statement 33). The public researchers and the biotechnology critics are more negative about this. They have a considerably higher percentage of "no realization" voters.

In the "health" domain, statement 40 shows remarkable differences between the industry experts and the other three groups. The industry group does not expect any unintended health impacts of food products due to the use of genetic engineering during processing. The other groups have a less strongly articulated expectation on this subject. Finally, in the domain "scientific and technological development" some remarkable differences in statements 65 and 69 are found. The biotechnology critics have a considerable number of "no realization" voters in its group as it comes to the use of cloned embryos (65); the others do not have them. The routine genetic engineering of monogenetic traits in fish breeding is not expected to be realized by a considerable number of the farmers and the biotechnology critics, while the other two groups see realization in six to ten years as an option.

Table 3.30: Time of realization for selected statements with considerable differences between the expert groups in the Netherlands

State-ment No.	Content	0-5	6 - 10	11 - 15	16 - 20	> 20	No reali-zation	Expert group
4	**Acceptance** Research on genetic engineering of farm animals is not funded by public institutions in the Netherlands due to ethical reasons (e. g. animal rights, animal welfare, preservation of the Creation).	26 22 38 42	15 17 25 5	15 2 0 2	0 0 0 0	0 2 4 0	41 50 29 47	Industry Research Farmers Biotech-nology critics
13	**Regulation** The uniform implementation of the EU biosafety directives in all EU countries has led to a higher attraction of the EU for companies based outside the EU.	46 37 15 39	42 37 46 35	0 0 15 3	0 0 0 0	0 0 0 0	8 14 15 13	Industry Research Farmers Biotech-nology critics
18	**Economy** The widespread use of modern biotechnology in agriculture reduces the production costs for agricultural bulk products (e. g. cereals, milk) by approximately 30 %.	9 0 10 3	59 48 33 55	13 26 10 5	0 2 5 3	0 0 0 3	19 24 38 26	Industry Research Farmers Biotech-nology critics
30	**Environment** Despite the widespread use of methods and techniques related to modern biotechnology, all in all, no negative effect on maintenance of biodiversity in agriculturally influenced ecosystems can be discovered.	52 26 32 27	10 36 12 22	7 2 4 3	10 0 4 0	0 0 8 3	7 14 20 38	Industry Research Farmers Biotech-nology critics
33	**Environment** Organic farmers in the Netherlands are allowed to integrate specific genetic engineering approaches (e. g. use of genetically engineered biocides or animal vaccines) in their production process.	12 12 43 19	59 42 36 35	18 4 0 4	0 0 0 0	0 0 0 0	12 31 14 38	Industry Research Farmers Biotech-nology critics

State-ment No.	Content	0-5	6 - 10	11 - 15	16 - 20	> 20	No reali-zation	Expert group
40	**Health** Unintended impacts on the health of consumers occur in some products or manu-facturing processes due to the use of genetic engineering during food processing.	10 20 30 29	15 23 20 29	0 14 10 0	0 0 0 0	0 0 0 0	70 37 20 29	Industry Research Farmers Biotech-nology critics
65	**Scientific and Technological development** The use of cloned embryos of cattle, sheep and goats is a widespread technique for the reproduction of farm animals.	50 61 63 43	27 15 15 18	10 7 4 5	3 0 0 0	0 0 4 0	10 15 15 35	Industry Research Farmers Biotech-nology critics
69	**Scientific and Technological development** Monogenic traits are speci-fically and stably altered in fish important for aquaculture by genetic engineering as a matter of routine (e. g. anti-freeze gene, growth hormone gene).	7 35 0 17	50 39 75 17	14 13 0 11	7 4 0 11	0 0 0 0	7 4 25 33	Industry Research Farmers Biotech-nology critics

Influential factors

The members of the Delphi panel were asked to indicate the most important influential factors for the realization of the situation described in the statement with a maximum of three factors. In table 3.31 the average percentage calculated over all statements is presented for each influential factor.

Over all statements, the Dutch panel appears to attach most value to acceptance, R&D infrastructure and markets as the most important influential factors. In table 3.32 the "top 10" list of statements with the highest scoring influential factors registered in the Netherlands are shown. In table 3.33 for each of the twelve influential factors the highest score among the 71 statements is identified and the corresponding statement for this score is added in the table.

Table 3.31:   Relevance of each influential factor in the Netherlands

Influential factor	Relevance over all statements (%)
Acceptance	18.5
R&D infrastructure	17.7
Market	14.1
Industrial innovativeness	10.7
Personnel: education and skills	9.9
Funding	8.0
Regulation	7.2
Information	4.7
Technology transfer	3.9
Ecology	3.0
International collaboration	1.3
Policy	1.0
Total	100 %

Table 3.32:   "Top 10" statements with the highest scoring influential factors in the Netherlands

Statement No.	Content	Influential factor	Relevance (%)
15	The prices of food products in specific market segments, of which the quality is significantly improved by the application of modern biotechnology are at least 30 % higher than that of corresponding conventional products.	Markets	47
7	A restaurant chain or catering service specialized in offering trendy meals with genetically engineered ingredients open branches in almost all large cities in the Netherlands.	Markets	43
63	Breeding procedures based on genetic engineering are practically used for those farm animals (cattle, pigs, poultry, sheep, goats) which are of major economic importance for food production in the EU.	Acceptance	41
4	Research on genetic engineering of farm animals is not funded by public institutions in the Netherlands due to ethical reasons (e. g. animal rights, animal welfare, preservation of the Creation).	Acceptance	40
44	The genomes of most economically important bacteria used in food production are completely sequenced.	R&D infrastructure	39

Statement No.	Content	Influential factor	Relevance (%)
7	A restaurant chain or catering service specialized in offering trendy meals with genetically engineered ingredients open branches in almost all large cities in the Netherlands.	Acceptance	39
19	Small farmers in the Netherlands are not able to afford new genetically engineered plants and animals.	Markets	38
51	The molecular basis of virus resistance mechanisms in economically important perennial plants like e. g. vine, olive and fruit trees is elucidated.	R&D infrastructure	38
65	The use of cloned embryos of cattle, sheep and goats is a widespread technique for the reproduction of farm animals.	Acceptance	37
38	Food enriched with specific microorganisms having positive health effects achieve approximately 25 % turnover share in their product group (e. g. yoghurt).	Markets	37

Table 3.33: All influential factors and their highest score in the Netherlands

Statement No.	Content	Influential factor	Relevance (%)
15	The prices of food products in specific market segments, of which the quality is significantly improved by the application of modern biotechnology are at least 30 % higher than that of corresponding conventional products.	Markets	47
63	Breeding procedures based on genetic engineering are practically used for those farm animals (cattle, pigs, poultry, sheep, goats) which are of major economic importance for food production in the EU.	Acceptance	41
44	The genomes of most economically important bacteria used in food production are completely sequenced.	R&D infrastructure	39
24	Due to the hesitant application of genetic engineering (compared to other EU member states) the Agro-Food industry cuts 10 % or more of the jobs in this sector in the Netherlands.	Policy	37
30	Despite the widespread use of methods and techniques related to modern biotechnology, all in all, no negative effect on the maintainance of biodiversity in agriculturally influenced ecosystems can be discovered.	Ecology	35

Statement No.	Content	Influential factor	Relevance (%)
58	The molecular basis of virus resistance mechanisms in economically important perennial plants (e. g. vine, olive and fruit trees) is elucidated.	Funding	35
51	Artificial polyfunctional enzymes are practically used for analytical applications in the Agro-Food sector (e. g. rapid quantitative analyses of metabolic activities or the composition of raw materials).	Industrial innovative-ness	33
9	All food and beverages which are produced with the help of genetic engineering are subject to an official case-by-case procedure for approval in the Netherlands.	Regulation/ standards	31
13	The uniform implementation of the EU biosafety directives in all EU countries has led to a higher attraction of the EU for companies based outside the EU.		
3	The widespread diffusion of information on genetic engineering from different sources (e. g. public and private institutions, several interest groups and associations, media) has increased the acceptance of food products made with the help of genetic engineering in the Netherlands.	Information	30
27	The public debate about environmental risks of the deliberate release and marketing of genetically engineered microorganisms (e. g. uncontrollable spread, disturbance of natural balances) has led to a broad rejection of such activities in the Netherlands.		
17	The practical use of modern biotechnology in the food industry creates a small number of new jobs (e. g. in specialized service companies and the supplying industry).	Personnel	25
16	The widespread use of modern biotech-nology enables small and medium-sized companies of the Agro-Food sector in the Netherlands to introduce a large variety of innovative processes and products.	Technology transfer	20
13	The uniform implementation of the EU bio-safety directives in all EU countries has led to a higher attraction of the EU for companies based outside the EU.	Internatio-nal colla-boration	12

Market is the factor achieving the highest score and this is realized in statement 15. Market can be found five times in the "top 10" list presented in table 3.32. As can be concluded from table 3.31, in the Dutch Delphi panel consensus seems to exist about the fact that acceptance, R&D infrastructure and market are the factors that most strongly influence the application of biotechnology in economy and society. For successful future market introduction of biotechnological products, market pull factors ("market" and "social and ethical acceptance") play an important role, but have to be counterbalanced by science and technology push "R&D infrastructure". Regulation will play a moderate role in the future. Technology transfer and international cooperation will play almost no role at all as influential factor for future implementation of biotechnology in the Dutch Agro-Food sector.

In order to draw conclusions about the differences between the four Dutch expert groups, a "top 10" list for each group of the influential factors with the highest score was made (table 3.34). It is interesting to see the patterns of distribution of factors for the groups: for industry "markets" scores six times, "acceptance" and "R&D infrastructure" take the second and third position. For the researcher in the publicly funded research institutes other factors belonging to other statements are in the "top 10". "Acceptance" has a higher priority and also "policy" is in their list.

In the farmer group "top 10" list, "acceptance" is present four times. They also have "funding" and "industrial innovativeness" in their list. It is remarkable that the biotechnology critics score the highest for "markets": five times. "Acceptance" and "R&D infrastructure" follow in second and third position.

Table 3.34:   "Top 10" "influencing" factors for the four groups in the
                Netherlands

Industry			Public researchers			Farmers			Biotechnology critics		
%	Factor	State-ment No.	%	Factor	State-ment No.	%	Factor	State-ment No.	%	Factor	State-ment No.
45	Market	15	47	Market	15	48	Market	15	47	Market	7
43	Market	67	46	Accept.	33	45	Funding	58	42	Market	15
42	Accept.	4	45	Market	7	44	Ind. In	51	42	Market	19
41	Market	60	42	Accept.	63	39	R&D in.	44	40	Accept.	63
39	Market	38	41	Accept.	4	38	Accept.	4	40	R&D in.	51
39	Accept.	63	41	Policy	24	37	Market	38	40	R&D in.	44
38	Market	7	39	Market	45	37	Accept.	14	39	Accept.	65
37	R&D in.	48	39	R&D in.	58	36	Accept.	69	38	Accept.	33
37	R&D in.	54	39	Accept.	66	36	Accept.	65	37	Market	6
36	Market	19	39	Market	51	36	Market	21	37	Market	38

All groups indicate for statement 15 "market" is a very important influential factor. In statement 4 three groups (industry, public researchers and farmers) indicate "acceptance" is an important influential factor. The same applies for statement 63, but here the biotechnology critics take the place of the farmers. The same set of three groups (industry, public researchers and biotechnology critics) gives a very high priority to the "market" as influential factor for statement 7 on the trendy restaurants.

Especially the environment statements show remarkable differences between the groups. For statement 29 on the widespread cultivation of genetically engineered field crops resistant to herbicides which results in an approximate 50 % reduction of environment pollution, acceptance is for all four groups the most influential factor, but their opinion about the other factors differs: public researchers, farmers and biotechnology critics also give a priority to R&D which is not at all prioritized by industry. The farmer group also gives policy some priority. In statement 33 on the organic farmers who are allowed to integrate specific genetic engineering approaches in their production process, industry gives some priority to information, while farmers do not score it at all.

Finally, two scientific and technological statements must be mentioned. For statement 58, dealing with the elucidation of the molecular basis of virus resistance mechanisms in economically important perennial plants, both industry and farmers give a third place priority to industrial innovativeness (R&D infrastructure and funding being first and second for all four groups). The biotechnology critics give a third priority to technology transfer. For the researchers there is no third priority. Statement 66, on the elucidation of the principles underlying the development processes of oocytes which leads to the genetic engineering of early stages of embryo cells from farm animals, shows that R&D infrastructure and acceptance are the two very important factors and - excluding the farmers - also the only two relevant. The farmers mention as a third factor policy, which must be interpreted as a first step to regulation.

Importance

In table 3.35 the most important statements for "knowledge creation in science and technology" are presented. To allow the possibility of finding relations between the assessment of the three fields, the score of the statements on the other two factors - "important for competitiveness of economy" and "important for protection of the environment/sustainable development" - are included. Similar lists for the other two factors are presented in tables 3.36 and 3.37.

The majority of the statements important for "knowledge creation in science and technology" are developments with relevance for food processing (see table 3.35)

Table 3.35:   "Top 10" statements for "importance for knowledge creation in science and technology" in the Netherlands

Statement No.	Content	Index Value		
		Knowledge creation science & technology	Competi- tiveness of economy	Protec- tion of environ- ment
44	The genomes of most economi- cally important bacteria used in food production are completely sequenced.	0.96	0.16	-0.60
54	Genetic engineering approaches are developed which allow the alteration of polygenic traits in most economically important plant species.	0.92	0.65	-0.12
48	A large variety of genetically engi- neered microorganisms which can be precisely controlled and regu- lated in their metabolic activities during food production processes has been developed.	0.91	0.77	-0.48
58	The molecular basis of virus resistance mechanisms in econo- mically important perennial plants (e. g. vine, olive and fruit trees) is elucidated.	0.90	0.37	0.15
43	Techniques based on modern biotechnology are practically used for on-line control of quality para- meters in food processing (e. g. content of micronutrients or harm- ful substances).	0.89	0.81	-0.27
25	Enzyme systems are specifically developed to improve the environ- mental performance of conven- tional food-processing procedures.	0.87	0.83	0.79
23	Pharmaceutical substances produced by genetically engi- neered animals and plants (e. g. proteins, enzymes, hormones, antibodies) achieve a turnover share of at least 5 % of the phar- maceutical market in the Netherlands.	0.87	0.87	-0.34
59	Plant cell cultures in large-scale bioreactors are widely used for the production of high value compo- nents (e. g. pharmaceuticals, fine chemicals, proteins).	0.86	0.73	-0.26

Statement No.	Content	Index Value		
		Knowledge creation science & technology	Competi- tiveness of economy	Protec- tion of environ- ment
62	Rapid test systems based on modern biotechnology are widely used for pathogen identification in plant production and animal husbandry.	0.85	0.76	0.10
56	To prevent resistance against biocides in pathogens, new genetic engineering approaches for plant defence mechanisms have been developed (e. g. combination of different resistance genes, increase in pathogen tolerance).	0.85	0.67	0.38

like the complete sequencing of genomes of important bacteria, the development of microorganisms which can be precisely controlled and regulated in their metabolic activities during food production processes, on-line control of food quality parameters, process-integrated biocatalysts food for a better environmental performance (statements 44, 48, 43, 25). Other relevant developments for knowledge creation in science and technology are developments in plant breeding and production, like the creation of basic knowledge on the alteration of polygenic traits in plants, on new biotechnological approaches for plant defence mechanisms and on the elucidation of the molecular basis of virus resistance mechanisms in perennial plants (statements 54, 56, 58). The Dutch panel expects important scientific and technological spin-offs from research in the field of gene farming, in plants as well as in animals (statements 23, 59). Finally, rapid test systems for pathogen identification in plant and animal production are considered important for scientific and technological knowledge creation (statement 62) as well.

The list of statements with the highest score for "importance for competitiveness of the national economy" (table 3.36) shows a number of statements which directly relate to traditional economic benefits, such as new products like pharmaceutical substances, renewable resources, optimized enzymes and on-line control techniques (statements 23, 22, 49, 43), reduction of productions costs (statement 18), or which are in favour of SMEs (statement 16). However, it is very remarkable - and illustrative for how environmental problems are dealt with in the Netherlands - that the Dutch panel also gives a high economic priority to the enzyme systems which are developed for the improvement of the environmental performance of production processes (statement 25).

Table 3.36:   "Top 7" statements for "importance for competitiveness of economy" in the Netherlands

Statement No.	Content	Index value		
		Knowledge creation science & technology	Competitiveness of economy	Protection of environment
16	The widespread use of modern biotechnology enables small and medium-sized companies of the Agro-Food sector in the Netherlands to introduce a large variety of innovative processes and products.	0.69	0.90	-0.32
23	Pharmaceutical substances produced by genetically engineered animals and plants (e. g. proteins, enzymes, hormones, antibodies) achieve a turnover share of at least 5 % of the pharmaceutical market in the Netherlands.	0.87	0.87	-0.34
22	Due to the application of modern biotechnology, farmers produce renewable resources (e. g. biofuel, starch, fatty acids) which are used outside the food sector on 20 % or more of the arable land in the Netherlands.	0.71	0.86	0.27
25	Enzyme systems are specifically developed to improve the environmental performance of conventional food-processing procedures.	0.87	0.83	0.79
18	The widespread use of modern biotechnology in agriculture reduces the production costs for agricultural bulk products (e. g. cereals, milk) by approximately 30 %.	0.34	0.83	-0.11
49	Enzymes optimized by protein engineering are practically used in specific sectors of the food industry (e. g. starch processing, bakeries, breweries, cheese/dairy production).	0.83	0.83	-0.41
43	Techniques based on modern biotechnology are practically used for on-line control of quality parameters in food processing (e. g. content of micronutrients or harmful substances).	0.89	0.81	-0.27

All statements in the list "importance for the protection of the environment/ sustainable development" are from the environment domain (see table 3.37). Significant positive impacts on the environment are expected if modern biotechnology contributes to the transformation or reduction of emissions and waste from animal production (statements 31, 32). The Dutch panel also expects a considerable contribution from genetically engineered herbicide-resistant crops; an important point of discussion between environmental groups and industry (statement 29). Biodiversity has been a subject on the Dutch research agenda since the UN Rio de Janeiro conference on global environmental issues: the Dutch panel assesses a non-negative effect of biotechnology on biodiversity (statement 30).

Table 3.37:   "Top 6" statements for "importance for protection of the environment/sustainable development" in the Netherlands

Statement No.	Content	Index value		
		Knowledge creation science & technology	Competi- tiveness of economy	Protec- tion of environ- ment
32	Modern biotechnology significantly contributes to the transformation of 40 % or more of the organic agricultural waste into marketable .products (e. g. energy, secondary raw materials).	0.75	0.66	0.87
31	Modern biotechnology is widely used to reduce emissions and waste from animal production (e. g. less manure due to enzymes as feed additives, improved anaerobic waste treatment, biofilter).	0.62	0.63	0.87
29	The widespread cultivation of genetically engineered field crops resistant to herbicides results in an approximate 50 % reduction of environmental pollution in plant production.	0.35	0.45	0.83
25	Enzyme systems are specifically developed to improve the environ- mental performance of conventional food-processing procedures.	0.87	0.83	0.79
30	Despite the widespread use of methods and techniques related to modern biotechnology, all in all, no negative effect on the maintenance of biodiversity in agriculturally influenced ecosystems can be discovered.	0.20	0.05	0.75
28	Genetically engineered microorganisms and plants producing biocides are widely used for biological pest control.	0.75	0.67	0.74

Finally, in table 3.38 the ten highest scores for all three "importance factors" are presented. Statements on the "top 6" list of sustainable development also have a positive score on the other two fields (see table 3.37). However, most of the statements with the highest scores for science and technology and competitiveness of economy (tables 3.35 and 3.36) have a negative score in the environmental field. In the case of situations described in three statements, a win-win situation for economy and ecology can be reached: the enzyme systems for improving the environmental performance of conventional food-processing procedures, the biotechnological transformation of organically agricultural waste into marketable products, and the use of biotechnology for emission and waste reduction from animal production. The research which has to be done to realize the described situations can contribute to future scientific and technological development in all cases.

Table 3.38:   "Top 10" statements for all three "importance" factors in the Netherlands

Statement No.	Content	Index value		
		Knowledge creation science & technology	Competi- tiveness of economy	Protec- tion of environ- ment
44	The genomes of most econo- mically important bacteria used in food production are completely sequenced.	0.96	0.16	-0.60
54	Genetic engineering approaches are developed which allow the alteration of polygenic traits in most economically important plant species.	0.92	0.65	-0.12
48	A large variety of genetically engineered microorganisms which can be precisely controlled and regulated in their metabolic activities during food production processes has been developed.	0.91	0.77	-0.48
16	The widespread use of modern biotechnology enables small and medium-sized companies of the Agro-Food sector in the Nether- lands to introduce a large variety of innovative processes and products.	0.69	0.90	-0.32

Statement No.	Content	Index value		
		Knowledge creation science & technology	Competi- tiveness of economy	Protec- tion of environ- ment
58	The molecular basis of virus resistance mechanisms in econo- mically important perennial plants like e. g. vine, olive and fruit trees is elucidated.	0.90	0.37	0.15
43	Techniques based on modern biotechnology are practically used for on-line control of quality parameters in food processing (e. g. content of micronutrients or harmful substances).	0.89	0.81	-0.27
25	Enzyme systems are specifically developed to improve the environmental performance of conventional food-processing procedures.	0.87	0.83	0.79
23	Pharmaceutical substances produced by genetically engi- neered animals and plants (e. g. proteins, enzymes, hormones, antibodies) achieve a turnover share of at least 5 % of the pharmaceutical market in the Netherlands.	0.87	0.87	-0.34
32	Modern biotechnology signifi- cantly contributes to the transfor- mation of 40 % or more of the organic agricultural waste into marketable products (e. g. ener- gy, secondary raw materials).	0.75	0.66	0.87
31	Modern biotechnology is widely used to reduce emissions and waste from animal production (e. g. less manure due to enzymes as feed additives, improved anaerobic waste treatment, biofilter).	0.62	0.63	0.87

Particularities of the results in the Netherlands

The public debate about the use of modern biotechnology in the Netherlands has a very long history. Already in the mid-70s, when the scientists themselves put a temporary ban on recombinant DNA research during the Asilomar conference (1973) in the USA, the first initiatives were taken by socially responsible scientists to question the social aspects and impacts of this new technology. Because the research community in the Netherlands, in universities and in industry, was also already very aware of the importance of this new technology for research and innovation, the debate in the Netherlands started also at that early period.

Dutch scientists and industry organized themselves in a number of professional organizations and were, and still are, a strong lobby group in The Hague (geographical basis of the Dutch government) and later in Brussels (EU government). They were very successful in organizing public money in the form of R&D programmes. Biotechnology has, especially at the moment, a strong basis in the Netherlands in the agricultural and food sector and in the environmental sector. Recently, the use of biocatalysis in fine chemicals has gained importance. The pharmaceutical industry - traditionally the most important biotechnology user and developer - has only a small industrial basis in the Netherlands.

But also the critics of biotechnology - thanks to the liberal culture in the Netherlands - were able to gain ground, often facilitated by government funding for their arguments and activities. However, there were also anonymous radical groups that destroyed experimental fields with genetically modified crops: "The Angry Bintjes" (Bintje is the name of a famous Dutch potato) and "The Raging Raiser" operated in the period, in which the first open field trials took place (end of 80s/beginning 90s).

Although the confrontation between the supporters and critics of biotechnology has sometimes been very tough (posters of the animal protection organization with a shocking photo montage of a naked mother with animal breasts feeding her baby, publication of more or less secret deals between companies and the government etc.), at this moment there is a rather stable status quo. The position of the political and social actors is rather clear and workable. To a great extent this can be ascribed to the regular meetings of these groups in the so-called "informal consent". Industry and social groups (consumers and environmental organizations) regularly deliberate about a number of issues dealing with biotechnological activities of industry and the consumers and environmental aspects.

Legislation is very well elaborated in the Netherlands and most social issues are more or less regulated. Health and safety aspects are as far as possible being taken care of by legislation. Ethical aspects are dealt with by an ethical committee that screens the research projects in the field of animal biotechnology. Biotechnological research with humans is forbidden, with humans material it is under strict regulation. At this moment labelling of gene food products is in the limelight, because although general (EU) rules have been set, it now comes down to daily practice, of labelling or not. Since last year Greenpeace - because soya and maize come by boat - has also entered the Dutch playing field and has found itself in good company of the very experienced and specialized environmental, consumers and other non-governmental organizations.

This position is also reflected in the Delphi survey: the Dutch panel in general has a very positive attitude towards regulatory activities with respect to biotechnology (labelling, case-by-case procedure for approval), towards the assessment of social impacts in this procedure, the participation of the public in this procedure, and towards the use of specific technical equipment and standardized methods for controlled purposes. Some of the situations described in the statements on regulation already have been or soon will be realized in the Netherlands: labelling, and procedures for market approval. Research institutes in the Netherlands are working on methods to identify genetically modified substances or traces from genetic engineering in food. The panel appreciates it when retailers and food companies develop specific labels for non-GMO products and they expect these products on the market in the next five years. The Dutch panel has no very pronounced opinion about the consumers attitudes: there is a relatively high indifference towards the described situation where most consumers in the Netherlands have quickly got used to all kinds of food and beverages made with the help of genetic engineering. This corresponds with the general Dutch attitude which shows great liberty to others, incl. consumers, opinions and performance.

Partly as a result of, but also as an important condition for, the Dutch debate for some years, there is the organization Foundation for Public Information on Science, Technology and the Humanities (PWT) sponsored by the Dutch government, informing the public (also) about biotechnology and provids the education system which informs on biotechnology mediated by a rich variety of media. The Dutch government regularly asked the Rathenau Institute to organize public debates about specific aspects of biotechnological developments and applications. These activities might be responsible for the fact that the Dutch panel has a very positive attitude towards initiatives taken by the government or social organizations to inform the public about modern biotechnology.

The Dutch panel is more or less negative about the government not funding biotechnological research of farm animals. In fact, the Dutch government has funded this kind of research. This research can be approved only after a research proposal has been discussed in an ethical committee - set up by the government - and this committee has advised about conditions for funding etc. That is why 41 % of the panel indicates that the situation described in the statement is not a realistic option.

The environment issues are receiving much attention in Dutch politics and biotechnology is considered an important instrument for sustainable development. The Netherlands, together with Germany and the USA, is the furthest ahead with environmental biotechnology and also has a strong market position in this growing market.

The Dutch panel is also most knowledgeable in this field and has a positive attitude towards the situations described in the "environment" statements, especially for all situations which are dealing with biotechnological techniques to help to solve the environmental problem (enzyme systems for cleaner food processing, biocides for pest control, herbicide-resistant crops, reducing of animal waste by enzymes, biofilters and anaerobic cleaning, biotransformation of organic waste). For most situations they expect a realization in the coming six to ten years.

The impacts of biotechnology on health seem to be a black box for the Dutch panel. The Dutch panel evaluates itself as less knowledgeable in this domain. Reasons for this could be that the Netherlands only has very few pharmaceutical companies, but also because that part of the medical sector that is working in this field does not present itself as workers in the field of biotechnology, but instead as medical professionals. Nevertheless, the Dutch panel is positive about the use of biotechnology in the health domain, such as producing food for allergy sufferers or food with superior nutritional value supporting dietary health requirements and also about using biotechnology for monitoring long-term health effects.

The economic impact of biotechnology is an important issue for the Dutch panel: a relatively large number of high scores can be observed for the statements in the economy domain. Market gets a number of the highest scores in the set of influential factor and as mentioned before, four items from the "top 10" of negative attitudes are in this domain. This negative attitude of the Dutch panel is towards the higher prices of food products in specific market segments, of which the quality is significantly improved by the application of modern biotechnology, the small farmers who are not able to afford new genetically engineered plants and animals, the 30 % decrease in traditional jobs in agriculture as a direct consequence of the widespread use of modern biotechnology in animal and plant production, and the cuts of 10 % or more of the jobs due to the hesitant application of genetic engineering in the Dutch Agro-Food industry (compared to other EU member states). The situations described in these statements are all also assessed as not realistic in almost all cases because of market arguments. The production of renewable resources (e. g. biofuel, starch, fatty acids) for non-food applications on 20 % or more of the arable land (statement 22) is expected to have an important impact on economic development in the Netherlands. This issue has already been addressed in a number of R&D programmes and the Dutch research community has been successful in producing biodegradable "plastics" from potato starch.

If it comes to acceptance of biotechnology in general, the Delphi survey shows the following pattern. On the whole, acceptance of biotechnology certainly is an issue in the Netherlands, but - also as a result of information campaigns, public debates and the informal consent between industry and a number of important social groups, there is a diversification with respect to the application area. The Dutch panel is rather indifferent towards genetically modified enzymes in food, especially towards beer; it gets the highest indifference score of the whole questionnaire! This attitude can also be observed for situations in which approximately 90 % of the enzymes used in the food-processing industry are produced by genetically engineered organisms, the production of industrial enzymes in genetically engineered field crops is developed for large-scale application and some other statements describing situations of GMOs and genetically modified enzymes systems used on a large scale. The Dutch panel does not have a pronounced positive or negative attitude towards these subjects. Nevertheless, the panel has a very pronounced opinion about the influential factors for these statements: entrepreneurship, R&D infrastructure and acceptance score very high.

For plants the attitude towards the use of genetic engineering techniques is positive/indifferent. The panel is more positive if environmental benefits or pathogen detection are described. The panel is getting more indifferent if alternatives for genetic engineering are mentioned (marker-assisted breeding). However, if it comes to the use of modern biotechnology in animal breeding, the panel's attitude switches from positive/indifferent to negative/indifferent with a very negative attitude towards the use of breeding procedures based on genetic engineering for those farm animals (cattle, pigs, poultry, sheep, goats) which are of major economic importance for food production in the EU. Acceptance gets its highest score of the whole questionnaire here. The described situation will not be realized according to the panel. On the contrary, the use of modern biotechnology in animal vaccines and animal feed gets a positive assessment. Both will have an important impact on economic development in the Netherlands.

3.5   Spain

Degree of knowledge

In general terms, the Spanish experts tend to consider themselves as being not very familiar with the issues they are faced with, while reading the questionnaire. "Not familiar" is the option chosen by around one third of the experts to define their own degree of knowledge, while the ones who define themselves as being "very familiar" form the smallest group (see figure 3.12). The self-assessed degree of knowledge on the subject of Agro-Food biotechnology shows a very interesting result when comparing this category between the different expert groups. The distribution of the different levels of knowledge does not change substantially between researchers, industry, consumers and critics.

The number of answers significantly differ between the 71 statements. This could be due to the design of the questionnaire, since more general statements had been interspersed with specific ones. There is not a clear pattern in the response rate of the different sections of the questionnaire, i. e. there is not a clear-cut tendency which divides the experts' knowledge depending on the thematic area, in which the statements are organized (regulation, acceptance, scientific and technological development etc.). Despite this, a decrease in the general level of expertise can be observed from half of the questionnaire which is mainly the section devoted to the scientific and technological development of biotechnology, an area which is quite unknown for people not working directly in it. The subject addressed in statements 26 and 27, which are about the negative impacts genetically modified organisms may have on the environment, is the one the Spanish sample is more familiar with.

Figure 3.12:   Degree of knowledge of the experts in Spain

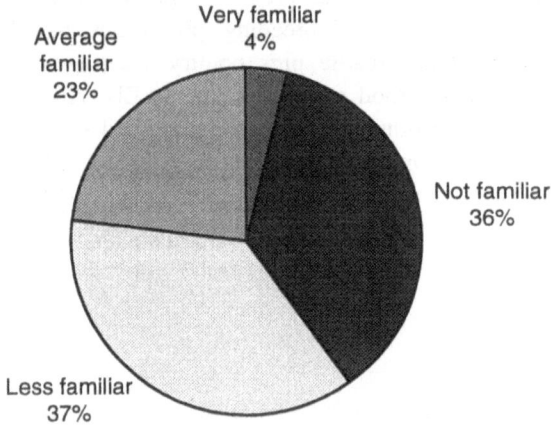

The highest response rate for the whole belongs to statements 2 and 27 sample (134 and 133 answers out of 191 experts, respectively):

- Statement 2: most consumers in Spain have quickly got used to all kinds of food and beverages produced with the help of genetic engineering.

- Statement 27: the public debate about environmental risks of the deliberate release and marketing of genetically engineered microorganisms (e. g. uncontrollable spread, disturbance of natural balances) has led to a broad rejection of such activities in Spain.

In general, the statements regarding information and consumers' issues have a relatively high response. The same applies to the statements relating to biotechnology and creation of new jobs, as well as some environmental effects. Those are big areas of interest appearing frequently on the mass media headlines - consumers' issues, employment and environmental degradation - and so, they form part of very familiar issues, susceptible to be a subject of everyday conversation. In contrast, developments described in statement 7 ("A restaurant chain or catering service specialized in offering trendy meals with genetically engineered ingredients open branches in almost all large cities in Spain") or statement 37 ("The increased use of modern biotechnology in food production and processing results in additional allergies in people actively involved in these processes") are very poorly responded to.

Personal attitude

The Spanish experts (over 70 % of the sample) show a very positive personal attitude towards the biotechnological developments expressed in almost two thirds of the statements. So, their opinion on such issues could be interpreted as quite enthusiastic.

Focusing on positive attitudes held by Spanish experts, table 3.39 only shows the "top 10" statements, as the whole list of statements considered positive by more than 50 % of the experts add up to 54 statements.

As a general comment, most of those "top 10" statements for positive attitude refer to the domain "scientific and technological development" and so quite technical issues; two refer to the improvement of the environmental impact of agriculture and animal husbandry, and only one statement is concerned with health issues (statement 41). This statement is not about the positive effect that genetically engineered food products may have on human health, but about the investigation of epidemiological studies on long-term effects of biotechnological foods. So, it is not the actual reduction of risks, but the better knowledge about such risks which is regarded as a very positive advance by more than 96 % of the sample.

The two developments/situations better considered by the Spanish sample are also expected to be implemented or to happen in the next five years: the development of analysis systems to prevent pathogens in plant production and animal husbandry and the improvement of hygiene monitoring methods in food processing.

In fact, the improvement of quality standards in food production processes is also present in two other statements, one regarding the development of on-line control of quality parameters and another dealing with the designing and application of biotechnological microorganisms which can be precisely controlled and regulated in their metabolic activities. As a result, it could be concluded that a better performance in food processing, in terms of quality control of the final product, is one of the trends that the Spanish sample considers more positive to reinforce.

Table 3.39: Statements with very "positive attitude" in Spain

Statement No.	Content	Proportion of "positive attitude" (%)
62	Rapid test systems based on modern biotechnology are widely used for pathogen identification in plant production and animal husbandry.	99
42	The hygiene monitoring of food processing is significantly improved due to the widespread use of modern biotechnological analytical methods.	99
31	Food enriched with specific microorganisms having positive health effects achieve approximately 25 % turnover share in their product group (e. g. yoghurt).	98
43	Techniques based on modern biotechnology are practically used for on-line control of quality parameters in food processing (e. g. content of micronutrients or harmful substances).	96
58	The molecular basis of virus resistance mechanisms in economically important perennial plants like e. g. vine, olive and fruit trees is elucidated.	96
32	Modern biotechnology significantly contributes to the transformation of 40 % or more of the organic agricultural waste into marketable products (e. g. energy, secondary raw materials).	96
41	The long-term health impacts of the use of Agro-Food products made with the help of genetic engineering on consumers, farmers and employees are investigated (e. g. by epidemiological surveys).	96
57	New monitoring and control techniques for genetically engineered plants in open fields have improved the reliability of risk assessment significantly.	95
55	Genetically engineered plant varieties resistant to salinity and/or drought have improved agricultural productivity in arid environments significantly.	95
48	A large variety of genetically engineered microorganisms which can be precisely controlled and regulated in their metabolic activities during food production processes has been developed.	95

It is remarkable here to point out that two developments considered positive by over 95 % of the Spanish experts refer to a very Mediterranean problem with a great economic relevance. Water stress and the salinity of the scarce water available make the designing of plants resistant to salinity and drought very interesting, as they would improve agricultural productivity in Spain. Furthermore, the elucidation of the molecular basis of virus resistance mechanisms in perennial plants (e. g. vine, olive and fruit trees), very important in Spanish national economy, is consequently seen as very positive as well.

In addition, the reduction of waste and emissions in farming due to the use of biotechnology is seen as very positive. In this sense, statement 32 is about one of the strategic research lines already highlighted by the Spanish national expert committee, as it deals with the transformation of organic waste into marketable products.

In summary, positive attitudes are more common regarding the realizations proposed by the statements belonging to the domains "scientific/technological development", "regulation", "acceptance" and "environment". The advances in those issues are considered rather positive by the Spanish experts, especially the ones regarding "regulation".

Only two statements reach over 50 % indifference response rate, so more than half of the experts who answered seem to be able to express an opinion in practically all issues addressed by the questionnaire. Those two statements which leave people more indifferent are 45 and 7 (64.6 % and 57.8 % of the respondents, respectively), followed by statements 47 and 14 (49.3 % and 40.6 % of the responses). All of them recreate situations, where genetically engineered organisms are widespread marketed products or, as is the case for statement 47, biotechnology fails in producing food and beverages in satisfying traditional taste preferences. Experts could either not be really interested - nor uninterested - in the commercial success of such products or not fully aware of its consequences, or maybe foresee them so far away in time that they do not have a formed opinion.

On the other hand, regarding the developments and situations which evoke a negative attitude by 50 % of the experts' sample, these are concentrated in twelve statements (adding statements 4 and 17 to the ones already presented in table 3.40). None of them belong to the domain "scientific and technological developments". That means, negative opinions are more frequent concerning the framework conditions of biotechnology applied to the Agro-Food sector than to the technology itself.

Those "negative" situations refer in many cases to economic damages caused by different impacts of the implementation of genetic engineering techniques. Such damages could affect either consumers, Spanish firms, farmers or the labour market itself. The reduction of jobs, either because of the delay in introducing biotechnological processes and techniques in Spain or, paradoxically, because of using them on a broad scale, is one of the risks which arouse most concern.

Table 3.40:   Statements with very "negative attitude" in Spain

Statement No.	Content	Proportion of "negative attitude" (%)
40	Unintended impacts on the health of consumers occur in some products or manufacturing processes due to the use of genetic engineering during food processing.	89
27	The public debate about environmental risks of the deliberate release and marketing of genetically engineered microorganisms (e. g. uncontrollable spread, disturbance of natural balances) has led to a broad rejection of such activities in Spain.	83
19	Small farmers in Spain are not able to afford new genetically engineered plants and animals.	83
26	The widespread deliberate release of genetically engineered organisms results in a significant transfer and recombination of the introduced genes with unintended negative impacts (e. g. acquisition of resistances to herbicides and pathogens in weeds).	81
37	The increased use of modern biotechnology in food production and processing results in additional allergies in people actively involved in these processes.	81
13	The uniform implementation of the EU biosafety directives in all EU countries has led to a higher attraction of the EU for companies based outside the EU.	77
34	The widespread use of genetically engineered organisms in food production significantly aggravates the loss of organisms traditionally used in Spain.	75
24	Due to the hesitant application of genetic engineering (compared to other EU member states) the Agro-Food industry cuts 10 % or more of the jobs in this sector in Spain.	73
20	The widespread use of modern biotechnology in animal and plant production leads to an approximate 30 % decrease in traditional jobs in agriculture.	72
15	The prices of food products in specific market segments of which the quality is significantly improved by the application of modern biotechnology are at least 30 % higher than that of corresponding conventional products.	67

Another economic disadvantage for Spain would be that, as Spanish farmers have to get genetically improved seeds from multinationals at an "international" price, the national farming sector would lose market rate. This would promote a dependence situation, encouraged by the fact that there is not a strong biotechnological industry in the country. Finally, another worry regards the possibility that genetically engineered food products are more expensive than the other ones, so this "high quality food" is restricted to well-off domestic economies.

Some other undesirable effects would refer to the impact of biotechnology on public health (almost 89 % of the sample see it as very negative, logically, unintended health impacts on people involved in production processes using biotechnological products), loss of biodiversity, the building up of insect resistance to biocides etc. It is important to remark that these situations are considered as being not very likely to happen, at least in the short run, as the estimations in the category "time of realization" show.

In contrast to the other answering categories, there are some differences in the "personal attitude" between the different expert groups in Spain whose analysis brings some information on the opinion which the different social sectors hold regarding biotechnology applications. A first conclusion coming from the comparison between expert groups is that the domain "scientific/technological developments" is never considered among the "top 10" statements for negative attitude. So, the problems deriving from modern biotechnology applied to the Agro-Food sector in Spain are more related to the sort of management of this subject than to the research in this technical field itself. In fact, for most Spanish experts framework conditions are the real challenge for the development of modern biotechnology.

In the case of researchers the domain "scientific/technological developments" is the one which arouses a better personal attitude: nine of the "top 10" statements for positive attitude are under that headline (see figure 3.13). The perception of the benefits of the scientific development decreases in the other expert groups, although all three groups agree in showing a very positive attitude regarding statement 62: "rapid test systems based on modern biotechnology are widely used for pathogen identification in plant production and animal husbandry".

Scientific developments referring to plant production are the ones seen as more important by research institutions and experts from the industrial field, while consumers and critics give more importance to the approaches which improve the hygienic standard of food processing.

Figure 3.13:   "Positive attitude" by groups and domains in Spain

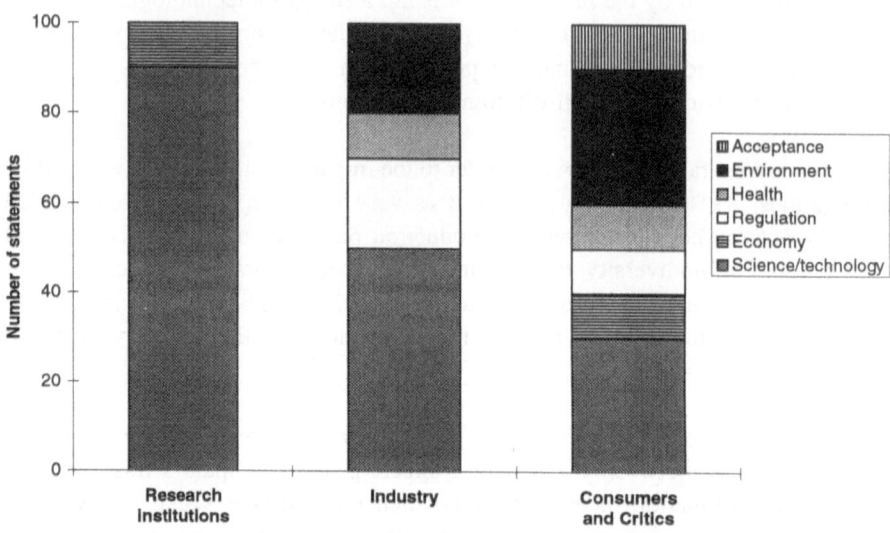

Industry experts as well as consumers and critics also highly appreciate the implementation of a case-by-case procedure to approve the marketing of food and beverages produced with the help of genetic engineering. Both expert groups also highlight the positive effects that modern biotechnology may have on the environment as, for instance, the reduction of the pollution originated by the use of pesticides and agricultural waste, an aspect which is not regarded as that positive in the researchers' opinion.

On the other hand, "economy", "health" and the "environment" are the areas of more concern, while dealing with the possible negative impacts of biotechnology, the economic issues are the most often mentioned. This trend is valid for all three expert groups (see figure 3.14). They also agree in considering the development of statement 40 ("Unintended impacts on the health of consumers occur in some products or manufacturing processes due to the use of genetic engineering during food processing") as very negative, as the three groups have it on the list of the "top 10" for negative attitude: 87 % of the experts of research institutions and industry and 96 % of the consumers and critics reject this development.

Finally, regarding the environment, the three groups show a very negative attitude (between 66.7 % and 87.5 % of the experts) towards statements 26, 27, 34:

• The widespread deliberate release of genetically engineered organisms results in a significant transfer and recombination of the introduced genes with unintended negative impacts (e. g. acquisition of resistance to herbicides and pathogens in weeds) (statement 26).

- The public debate about environmental risks of the deliberate release and marketing of genetically engineered microorganisms (e. g. uncontrollable spread, disturbance of natural balances) has led to a broad rejection of such activities in Spain (statement 27).

- The widespread use of genetically engineered organisms in food production significantly aggravates the loss of organisms traditionally used in Spain (statement 34).

In this context it is remarkable that even consumers and critics are aware that the public rejection of modern biotechnology can be due to the publicity given to its negative impacts and, furthermore, consider such phenomenon as not desirable. On the other hand, this expert group is the only one which seems to encourage strongly the establishment of "a governmental institution which provides information on modern biotechnology in the Agro-Food sector to all interested persons and institutions" in Spain (statement 1), as they include this statement among the "top 10" for positive attitude.

Many of the comments which the experts wrote in their questionnaires mention that a great problem in Spain is that the public does not have reliable information on biotechnological issues, so the only information available is the one released by ecologist organizations in their campaigns against transgenic food products. A good part of the experts, mainly scientists and researchers, but also some people related to the ecology movement, complained about this situation. It has to be said that the already mentioned attitude of Spanish industry, their reluctance to release any information about its activities or researches, does not help to solve this lack of information on genetic engineering.

Industry is the only expert group which shows a very negative attitude towards a statement of the "acceptance" area: "research on genetic engineering of farm animals is not funded by public institutions in Spain due to ethical reasons (e. g. animal rights, animal welfare, preservation of Creation)" (statement 4). People from the industrial sector appear very disappointed with the general social feeling against genetically engineered food in Spain and the limited activities that the government is carrying out to change the situation. As they stated in their comments, this is one of the main factors which prevents them investing more in biotechnological research and marketing the products that they have already developed.

Figure 3.14:   "Negative attitude" by expert group and by theme in Spain

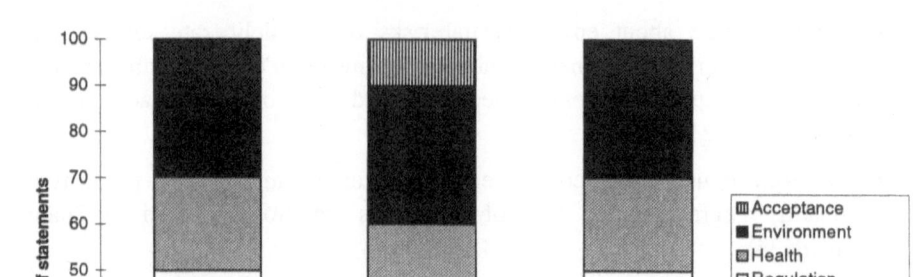

Time of realization

Most Spanish experts (83 %) believe that the developments included in the statements will be realized within the next ten years (see figure 3.15), 1 % believes that they could occur in more than 20 years and another 1 % thinks that they will never be realized.[4] In fact, only the sub-categories "up to five years" or realized in "six to ten years" reach in some statements more than 40 % responses. The other sub-categories of "time of realization" were not chosen by so many experts.

Regarding the statements more likely to happen in the short run, more than 80 % of the respondents stated that rapid analysis systems based on biotechnology will be widely used to identify pathogens in the next five years, food enriched with specific microorganisms having positive health effects achieve approximately 25 % turnover share in their product group (e. g. yoghurt), the labelling of all food products made with the help of genetic engineering is compulsory and fully implemented in Spain, the hygiene monitoring of food processing is significantly improved due to the widespread use of modern biotechnological analytical methods and, as well, the

---

4   By comparing the first and second rounds of the Delphi survey, it can be seen that the time of realization of the statements has shortened in general. For example, there is a shift from 60 % of respondents of the first round who thought that the statements will be realized in the next 10 years, to 83 % in the second round. Besides this, both in sub-categories which express a time scale of greater than eleven years and in those which express "will not be realized", the response decrease was uniform.

Spanish government will set up an institution to offer information on biotechnological issues related to Agro-Food (see table 3.41).

Figure 3.15:   Time of realization of the statements in Spain (according to the median class)

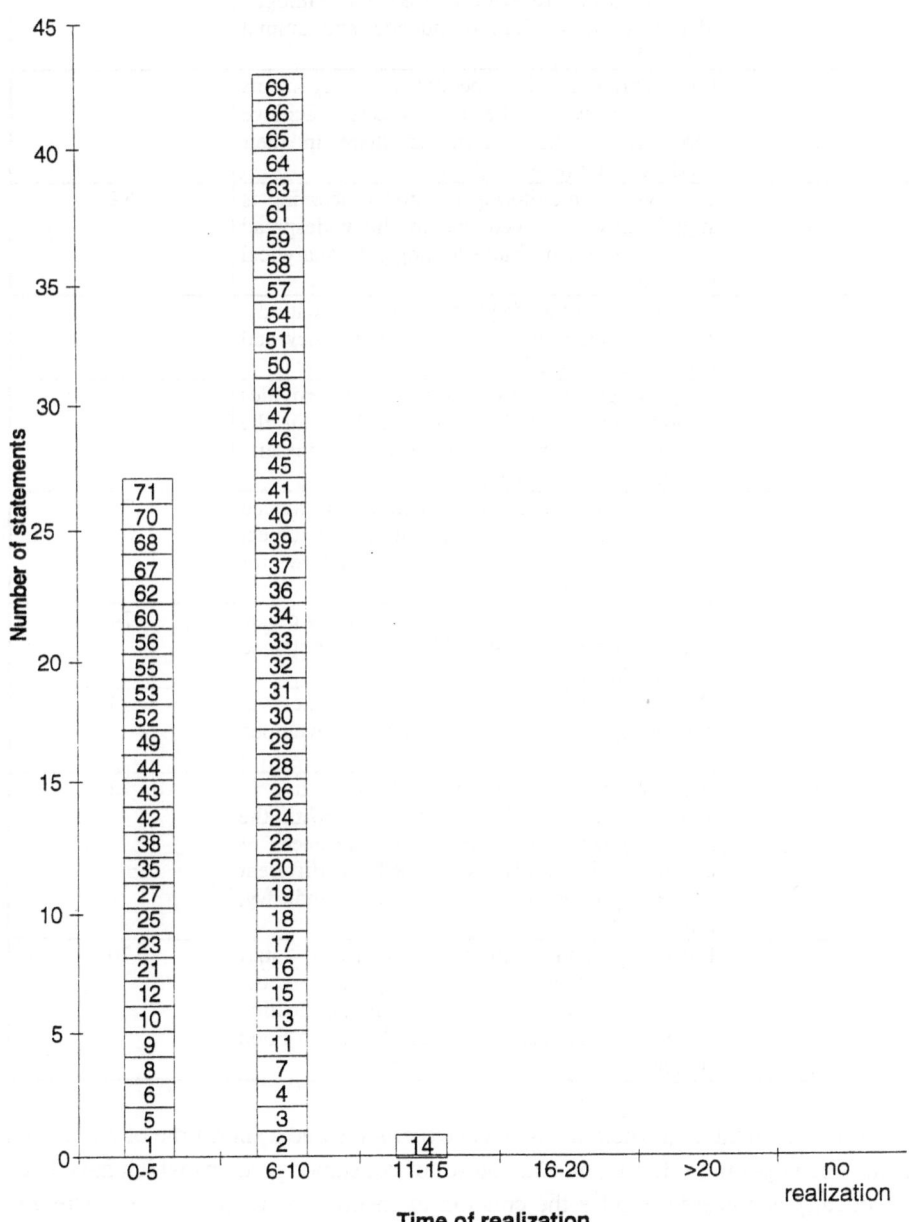

Table 3.41:   Short-term developments in Spain

Statement No.	Content	Proportion of "< 5 years" (%)
62	Rapid test systems based on modern biotechnology are widely used for pathogen identification in plant production and animal husbandry.	87
38	Food enriched with specific microorganisms having positive health effects achieve approximately 25 % turnover share in their product group (e. g. yoghurt).	84
42	The hygiene monitoring of food processing is significantly improved due to the widespread use of modern biotechnological analytical methods.	82
8	The labelling of all food products made with the help of genetic engineering is compulsory and fully implemented in Spain.	81
1	A governmental institution which provides information on modern biotechnology in the Agro-Food sector to all interested persons and institutions is established in Spain.	81
9	All food and beverages which are produced with the help of genetic engineering are subject to an official case-by-case procedure for approval in Spain.	78
6	Food companies and retailers proactively create specific labels for food and beverages produced without the help of genetic engineering.	74
52	The production of industrial enzymes in genetically engineered field crops is developed for large-scale application.	73
5	An advisory committee with the objective to assess whether food products made with the help of genetic engineering meet the needs of consumers is jointly established by different interest groups (e. g. consumers, industry, retailers) in Spain.	72
12	European and national authorities start initiatives which involve the public in the debate and decision-making on the application of modern biotechnology in the Agro-Food sector.	70

Another important area where the Spanish experts foresee a short time of realization is the one regarding information and control of consumer products which have been genetically engineered besides the creation of institutions to control, evaluate and monitor such products, both before and after being marketed. Despite this, in the opinion of a significant percentage of the Spanish experts (16 %) genetically engineered food products will not achieve a 30 % market share of the food market in the following 20 years (see table 3.42). As we detected in some of the comments made by the participants in the survey, this reluctance to accept a high penetration of such products in the Spanish market in the short term should be qualified. Probably, quantifying their popularity in a specific percentage poses a problem. In the case of Spain, 30 % was felt as being too high.

Table 3.42:   Long-term developments in Spain

Statement No.	Content	Proportion of "> 20 years" (%)
7	A restaurant chain or catering service specialized in offering trendy meals with genetically engineered ingredients open branches in almost all large cities in Spain.	22
14	Food made with the help of genetic engineering achieve a turnover share of 30 % or more of all food consumed in Spain.	16
20	The widespread use of modern biotechnology in animal and plant production leads to an approximate 30 % decrease in traditional jobs in agriculture.	5
61	After years of experimentation with genetic engineering, the majority of plant breeders strongly prefers the combination of marker-assisted breeding with traditional breeding methods, compared to the use of genetically modified plants.	4
34	The widespread use of genetically engineered organisms in food production significantly aggravates the loss of organisms traditionally used in Spain.	4

In table 3.42 the statements are included with a relatively high response rate for the sub-category "more than 20 years". It is remarkable that these statements for a longer time of realization are based on very small percentages. From those, the possibility of a restaurant chain selling genetically engineered food opening new branches through the country is the one seen as least feasible (statement 7). Despite this, and despite the fact that 16 % of the sample think it will not occur in the next 20 years and 9 % affirm it will not happen, almost half of the respondents (49 %) still believe it will be realized in the next five years.

Regarding the three expert groups, differences between them appear to occur in the assessment of long-term developments as well as unrealistic future options, while they are rather rare between the three groups of experts in the assessment of short-term developments. Despite this, it can be observed that consumers' and critics' mean is lower than the other groups' in the time category "up to five years", i. e.

consumers and critics delay the realization of the biotechnological developments mentioned in the questionnaire. As a result, the sub-categories "more than 20 years" and "will not be realized" achieve the highest relevance in this expert group.

Influential factors

The most important factor influencing biotechnology development in Spain is "infrastructure in research and development" (R&D). For example, between 21 % and 30 % of the sample consider "R&D infrastructure" one of the three most influential factors for the realization of over half of the statements in the questionnaire (37 out of 71). "Technology transfer" from research institutions to private firms goes very close together with R&D infrastructure as a factor conditioning the development of genetic engineering in the Agro-Food sector in Spain, followed by "markets" and "information".

On the other hand, "policy", "funding", "ecology" and "international cooperation" are not considered very influential in the Spanish case, with the last two mentioned here seen as least relevant ("international cooperation" reaches the 20 % level of the answers for no statement, while 25 % and 22.8 % of the sample believe "ecology" is a very strong conditioning factor for the realization of statements 26 and 30, both belonging to the domain "environment").

By statements, the highest percentual values for each influential factor are related, logically, to the themes the statements belong to (see table 3.43). So, investments in R&D is a very influential factor for the statements about the research developments in science and technology. The same applies to industrial innovativeness, which seems to be quite relevant to the fostering of research and, as well, to make a success of biotechnological products and processes. The most obvious examples here are the ones relating to the influential factors "social/ethical acceptance" and "policy" which are especially relevant for the statements on acceptance and regulation issues.

In this sense, no significant differences are noticeable by expert group. That is, researchers, industry, consumers and critics present a similar distribution, while defining the factors which influence the development of biotechnology in future. Despite this, it has to be said that R&D infrastructures and technology transfer are mentioned more often and in a better position in the industry group. This is also the only group in which the factor "policy" is represented to a higher extent, while consumers and researchers appear to prefer the factors "market" and "information".

Table 3.43:   Most influential factors in Spain

Statement No.	Factor	Highest value in each sub-category (%)
15	Markets	37
44	R&D infrastructure	35
3	Information	35
7	Social/ethical acceptance	34
17	Industrial innovativeness	34
51	Technology transfer	33
12	Policy	28
9	Regulations/standards	27
26	Ecology	25
7	Personnel	22
19	Funding	21

Importance of the statement

The contents of the statements of the questionnaire seem to be especially important for improving the economic competitiveness, while they do not appear to be crucial at all for the protection of the environment and the implementation of sustainable production strategies in Spain (see figure 3.16). In general, this sub-category is the one which produces most indifference.

It is important to point out that, although the three expert groups coincide in placing more importance on the factor "economic competitiveness" than on the rest (as in the global sample), there is apparently a tendency on the part of the consumers and critics to score higher in each sub-category. That could mean that this group gives more importance than the others to the general impacts of biotechnological developments.

Although the statements considered important for each one of the sub-categories are diverse, there are some which appear to be relevant for all three sub-categories or, at least, for two of them, as is shown in table 3.44. It could be said that those statements, especially 55 and 56, could probably be the most relevant biotechnological issues in Spain. It is quite significant that both of them, as well as statement 57, refer to biotechnology applied to plants, which is probably the most developed area within the Agro-Food sector in Spain.

Figure 3.16:   General importance of the statements in Spain (mean of index of all statements)

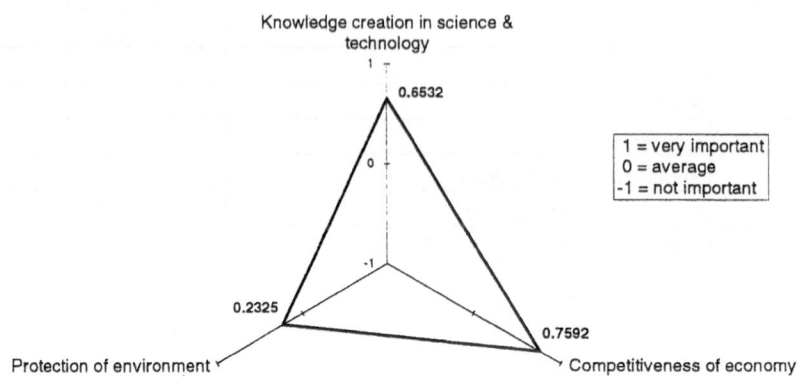

Table 3.44:   Statements classified as "very important" in Spain

Statement No.	Content	Index value		
		Science & tech-nology	Eco-nomy	Environ-ment
56	New genetic processes have been developed improving plants defence mechanisms (i. e., combinations of resistant genes, improvement of pathogen tolerance) in order to prevent pathogens resistance to biocides.	0.95	0.93	0.68
55	Genetically engineered plant varieties resistant to salinity and/or drought have significantly improved agricultural productivity in arid environments.	0.94	0.93	0.68
57	New monitoring and control techniques for genetically engineered plants in open fields have improved the reliability of risk assessment significantly.	0.97	-	0.81
25	Enzyme systems are specifically developed to improve the environmental performance of conventional food-processing procedures.	0.96	-	0.77

The tendency observed in the relation between the theme of the statement and the type of factor which influence them most is also evident in this section. For instance, the statements classified as "very important" for "knowledge creation in science & technology" are the ones on the subject "scientific and technological developments" and not the ones under the domain "regulation". So, the "top 10" statements for each one of the three sub-categories are the following:

- Knowledge creation in science and technology: statements 57, 44, 41, 25, 56, 67, 62, 68, 55, 66

- Competitiveness of economy: statements 18, 16, 46, 70, 59, 52, 49, 50, 55, 56

- Environmental protection: statements 31, 32, 30, 29, 28, 57, 25, 56, 55, 33

Having said this, it is remarkable that most of the important developments to improve the economic competitiveness of Spanish firms do not belong to the domain "economy" but to "scientific and technological developments"[5]. Two groups of statements seem interesting from an economic point of view:

- Enzymes: widespread application of genetically engineered enzymes to food production and processing (statements 49, 50 and 52)

- Plants: application of modern biotechnology to improve plants resistance to drought, salinity and plagues as well as the use of plants to produce non-food products (statements 55, 56, 59)

The results of the Delphi survey support the estimation of the Spanish national expert committee, which assessed these two research fields as being of strategic interest in Spain in future. The results for the sub-category "environmental protection/sustainable development" reinforce this argument as six of the statements the experts considered as "very important" in this sense refer to biotechnology applied to plant production (statements 28, 29, 30, 55, 56, 57) and one to enzymes specifically designed to improve the environmental performance of the conventional methods for food processing (statement 25).

The other three statements "very important" to environmental protection refer to the use of genetically engineered organisms to process farm and agricultural waste (statements 31 and 32) and to the integration of genetic engineering approaches in organic farming (statement 33).

---

5 Despite this, statements belonging to the themes "regulation" and "acceptance" have the greatest increases regarding their importance to economic competitiveness, with average values of approximately 0.2 of the index between the first and second round of the Delphi survey. This means that, after the feedback, Spanish experts believe more strongly that the establishment of more regulations and the promotion of public acceptance are two key factors to improve national competitiveness in biotechnology applied to the Agro-Food sector.

On the other hand, the most relevant issues for the development of science and technology refer to diverse types of research lines, from plant monitoring and control to the decoding of the sequence of economically important bacteria. Because of its particular climate conditions, the production of plants resistant to salinity (statement 55) is, again, included in the "top 10" of the Spanish scientific activity.

It is interesting to point out here that one of the developments considered very important to the advancement of science and technology is statement 41: "The long-term health impacts of the use of Agro-Food products made with the help of genetic engineering on consumers, farmers and employees are investigated (e. g. by epidemiological surveys)." These health impacts seem to indicate that such effects are still unknown, although the common attitude that the Spanish scientific community has towards the lay public is to deny risks in biotechnology application.

Finally, statements about the progress in basic research (statements 62, 66, 67 and 68) are considered amongst the "top 10" only for "knowledge creation", but not for the other two sub-categories. Probably, experts not working in the field are not able to judge the scientific relevance of such advances as the widespread use of genetic maps to identify some economically important trait in farm fishes, for instance (statement 68).

While looking at the correlations between sub-categories, there seems to be a quite clear correlation between the sub-categories "competitiveness of the economy" and "knowledge creation in science and technology". Accordingly, a deeper knowledge of biotechnological issues should have a positive impact on the national economy. In this sense, the advance in the knowledge of plant biotechnology is one of the areas which appear to be more attractive to Spanish experts, and so more likely to have a positive impact on the national economy.

Besides this, the creation of knowledge does not involve, as a general tendency, an improvement in environmental protection or in the sustainability of human activities. There are some specific cases, regarding again the improvement of plant production by using genetic engineering, in which scientific interests go in the direction of environmental protection (statements 28, 29, 30, 31, 32). These particular statements are also very significant in the analysis of the correlation "environmental protection" and "economic competitiveness". As a general trend, the economy is much more important than environmental issues, but the improvement of plant resistance and productivity satisfies both parameters in the Spanish experts' opinion.

Summary of Spanish results

- The level of knowledge on biotechnological issues seems to be quite low for the Spanish experts.[6] This data, though, has to be qualified. There is not a clear trend in the distribution of the category "degree of knowledge" among the different expert groups although, obviously, scientists would know more than journalists, for instance. Self-definition of knowledge always brings this problem, which needs further study.

- There is no trend on the level of response through the questionnaire. No domain is highlighted as being better known by the experts than others.

- The biotechnological developments treated by the statements will mainly be realized within the coming ten years, as most of the Spanish experts believe (83 %). This will especially relate to improvements in the quality of food products, development of pharmaceuticals and control systems to monitor the quality of food processing and, as well, to the development of genetically engineered enzymes.

- The formulation and application of new regulations on biotechnological applications, and on the consequent commercialization of its products, is the issue which is expected to be realized rather in the short term. Clear regulations are seen as a very influential and necessary element for the promotion of modern biotechnology in Spain.

- The general attitude of Spanish experts to biotechnology is more positive than negative. The statements which arouse the most positive attitudes are about scientific and technological developments rather than dealing with framework conditions.

- This generally positive attitude is standard in all expert groups, which do not show any significant differences from each other.

- Experts' negative opinion on biotechnology is quite concentrated on certain issues regarding on negative impact it may have to the Spanish economy and labour market, as well as the possible link between using genetically modified organisms and increasing the risk of suffering from allergies or causing environmental damages.

- R&D infrastructures are considered the most influential factor for the development of modern biotechnology in Spain, while international cooperation

---

6 The feedback, characteristic of a Delphi survey, had a clear impact on the self assessed level of knowledge of the participants: it increased in the second round. This result could have something to do with two news stories widely covered by the Spanish mass media while the second round was being carried out: the unloading of Monsanto's transgenic soya bean in Barcelona, and the cloning of "Dolly".

is seen as negligible. Markets and technology transfer also have a high importance.

- The future visions included in the questionnaire are especially important to foster economic competitiveness, as the Spanish experts see it, and their contribution to environmental protection or to sustainability is quite low. Apart from that, the expert group "industry" considers policy as a very influential factor, the only group to do so.

- In the consumers' and critics' view market conditions and (the release of reliable) information are the main factors that determine the development of modern biotechnology in Spain.

- Plant biotechnology is the main area of interest in Spanish research, and its advancement is seen as very important for knowledge creation, the economy and the protection of the environment.

- Genetic engineering applied to plants and enzymes are the two strategic fields for applied research in Spain at the moment, as they seem to be the ones more likely to have an economic impact in the short term.

- There is a clear correlation between economic competitiveness and biotechnological research, they progress together. This does not happen to environmental protection/fostering of sustainable development because, as the results indicate, economic or scientific progress does not necessarily involve a better treatment of the environment according to the opinion of the Spanish sample.

4.   Comparative analysis

4.1   Differences between countries

In this chapter common features and differences between the five countries are analysed. The presentation of the results is done according to the structure of the questionnaire. The target is to elaborate major differences between the five countries involved but not to analyse their response behaviour in detail. In order to filter out main differences concerning the personal attitude and the time of realization, a specific statistical test has been used which compares the distribution of answers between two samples. The respective statistical analysis has been carried out with the U-Test of Mann-Whitney for non-parametric data (Kähler 1995).

4.1.1   Degree of knowledge

In general, the experts in the five countries assess their individual knowledge of biotechnology applied to the Agro-Food sector as being quite low. Summing up the given self-estimation of all experts to all 71 statements the overall distribution of knowledge can be compared between the different countries. The results of this analysis are shown in table 4.1.

Table 4.1:   Overall degree of knowledge of all experts asked in the five
               European countries

Degree of knowledge	Germany	Greece	Italy	The Netherlands	Spain
Very familiar	12 %	4 %	9 %	7 %	4 %
Average familiar	45 %	12 %	24 %	37 %	23 %
Less familiar	29 %	54 %	34 %	31 %	37 %
Not familiar	14 %	30 %	34 %	26 %	36 %
Number of interviewees	522	192	149	204	151

Overlooking table 4.1, two groups of countries can be distinguished in relation to the category "degree of knowledge". The German and Dutch experts have higher self-estimation of their expertise. The majority of these samples estimate their personal expertise as "average familiar" for each statement. Especially in Germany, 57 % of the answers are " average familiar" or "very familiar". The percentage of answers for "very familiar" among the German sample is double the Dutch one.

Conversely, the experts of the involved Mediterranean countries assess their degree of knowledge as considerably low. Again, among these three countries the Greek experts show the lowest level of knowledge, i. e. only 16 % of the answers are "very familiar" or "average familiar" with the issue addressed by the questionnaire. The Spanish and Italian samples have a very similar structure regarding the self-estimation of knowledge. Despite this, the percentage of the answers for "very familiar" among the Italian samples is double the one for the Spanish sample.

In all countries more than 25 % of the total answers are defined as "not familiar", except Germany, whose panel only considers itself unfamiliar in 14 % of the answers. In the case of the Netherlands the category "not familiar" appears in one out of four answers, while about one third of the answers are "not familiar" in the Mediterranean countries.

The most "unknown" statements are the same for all countries: these are the statements 21, 67, 68 and 69, dealing with biotechnological approaches in fish production, and also statement 60 which focuses on the content of erucic acid in rapeseed. On the other hand, the issues that seem to be best known by the samples are the ones describing general framework conditions.

4.1.2   Personal attitude

Figure 4.1 gives a general overview of the personal attitude of the experts in the five countries. For this purpose the relative frequencies of the first three sub-categories of "personal attitude" are classified in the following way:

- Highly positive: first sub-category higher than 60 %
- Slightly positive: first sub-category between 50 % and 60 %
- Indifferent: all three sub-categories below 50 % or second sub-category between 50 % and 60 %
- Slightly negative: third sub-category between 50 % and 60 %
- Highly negative: third sub-category higher than 60 %

More than half of the statements are appreciated by the experts of the analysed European countries (see figure 4.1). In general, the Spanish experts have the strongest positive stance concerning the application of biotechnology and genetic engineering in the Agro-Food sector, followed by the Italian and Greek experts. In comparison to the Mediterranean countries, the experts of the Central European countries (Germany, the Netherlands) react in a more reserved manner with a lower percentage of "highly positive" and a rather high percentage of "indifferent" personal attitude. Less than a fifth of all statements are opposed by the experts of the five European countries. All in all, the experts of the Central European countries are more critical than the experts of the Mediterranean countries, whereby the Spanish experts are by far the most positive ones.

Figure 4.1: Personal attitude of the experts in the five European countries - tendencies of answering

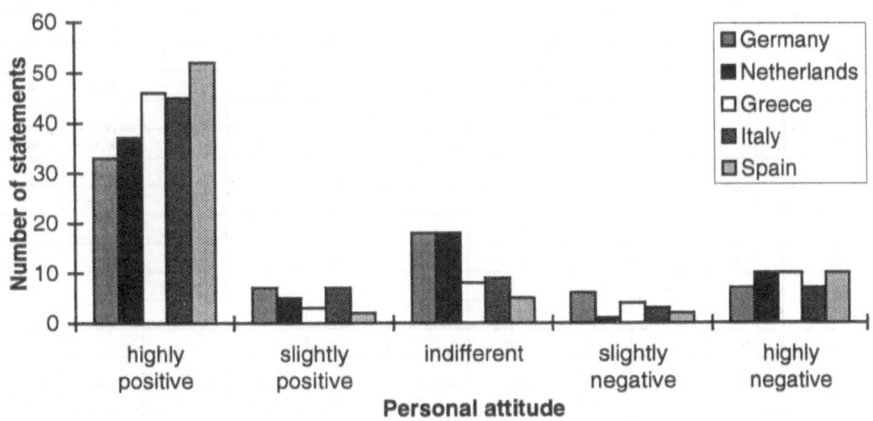

The U-Test of Mann-Whitney of non-parametric data has been used to filter out significant differences concerning the personal attitude of the experts in the five analysed countries. For this purpose the significance level of the Mann-Whitney test was set to 0.01. In table 4.2 the results of this analysis are registered.

Overlooking table 4.2, only in ten statements dono highly significant differences occur in the statistical analysis. The experts of all the five countries appreciate biotechnological approaches which contribute to reducing health problems like allergies (statement 36) and to developing market segments outside the food chain with the help of modern biotechnology (statements 22, 60). Moreover, they are in favour of extending the use of monitoring systems based on modern biotechnology (statements 43, 57). On the other hand, they oppose statements dealing with negative impacts on health (e. g. additional allergies) (statements 37, 40) and economy (e. g. reduction of employment) (statement 24, 20). A slightly different picture arises concerning the statements related to the environmental impacts of modern biotechnology. Even if in table 4.2 some singular differences are marked between the analysed countries, a more detailed analysis shows that the majority of the experts in all countries on the one hand appreciate developments with positive impact on the environment (statements 25, 31, 32), and on the other hand reject those statements dealing with negative impacts on the environment (statements 26, 34).

Table 4.2:   Differences in the personal attitude between the five countries

Statement No.	Significance level [1] of the differences of the answers of the compared countries									
	DE/NL	DE/IT	DE/ES	DE/GR	NL/IT	NL/ES	NL/GR	IT/ES	IT/GR	ES/GR
1	★			★★					★	
2	★★	★	★★				★★	★	★	★★
3	★★	★★	★				★★		★★	★
4		★		★★	★		★★		★★	★★
5	★			★★						
6		★		★★	★		★★		★★	★★
7										
8				★★			★★		★★	★★
9	★★	★★	★★	★★					★	
10	★★		★	★★					★★	
11				★★			★		★	★
12	★★			★★	★	★★			★	★★
13		★	★★	★	★	★★	★	★★		★★
14		★		★						
15	★		★	★						
16	★		★★			★		★		★
17		★★	★★	★★	★★	★★	★★	★★		★★
18			★★	★		★★		★★		★
19	★★		★		★					
20										
21		★★	★★	★★	★★	★★	★★	★		★
22										
23	★		★★					★		★
24										
25	★	★							★	
26				★						
27			★★	★★	★	★★	★★	★★	★★	★★
28		★	★★	★★		★				
29		★	★			★		★		
30	★	★		★★	★★		★★			★
31	★		★★				★		★	★★
32		★	★							
33	★		★★	★★		★				
34							★			
35				★			★		★	★★
36										
37										
38		★★	★★	★★	★★	★★	★★			★
39	★	★★	★★	★						
40										
41		★								
42			★							
43										
44		★★	★★	★★		★				
45	★★	★★	★★			★	★		★	★★
46	★		★★			★★		★★		★★
47	★★			★★	★	★				

Statement No.	Significance level [1] of the differences of the answers of the compared countries									
	DE/NL	DE/IT	DE/ES	DE/GR	NL/IT	NL/ES	NL/GR	IT/ES	IT/GR	ES/GR
48	★	★	★★	★		★★		★		★
49	★	★	★★			★★		★		★★
50		★★	★★		★	★★		★	★	★★
51	★★	★★	★★	★	★					★
52		★	★★	★	★★	★★		★	★★	★★
53	★	★★	★★	★	★	★★				★
54	★		★★	★		★★		★★		★
55		★				★				
56			★★			★		★		★
57										
58					★					
59	★	★	★							
60										
61				★		★★				★
62		★								★
63		★★	★★	★★	★★	★★	★★	★		★
64		★	★★	★	★	★★	★	★★	★★	★★
65		★★	★★		★	★★	★	★★	★★	★★
66		★★	★★	★	★★	★★	★	★★		★★
67		★	★★			★★	★	★	★	★★
68		★★	★★		★★	★★			★	★★
69		★★	★★		★★	★★		★	★	★★
70	★★	★	★★			★★	★★	★★	★	★★
71			★	★						
Number of statements [1]										
★	16	18	11	16	12	10	10	13	14	16
★★	10	16	33	19	9	22	13	10	9	22

1) The distribution of the answers of the different countries has been analysed by using the U-Test of Mann-Whitney. In the table the results are documented in the following way:
★ : Differences are highly significant (0.01 level)
★★ : Differences are more significant than 0.001

Summing up, this indicates that negative economic and health impacts as well as negative and positive environmental impacts of modern biotechnology are assessed rather homogeneously in all involved countries. The same occurs concerning the development and diffusion of renewable resources and monitoring techniques due to modern biotechnology.

Apart from these areas there are rather clear differences between the analysed countries. In all, one can distinguish between a critical, more cautious country group, comprising the German, Dutch and in some areas Greek experts, and an optimistic, rather technology enthusiastic group comprising the Italian and Spanish experts. The most obvious differences between these two groups emerge concerning statements which deal with the application of modern biotechnology in animal husbandry and animal breeding (statements 63, 64, 65, 66, 67, 68, 69). The German,

Dutch and Greek panels estimate the use of modern biotechnological approaches and genetic engineering in animal production rather critically, i. e. they have an indifferent or negative personal attitude, while the Spanish and Italian panel members regard these developments more positively.

In addition to this country division, national particularities can be figured out related to the assessment of specific statements.

The German experts are rather critical regarding case-by-case procedures for approval of food and beverages which are produced with the help of genetic engineering (statement 9) and a widespread integration of genetic engineering approaches in the production process of organic farmers (statement 33). These findings correspond with the results of a German population survey, showing that there is still a strong opposition to the application of biotechnology and genetic engineering in food processing (Hampel et al. 1997). The German experts are also rather reserved concerning the use of artificial polyfunctional enzymes for analytical applications in the Agro-Food sector (statement 51).

The Greek experts highly appreciate compulsory labelling of food and the introduction of control mechanisms for identification of food made with the help of genetic engineering, as well as detailed information about this kind of food (statements 8, 11, 35). These estimations are along the same lines as the cautious attitude of the Greek experts concerning the use of modern biotechnological approaches in animal breeding and production.

In Spain the national particularities emerge mostly in statements dealing with scientific and technological developments. Corresponding to the already mentioned rather technology enthusiastic stance of the Spanish experts, they show an extremely positive attitude towards genetic engineering approaches to improve plant resistance (statements 54, 55, 56) and the use of modern biotechnology for animal breeding and production (statements 63, 64, 65, 66, 67, 68, 69, 70), in contrast especially to the German and Greek panel. In the economic field the Spanish experts react rather positively towards the introduction of innovative processes and products in SMEs, the creation of new jobs and the decrease of production costs due to the use of modern biotechnology in the Agro-Food sector (statements 16, 17, 18, 23).

Both Spanish and Greek particularities can be figured out in case of statement 27, describing that the widespread public debate about environmental risks of the deliberate release of genetically engineered microorganisms has led to a broad rejection of such activities. The Spanish experts take the most negative position towards the stated situation while the Greek experts show by far the most positive stance. The comments of the experts in the five countries document that the majority of them appreciate a widespread public debate in general. However, different positions arise concerning the mode of discussion, i. e. especially the

Spanish experts followed by the Dutch, Italian and German experts have doubts about an objective discussion. They partly point out that there has been no public debate so far or only an emotional but less factual one. In contrast, the Greek experts do not doubt the possibility and objectivity of such discussions.

Taking this example into account it is understandable why the U-test shows a high amount of significant differences between Greek and Spain besides the expected clear differences between Germany and Spain (see table 4.2).

4.1.3   Time of realization

Figure 4.2 gives a general overview of the time period necessary to realize biotechnology-related scientific and technical developments as well as future visions related to the framework conditions in this field assessed by the experts of the five European countries.

Figure 4.2:   General overview of the time frame related to all statements assessed by the experts of the five European countries

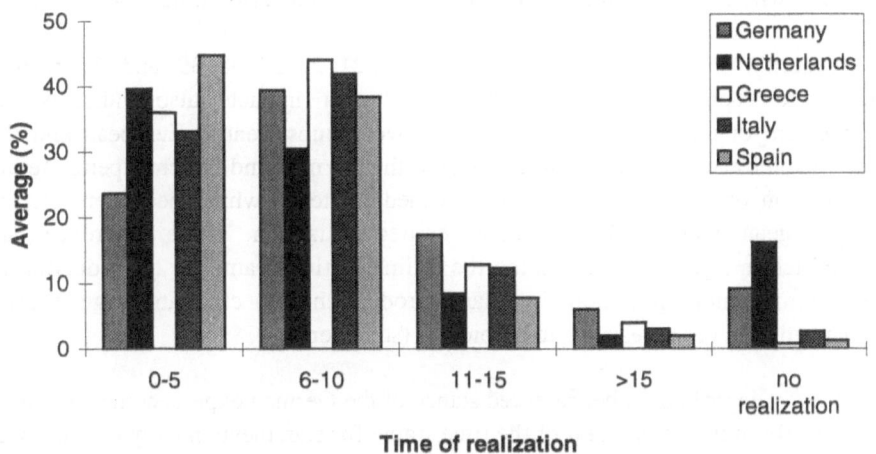

Time of realization

As illustrated in figure 4.2, more than 60 % of the statements are seen as realizable within the following ten years by the majority of the experts involved in the survey. Similar to the category "personal attitude", two poles can be observed. On one hand the Spanish experts, who believe in a short-term realization regarding most of the biotechnological developments addressed in the questionnaire, and on the other hand the German experts. Most of them predict a realization of the statements in medium term. In addition, even around 10 % of the German experts regard the statements as not realizable at all. The response behaviour of the Dutch, Italian and

Greek experts is positioned between these two poles. However, with respect to the Dutch estimation, it is important to point out that a considerable number of statements is seen as not realizable at all in this country (see table 4.4).

Similar to the category "personal attitude", the U-Test of Mann-Whitney of non-parametric data has been used to filter out significant differences concerning the estimated time of realization of the experts in the five analysed countries. For this purpose the significance level of the Mann-Whitney test was set to 0.01. The results of this statistical test are documented in table 4.3.

Overlooking table 4.3, only in statement 14 are no highly significant differences between the five countries registered. This statement relates to food made with the help of genetic engineering that achieves a turnover share of 30 % or more of all food consumed. The majority of the experts in the five countries uniformly expect that this development will take place within the next ten to 15 years.

All in all, a higher number of significant differences between the analysed countries occur in the category "time of realization" compared to "personal attitude". The highest number of differences are registered between Germany and Spain, similar to the category "personal attitude". Less differences arise between the Mediterranean countries, whereby the greatest similarities can be found between Italy and Spain.

The experts' estimation of the statements 18, 21, 28, 29, 38, 39 and 51 mainly relating to economic, environmental and health impacts also indicates the arrangement of the five countries into the two groups Central European and the three Mediterranean countries. In every case the German and Dutch experts predict a realization of the developments in the medium term, while the experts of the Mediterranean countries believe in an earlier realization. These differences are exemplified in figure 4.3, illustrating the estimated time frame for the reduction of the environmental pollution in plant production by cultivating genetically engineered field crops resistant to herbicides (statement 29).

The already described, rather reserved stance of the German experts comes out more significantly in the assessment of the time frame for statements mostly dealing with scientific and technological developments. According to the German expert panel, the use of modern biotechnological approaches for control and improvement of food and processing (statements 43, 46), the use of genetic engineering approaches in plant breeding and the prevention of resistance against biocides in pathogens (statements 52, 53, 56)) is assessed as being realizable in the medium or even long term. The same applies to the use of modern biotechnology techniques for animal breeding and production (statements 65, 66, 70, 71). In contrast, the experts of the other countries believe in an earlier realization of these developments.

Table 4.3:   Differences in the time of realization between the five countries

Statement No.	Significance level [1] of the differences of the answers of the compared countries									
	DE/NL	DE/IT	DE/ES	DE/GR	NL/IT	NL/ES	NL/GR	IT/ES	IT/GR	ES/GR
1	★★	★★		★	★★	★	★★	★★		★
2	★★			★	★★	★	★			
3	★★				★★	★★	★★			
4			★			★★	★			★
5	★★	★★		★	★★	★	★★	★★		★
6		★			★					
7	★★			★★	★★	★★	★★		★	
8	★★	★	★★		★★		★★	★	★	★★
9	★★	★★	★★	★★	★★	★★	★★	★		★★
10	★★	★★	★★	★★	★★	★	★★	★★		★★
11	★★		★		★		★★			
12	★★	★	★		★★	★★	★★			
13	★		★	★						
14										
15	★★	★★	★★	★★	★★	★★	★★		★★	
16	★★	★	★★	★★						
17	★★			★	★★	★★	★		★	
18		★★	★★	★★	★	★★	★★	★		
19	★★	★	★	★	★★	★★	★★			
20	★★		★	★★	★★	★★	★★			
21		★★	★★	★★	★★	★★	★★			
22	★★		★★							★
23	★★	★	★★		★★		★★			★★
24	★★				★★	★★	★★	★		
25	★★				★★	★★	★★			
26	★★				★★	★★	★★			★
27	★★		★	★★	★★	★★	★★		★★	★★
28		★★	★★	★★	★★	★★	★★	★★	★★	
29		★★	★★	★★	★★	★★	★★		★	★
30	★★	★★	★★	★★						
31	★★	★★	★★	★	★★	★★	★★	★	★	★
32	★★	★★	★★	★★		★★				
33		★★	★★	★★		★		★		★
34	★★		★★	★★	★★	★★	★★			
35				★★			★★	★	★★	★★
36			★★			★				★
37	★★				★	★★	★★			
38		★★	★★	★★	★	★★	★★	★		★
39		★★	★★	★★	★★	★★	★★			
40	★★					★	★			
41	★★		★★	★★	★★			★★	★	
42	★★	★	★★	★★	★		★	★		★
43	★★	★★	★★	★			★			
44	★★	★★	★★	★★		★★		★★	★	
45		★★	★★	★★	★	★	★★			
46	★	★	★★	★★		★	★			
47	★★		★★	★★	★★	★★	★★		★	

Statement No.	Significance level [1] of the differences of the answers of the compared countries									
	DE/NL	DE/IT	DE/ES	DE/GR	NL/IT	NL/ES	NL/GR	IT/ES	IT/GR	ES/GR
48		★★	★★	★★	★	★★	★★			
49	★★	★	★	★★	★	★	★			
50			★★	★		★		★		
51		★★	★★	★★	★★	★★	★★			
52	★	★★	★★	★★		★★		★★		★★
53	★★	★★	★★	★★						
54	★★	★★	★★	★★		★★	★★	★★	★★	
55	★★	★★	★★	★★	★★	★★	★★	★★	★	
56	★★	★★	★★	★★		★★		★★		★
57		★	★		★	★			★	★
58	★	★★	★★	★★		★	★	★	★	
59		★	★★	★★		★	★			
60		★	★						★	★
61	★			★	★	★	★★			
62	★★	★★	★★	★	★★		★★	★		★★
63	★★	★★	★★	★★	★★	★★	★★		★	
64		★★	★★	★★	★★	★★	★★			
65	★	★★	★★	★						
66	★	★★	★★	★★			★			
67	★★	★★	★★				★★		★	★★
68		★	★★			★★				★
69			★							
70	★★	★	★★	★★						
71	★★	★★	★★	★★		★		★		
Number of statements [1]										
★	7	14	11	12	11	16	11	13	14	15
★★	42	33	45	39	31	34	38	10	15	9

1) The distribution of the answers of the different countries has been analysed by using the U-Test of Mann-Whitney. In the table the results are documented in the following way:
★:   Differences are highly significant (0.01 level)
★★:   Differences are more significant than 0.001

Additionally, a specifically Dutch response behaviour emerges in a large number of statements. Overlooking the assessment of these statements, mostly dealing with acceptance, regulation and impacts of modern biotechnology on environment, health and economy (statements 3, 4, 5, 7, 9, 10, 12, 15, 17, 19, 20, 24, 25, 26, 27, 34, 37, 40, 47, 61 and 63), the Dutch panel members show a very pronounced stance, i. e. they expect a rather early or a very late realization of the respective development. Those statements which are seen as not realizable at all by a considerable part of the Dutch panel are registered in table 4.4.

Figure 4.3:   Assessment of the "time of realization" by the experts of the five
               European countries

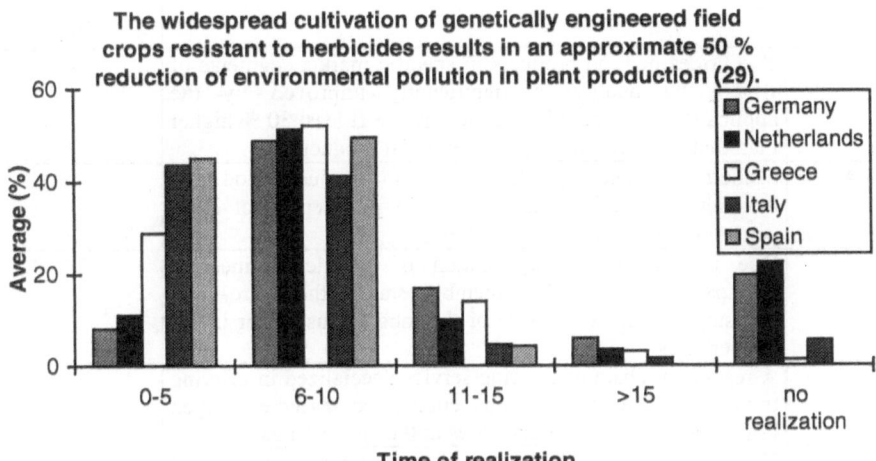

**The widespread cultivation of genetically engineered field
crops resistant to herbicides results in an approximate 50 %
reduction of environmental pollution in plant production (29).**

Time of realization

The Dutch panel does not expect higher prices for products of modern
biotechnology. In addition, they express the opinion that a strong job decrease in
agriculture will not take place due to the widespread use of biotechnology. They
also do not find a situation realistic in which the hesitant application of genetic
engineering leads to a loss of 10 % of the jobs in the Agro-Food sector
(statements 15, 20, 24) (see table 4.4). In addition, the Dutch panel has confidence
to prevent some possible negative impacts of modern biotechnology. This relates
e. g. to negative health effects of this technology. Almost two thirds of the Dutch
experts believe that additional allergies due to the increased use of modern
biotechnology in food production and processing will not occur in future (see
table 4.4). The same applies to possible negative environmental impacts
(statements 27, 26).

Table 4.4:   Statements scoring high for "will not be realized" in the Netherlands

Statement No.	Content	No realization (%)
15	The prices of food products in specific market segments of which the quality is significantly improved by the application of modern biotechnology are at least 30 % higher than that of corresponding conventional products.	84
47	Modern biotechnology has failed to produce food and beverages satisfying traditional taste preferences of large consumer groups in the Netherlands.	75
24	Due to the hesitant application of genetic engineering (compared to other EU member states) the Agro-Food industry cuts 10 % or more of the jobs in this sector in the Netherlands.	74
7	A restaurant chain or catering service specialized in offering trendy meals with genetically engineered ingredients open branches in almost all large cities in the Netherlands.	69
20	The widespread use of modern biotechnology in animal and plant production leads to an approximate 30 % decrease in traditional jobs in agriculture.	65
19	Small farmers in the Netherlands are not able to afford new genetically engineered plants and animals.	65
27	The public debate about environmental risks of the deliberate release and marketing of genetically engineered microorganisms (e. g. uncontrollable spread, disturbance of natural balances) has led to a broad rejection of such activities in the Netherlands.	64
37	The increased use of modern biotechnology in food production and processing results in additional allergies in people actively involved in these processes.	61
34	The widespread use of genetically engineered organisms in food production significantly aggravates the loss of organisms traditionally used in the Netherlands.	49
26	The widespread deliberate release of genetically engineered organisms results in a significant transfer and recombination of the introduced genes with unintended negative impacts (e. g. development of resistances to herbicides and pathogens in weeds).	43
63	Breeding procedures based on genetic engineering are practically used for those farm animals (cattle, pigs, poultry, sheep, goats) which are of major economic importance for food production in the EU.	42
4	Research on genetic engineering of farm animals is not funded by public institutions in the Netherlands due to ethical reasons (e. g. animal rights, animal welfare, preservation of Creation).	41
40	Unintended impacts on the health of consumers occur in some products or manufacturing processes due to the use of genetic engineering during food processing.	38

Figure 4.4:   Statements with Dutch specific response behaviour in assessing the "time of realization"

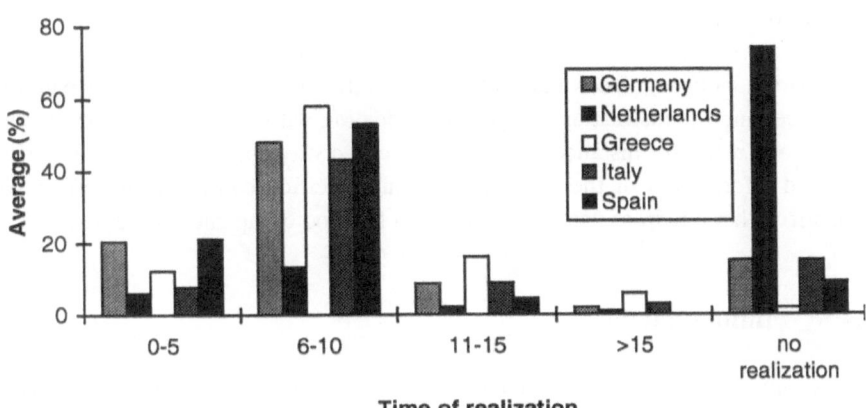

Due to the hesitant application of genetic engineering (compared to other EU member states) the Agro-Food industry cuts 10 % or more of the jobs in this sector in your country (24).

Enzyme systems are specifically developed to improve the environmental performance of conventional food processing procedures (25).

On the other hand, the Dutch experts predict some statements as being realizable in the short term in a very pronounced way (see figure 4.4). These statements mainly relate to the improvement of the environmental situation due to the application of biotechnological approaches (see figure 4.4), i. e. the use of environmentally friendly enzyme systems in food-processing procedures (statement 25) and some specific aspects of the regulation of gene food. This relates in particular to official case-by-case procedures for the market approval of genetically engineered food (statement 9), the involvement of social impacts and ethical conflicts during this process, as well as the involvement of the public in the decision-making on the application of modern biotechnology in the Agro-Food sector (statements 10, 12). The pronounced stance concerning different statements is probably caused by the fact that in the Netherlands an intensive discussion on the application of biotechnology in the Agro-Food sector has been carried out since the 70s (Menrad et al. 1996a).

Similar to the category "personal attitude", specifically Greek answering patterns arise, concerning the realization of statements dealing with information aspects. The Greek experts react, for instance, in a rather reserved way concerning the realization of detailed information of the consumers about specific health relevant effects of food made with the help of genetic engineering by food companies (statement 35).

4.1.4   Influential factors

For the future application of modern biotechnology the differing framework conditions in the five countries have to be taken into account. The results of the aggregated assessment of the influential factors are shown in figure 4.5. In this context an arrangement of the five countries in the two groups Central European and Mediterranean countries is also possible, whereby the differences mainly relate to the estimation of the factors information, social and ethical acceptance, policy, markets, technology transfer and personnel.

In particular, the demand-driven factors social and ethical acceptance of gene technology as well as the conditions on the relevant markets have got almost twice the weight in Germany and the Netherlands as in the three Mediterranean countries. In addition, activities and measures of the government are regarded as very important in the Central European countries. On the other hand, the German and Dutch experts are not concerned about the availability of qualified and skilled personnel.

In contrast, in the Mediterranean area R&D infrastructure, technology transfer as well as the availability of qualified and skilled personnel, i. e. all important "technology influenced" factors, are seen as rather important constraints. Moreover, the diffusion of background information on modern biotechnology is also regarded as more relevant in the Southern countries.

Considering discussions about an increasing significance of the international dimension in innovation activities, the factor "international collaboration" is rated surprisingly low in all five countries. Obviously, most of the experts of the Delphi panel do not share the view of a growing internationalization of biotechnology in the Agro-Food sector.

A second rather unexpected result is the low rating of "ecology" as an influential factor. Obviously, most of the experts have the opinion that potential environmental impacts of Agro-Food biotechnology should be handled in a rather pragmatic and factual way instead of a theoretical, rather ideologically coloured discussion. This includes that regulations existing in this area should aim at preventing or minimizing such effects.

Overlooking figure 4.5, some additional national particularities emerge. In the Netherlands the factors "R&D infrastructure" and "industrial innovativeness" score very high, obviously counterbalancing the mentioned demand-driven factors. Besides, the Spanish experts also estimate "aggregated industrial innovativeness" as more important than the German, Italian and Greek ones. In comparison to the other countries, the Greek experts regard the availability of funding as very important for the application of modern biotechnology in the Agro-Food sector, while the Italian experts lay more emphasize on regulation and standards.

Taking into account the content of the statements, the relevance of the influential factors varies between the different domains. In table 4.5 those domains are mentioned, in which the respective influential factor shows its greatest importance.

Table 4.5:   Relevance of different influential factors on the domain level

Influential factor	Highly important for the domain
R&D infrastructure	scientific/technological development
Personnel	scientific/technological development
Technology transfer	scientific/technological development
Industrial innovativeness	food production, plant production, economy
Markets	scientific/technological development, economy
Funding	scientific/technological development, economy
Regulation/standards	regulation
Policy	regulation, acceptance
Ecology	environment
Acceptance	animal husbandry/breeding
Information	environment, health, acceptance
International collaboration	regulation

Figure 4.5:　　Importance of influential factors in different countries

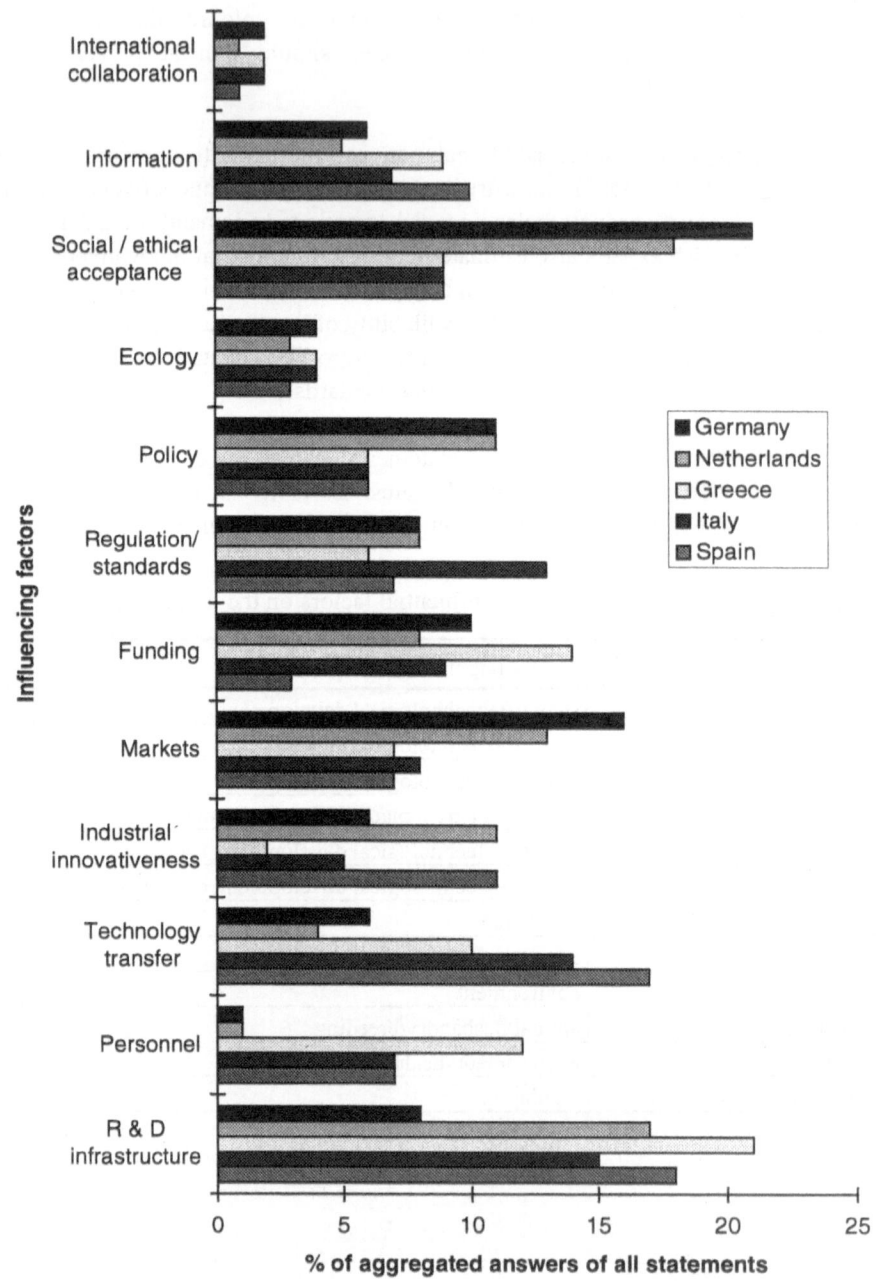

The experts of all involved countries estimate "R&D infrastructure" to be most important for scientific and technological development, whereat the highest relevance of the quantity and quality of R&D institutes is seen in plant biotechnology compared to the application of modern biotechnology in food processing and animal husbandry. In the involved five countries, especially for basic research activities like the elucidation of the molecular basis of virus resistance mechanisms in perennial plants, the development of genetic engineering approaches to alter polygenic traits in economically important plants, as well as the complete sequencing of the genomes of the most important bacteria in food production (statements 58, 54, 44), the availability and quality of respective scientific and technological institutions is regarded as an essential precondition for the future development. Additionally, corresponding to the technology enthusiastic stance of the Spanish experts in general, the latter also figure out R&D infrastructure as very important for genetic engineering of early stages of embryo cells and the use of genetic engineering in fish breeding and aquaculture (statements 66, 67, 68, 69).

In relation to the food industry, this estimation corresponds to the long tradition of using biotechnological approaches which form a fundamental basis for new approaches. Moreover, several research institutions already exist in this field. Concerning animal production and husbandry one might expect an even higher relevance of R&D infrastructure in general, not least because a lot of the basic principles have not been elucidated in this field so far. On the other hand, mainly in Germany and the Netherlands there is a fundamental debate on the social acceptance and ethical principles, especially of genetic engineering of animals. This might result in a certain hesitation of scientists and industry to apply these technologies.

The availability of educated and skilled staff in the biotechnology field is not seen as a very important constraint for the future development. It scores in general relatively low, whereat it got almost twice the weight in the Mediterranean countries compared to the Central European countries. In the Mediterranean countries statements related to scientific and technological developments have achieved the highest relevance. Examples in this direction are the innovation activities in plant breeding companies and the use of rapid test systems for pathogen identification in plant production (statements 61, 62). Besides, the experts of all countries uniformly point out personnel as crucial factor for the identification of gene food products with the help of standardized testing methods and the possible job creation in the food industry due to the use of modern biotechnology (statement 11, 17).

The organization of an efficient "technology transfer" between universities, research institutions and industry is seen as essential for the future application of modern biotechnology in the food industry, plant production and animal production in all countries except the Netherlands, whereat it has more weight in the Mediterranean countries. This relates in particular to the use of specific monitoring and control techniques in these fields, for instance, the use of rapid test systems for pathogen identification in plant production and animal husbandry, as well as new biotechnological approaches for on-line control of quality parameters in food processing (statements 62, 51).

"Industrial innovativeness" is regarded as a crucial factor for the application of modern biotechnology in the food industry and plant production, especially by the Dutch and Spanish panels. Additionally, specific innovative developments are influenced by the ability of industrial company to make a commercial success out of new scientific and technological findings. This relates in particular to the development of specific enzyme systems in order to improve the environmental performance of conventional food-processing procedures (statement 25), to the use of improved starter cultures and enzyme systems in the food industry (statements 48, 49) and to the development of new market opportunities outside the food chain, e. g. transformation of organic agricultural waste into marketable products (statement 32).

The conditions on the "markets" are regarded as highly relevant for economic aspects by the involved experts of this survey, whereby this factor has less weight in the Mediterranean as compared to the Central European countries. In the experts' opinion this factor mainly influences the pricing and import policy related to gene food products (statements 25, 21), the introduction or diffusion of new or improved products based on modern biotechnology (statements 14, 22, 38, 39), as well as the possible job creation in the food industry due to these techniques (statement 17).

The availability of private and public "funds" are regarded as a key factor for the future scientific and technological development, particularly in plant production, by the involved experts with exception of the Spanish panel. Especially, the elucidation of the molecular basis of virus resistance in perennial plants, the development of plant varieties resistant to salinity or drought, as well as the development of new monitoring and control techniques for genetically engineered plants in open field trials, highly depend on the availability of research funds and investment capital (statements 58, 55, 57). Additionally, the Greek experts point out that the current market situation and its future development also depend to a high extent on funding. This relates e. g. to innovation activities of small and medium-sized companies and the possibilities of small farmers to afford new genetically engineered plants and animals (statements 16, 19).

"Regulation and standards" have a comparatively high relevance for issues of regulation, itself, mostly for the Italian experts. The latter regard regulatory framework as most important for environmental and health aspects, as well as for the public acceptance of modern biotechnology. This relates, for example, to the investigation of long-term health impacts of Agro-Food products made with the help of genetic engineering on consumers, farmers and employees, the involvement of the public in debate and decision-making on the application of modern biotechnology in the Agro-Food sector and the implementation of EU biosafety directives in the different member countries (statements 41, 12, 13).

"Political activities" are seen as crucial for issues of regulation and acceptance by the experts of the five countries. For instance, they regard the market approval procedures of gene food products, the compulsory labelling of these products, an increased public participation in the debate and decision-making on the application of modern biotechnology as well as the implementation of EU biosafety directives in the different member countries (statements 9, 10, 8, 12, 13), as highly influenced by politics.

Considering the influential factor "ecology", a clear cross-relation occurs to statements of the environment domain in all countries. The main impact on the environmental situation in the respective country is seen with respect to the deliberate release of genetically engineered organisms resulting in a transfer and recombination of introduced genes with negative impacts, an increase of the loss of traditionally used organisms as well as the rejection of deliberate release and marketing of genetically modified organisms caused by the public debate about environmental risks of these activities (statements 26, 34, 27). In addition, the maintenance of biodiversity in agriculturally influenced ecosystems and the reduction of environmental pollution by cultivation of herbicide-resistant field crops (statement 30, 29) achieve a certain relevance in this context.

"Social and ethical acceptance" of modern biotechnology is regarded as one of the most important influential factor for its future application in the Agro-Food sector, especially in Germany and the Netherlands. This relates in particular to the field of animal breeding and husbandry. Especially the genetic engineering of early stages of embryo cells, the genetic alteration of polygenic traits in farm animals, the use of cloned embryos, and the practical use of breeding procedures based on genetic engineering in EU agriculture are regarded as highly depending on social and cultural opinions, as well as ethical concerns (statements 66, 64, 65, 63). The same applies to the public funding of research activities on genetic engineering of farm animals and the use of genetic engineering in fish breeding and aquaculture (statements 4, 67, 69). In the environmental field social and ethical acceptance of modern biotechnology is a major issue in the integration of these approaches in the production process of organic farmers and the deliberate release and marketing of genetically engineered microorganisms (statements 33, 27).

"Information" of consumers and the public, related to modern biotechnology in general and its environmental and health effects in particular, are seen as an important prerequisite to influence the acceptance of this technology by the involved experts. Accordingly, information is estimated to be crucial for an increased acceptance of gene food products and the habituation of consumers to this type of products (statement 3, 2). On the other hand, the majority of experts in the five countries emphasizes the necessity of information concerning the environmental risks of modern biotechnology (statements 27, 30) as well as possible health impacts of these technologies (statements 35, 37, 38, 40).

Comparing the estimation of the five European countries the factor "international collaboration" scores very low in general (see above). A certain relevance can be pointed out with respect to statements in the regulation domain whereas it is of minor importance in the other fields. Accordingly, the majority of the experts in the five countries see a necessity for international collaboration in the implementation of EU biosafety directives in the member countries, the labelling practice related to these products as well as initiatives to involve the public in the debate and decision-making on the application of modern biotechnology (statements 13, 9, 8, 12). In addition, international collaboration is seen as necessary in basic research activities related to modern biotechnology, like the complete sequencing of the genomes of important bacteria, the elucidation of the molecular basis of virus resistance in perennial plants and the development of plant varieties resistant to salinity or drought (statements 44, 58, 55).

4.1.5   Importance

The experts of the five European countries assessed the importance of the situation described in the statements for three fields: "knowledge creation in science and technology", "competitiveness of the national economy" and "protection of the environment/sustainable development".

In table 4.6 those statements are presented which are classified as most important for "knowledge creation in science and technology" in the five European countries. In table 4.6 developments in plant breeding and plant production are assessed as most important in this respect. This relates in particular to creation of basic knowledge in this field, the use of new monitoring and control techniques, (statements 58, 57, 62) and the genetic modification of complex regulated characteristics in plants (statements 54, 55, 56). Moreover, a significant extension of scientific knowledge is expected from the finishing of basic research activities with relevance for food processing (statements 44, 48), the development of enzyme systems with higher environmental performance (statement 25), as well as the investigation of long-term health impacts of genetically engineered products on consumers, farmers and employees (statement 41).

In addition, especially the Spanish experts also regard developments in animal production as essential for future knowledge creation. For instance, they classify progress in basic research activities like the elucidation of the principles underlying the developmental processes of oocytes and the development of new cell-biological methods for the production of polyploid fish (statements 66, 67) as very important. In addition, the application of genetic maps and molecular test systems for the identification of economically important traits and marker-assisted breeding of fish (statement 68) is pointed out as crucial as well. However, with respect to the statements relating to fish, it has to be taken into account that these statements are figured out as the statements with the lowest expertise of the experts in the five countries (see chapter 4.1.1). Correspondingly, the Spanish estimation has to be put into perspective.

Table 4.6: Statements classified as important for "knowledge creation in science and technology" in the five countries

Statement No.	Germany	Greece	Italy	The Netherlands	Spain
23		0.85		0.87	0.89
25	0.78	0.88	0.77	0.87	0.96
41	0.86	0.86	0.93	0.81	0.96
43	0.80			0.89	0.88
44	0.86	0.91	0.93	0.89	0.96
48	0.80		0.85	0.91	0.94
51	0.82			0.83	0.93
53	0.77		0.78	0.84	0.91
54	0.84	0.80	0.77	0.92	0.92
55	0.81			0.82	0.94
56	0.86	0.88	0.77	0.85	0.95
57	0.87	0.85	0.77	0.82	0.97
58	0.93	0.93	0.96	0.90	0.93
59	0.77	0.83		0.86	0.92
62	0.87	0.81	0.79	0.85	0.95
66	0.78				0.94
67		0.77			0.95
68			0.77		0.95
71	0.79	0.86		0.78	0.93

In table 4.7 those statements are presented which are classified as being very important for the "competitiveness of the economy" in the five European countries. As expected, the majority of experts of all involved countries point out statements describing positive impact on the economy as important for this field. They expect, for instance, strong impacts on the economy if modern biotechnology contributes to the extension of the industrial innovativeness of small and medium-sized companies in the Agro-Food sector or to the reduction of production costs for agricultural bulk products (statements 16, 18). Additionally, high economic impacts are expected if new markets for agriculture and food industry can be made accessible with the help of modern biotechnology. The latter relates e. g. to the production of pharmaceutical substances, optimized renewable resources or genetically engineered vaccines for farm animals (statements 23, 22, 71).

Furthermore, scientific and technological developments mainly in food processing and plant production are seen as essential for the competitiveness of the national economy. In particular, this relates to an increase in productivity in food processing due to the genetically engineered enzymes and to the use of specific monitoring and control systems in food production (statements 49, 50, 43). Besides, the application of genetic engineering in the field of enzyme modifications and in plant breeding are interesting from the economic point of view (statements 46, 52, 55, 56, 59).

Table 4.7: Statements classified as important for "competitiveness of the economy" in the five countries

Statement No.	Germany	Greece	Italy	The Netherlands	Spain
16	0.86	0.64	0.83	0.90	0.96
17	0.66	0.66	0.75		0.84
18	0.82	0.77	0.84	0.83	0.97
21	0.67	0.79	0.76		0.92
22	0.82		0.75	0.86	0.92
23	0.85		0.76	0.87	0.92
25	0.81		0.71	0.83	0.87
38	0.73	0.66	0.83	0.64	0.88
39	0.77		0.87	0.76	0.90
43	0.76		0.68	0.81	0.92
46	0.79		0.84	0.72	0.96
48	0.78		0.86	0.77	0.93
49	0.82	0.69	0.85	0.83	0.95
50	0.80		0.72	0.77	0.94
52	0.74		0.87	0.71	0.96
53	0.76		0.85	0.75	0.92
54	0.66	0.64	0.77	0.65	0.90
55	0.72	0.81	0.73		0.93
56	0.78	0.64	0.68	0.67	0.93
59	0.86	0.64	0.88	0.73	0.96
63	0.66	0.72	0.77		0.90
69			0.84		0.89
70	0.71		0.82	0.74	0.96
71	0.82		0.74	0.79	0.89

In table 4.8 those statements are presented which are classified as being very important for the "protection of the environment" in the five European countries. At a first glance it is obvious that there are only less differences between the five countries in the assessment of this sub-category. Statements classified as important to improve the environmental situation mainly belong to the domain "environment" (statements 25 to 33). This relates in particular to the application of modern biotechnology approaches in animal husbandry in order to reduce emission or organic waste problems (statements 31, 32). A significant positive environmental impact is seen as well if environment-friendly enzyme systems are developed for food processing (statement 25). Clear positive impacts for the environment are also expected if genetically engineered microorganisms and plants producing biocides are used for biological pest control, if salt- or drought-resistant plants varieties are cultivated in arid areas and if new monitoring and control techniques for genetically engineered plants are used in open fields to improve the reliability of risk assessment significantly (statements 56, 55, 57)

At least in three statements presented in table 4.8 it has to be asked whether the majority of the experts has assessed the sub-category "importance of the statement to protection of environment/sustainable development" according to the instructions given in the questionnaire. In this context it has to be taken into account that in the introduction of the questionnaire this sub-category was defined as a contribution to improve the environmental situation in the included countries. The respective statements describe a 50 % reduction of environmental pollution in plant production due to the cultivation of herbicide-resistant field crops as well as statement 30, in which no negative effects of the maintenance of biodiversity in agriculturally influenced ecosystems, despite the use of modern biotechnology, is mentioned. In addition, statement 26 (describing the deliberate release of genetically engineered organisms results in the transfer and recombination of the introduced genes with unintended negative impacts) has to be regarded rather critically in this context. In the latter case a negative impact of genetic engineering on the environment is described, which obviously has a certain impact on the environmental situation, but not in a positive way. Related to the other two statements more than 20 % of the German and Dutch experts expect that the described developments will not be realized at all. In addition, the comments of the experts document that at least some of them doubt that there will be a positive environmental impact of herbicide-resistant plants and that modern biotechnology will not damage biodiversity in agriculturally influenced ecosystems. Having these aspects in mind, it seems to be likely that most of the German respondents see high relevance of these statements for the environmental situation, but at least some expect an impact opposite to the description in the questionnaire.

Besides the consideration of the 71 statements concerning their importance for the single sub-categories, it is interesting to analyse the importance of the statements for two sub-categories. To find relations between the sub-categories "competitiveness of the national economy" and "protection of the environment/sustainable development" as well as "competitiveness of the national economy" and "knowledge creation in science and technology", the project team prepared a scatterplot of the respective indexes for all 71 statements in the five countries.

Table 4.8: Statements classified as important for "protection of the environment/sustainable development" in the five countries

Statement No.	Germany	Greece	Italy	The Netherlands	Spain
12		0.61			
25	0.73	0.52	0.77	0.79	0.77
26	0.75		0.62	0.57	0.54
27	0.59	0.76	0.63	0.47	0.54
28	0.80	0.53	0.76	0.74	0.85
29	0.85	0.72	0.72	0.83	0.89
30	0.83	0.67	0.78	0.75	0.90
31	0.89	0.62	0.91	0.87	0.94
32	0.89	0.80	0.91	0.87	0.91
33					0.67
55	0.63		0.66		0.68
56	0.62		0.62	0.38	0.68
57	0.78	0.62	0.84	0.54	0.81

Concerning the importance of the statements with respect to the "protection of the environment" and the "competitiveness of the national economy", it can be pointed out that in general only some statements of the questionnaire are assessed as being important for both sub-categories by the experts in the five countries. Those statements seen as being crucial for both fields (see table 4.9) relate to the improvement of the environmental performance of conventional food-processing procedures, the use of genetically engineered microorganisms and plants producing biocides for biological pest control, the biotechnological reduction of emissions and waste from animal production as well as the biotechnological transformation of organic agricultural waste into marketable products (statements 25, 28, 31, 32). In addition, the experts of all five countries estimate the production of renewable resources as well as the development of specific resistance mechanisms for plants, like genetically engineered field crops resistant to herbicides or plant varieties resistant to salinity and drought, as very important for the environment and for the respective national economy (statements 22, 29, 55, 56, 58). These areas represent suitable opportunities to promote modern biotechnology approaches in the included countries, because on the one hand it can be expected that such initiatives will not meet heavy social resistance due to anticipated environmental damage, and on the other hand represent areas with high economic perspectives.

With the exception of Greece, in all the other countries these statements achieve indexes higher than 0.5 for both sub-categories. In Greece the same statements are also assessed as most important for both fields, however, with less weight (indexes between 0.0 and 0.5). Obviously, the Greek experts are more reserved concerning positive interaction between "economy" and "environment". On the other hand, the Spanish experts seem to be very convinced about a positive relation between "economy" and "environment", as the majority of the other statements included in the questionnaire also gets positive indexes for both fields.

Table 4.9:   Statements classified as important for the "protection of the environment" and for the "competiveness of the economy" in the five countries

Statement No.	Content
22	Due to the application of modern biotechnology, farmers produce renewable resources (e. g. biofuel, starch, fatty acids) which are used outside the food sector on 20 % or more of the arable land in your country.
25	Enzyme systems are specifically developed to improve the environmental performance of conventional food-processing procedures.
28	Genetically engineered microorganisms and plants producing biocides are widely used for biological pest control.
29	The widespread cultivation of genetically engineered field crops resistant to herbicides results in an approximate 50 % reduction of environmental pollution in plant production.
31	Modern biotechnology is widely used to reduce emissions and waste from animal production (e. g. less manure due to enzymes as feed additives, improved anaerobic waste treatment, biofilter).
32	Modern biotechnology significantly contributes to the transformation of 40 % or more of the organic agricultural waste into marketable products (e. g. energy, secondary raw materials).
55	Genetically engineered plant varieties resistant to salinity and/or drought have improved agricultural productivity in arid environments significantly.
56	To prevent resistance against biocides in pathogens, new genetic engineering approaches for plant defence mechanisms have been developed (e. g. combination of different resistance genes, increase in pathogen tolerance).
58	The molecular basis of virus resistance mechanisms in economically important perennial plants like e. g. vine, olive and fruit trees is elucidated.

All in all, this result is understandable, as all of these statements describe developments with positive environmental effects and in addition, with obvious economic benefit. For instance, the improvement of plant resistance mechanisms enables farmers to achieve high yields by using less biocides. However, depending on the approach, e. g. the development of new genetic engineering methods for plant defence mechanisms in general (statement 56) or more specifically the elucidation of virus resistance mechanisms for economically important perennial plants (statement 58), the economic effects are different. In case of the more general approach (statement 56), it is evident that there would be very far-reaching consequences, i. e. besides an increase of productivity, this would also influence several steps in the respective product chain. In case the virus problems in fruit trees (considered among others in statement 56) are solved, this would have high economic benefits for fruit growers, because they would not be forced to replace their fruit trees after ten to twenty years, saving the respective costs for establishing a new orchard as well as the costs for crop failure during this process.

On the other hand, an increased productivity can cause additional or new problems. For example, in case of plant varieties resistant to salinity or drought (statement 55), a significant extension of more or less intensive farming activities in arid areas which often represent rather sensitive ecological systems can also induce a stronger exploitation of the soil and in consequence a long lasting decrease of the soil fertility.

Additionally, it has to be taken into account whether farmers can afford these plants (described in statement 19). Concerning this aspect, especially the experts of the Mediterranean countries are sceptical, probably caused by a higher proportion of small farmers in their countries. In contrast, the Dutch experts do not doubt about a widespread use of genetically engineered plants.

Regarding the statements important for both, "knowledge creation in science and technology" and "competitiveness of the national economy" one can point out that the experts of all countries uniformly consider the majority of the statements as important for both fields, i. e. generally they believe that scientific research has a significant impact on economic competitiveness. Those statements, identified as most important (with indexes higher than 0.7) for both fields, relate to application of biotechnology in the non-food sector like the production of pharmaceuticals and high value components (statements 23, 59), for control techniques (statements 43, 62) and for food production processes (statements 25, 48). Moreover, the production of enzymes in genetically engineered field crops, the improvement of resistant mechanisms and the use of genetic engineering in plant breeding (statements 52, 53, 55) are included as well (see table 4.10).

In Spain additionally around 25 statements are seen as very important (with indexes > 0.7) for both fields. These statements mainly relate to all areas of scientific and technological developments. This result also backs up the optimistic and technology enthusiastic picture that generally emerges by analysing the Spanish stance towards the application of biotechnology in the Agro-Food sector.

A common feature of the majority of the identified statements is that they describe rather broad approaches, as, for instance, the general improvement of the environmental performance of conventional food-processing procedures and the developmemt of genetically engineered microorganisms which can be precisely controlled and regulated in their metabolic activities during food production processes (statements 25, 48). In all cases the mentioned development induces far-reaching consequences for the food industry in general, like a reduction of the environment expenses, an enhancement of productivity and a general improvement of product quality. On the other hand, especially in Germany, a widespread use of genetically engineered organisms in the food industry is highly disputed with respect to social and ethical concerns, despite the predicted positive economic effects.

Table 4.10: Statements classified as "important for the knowledge creation in science and technology" and the "competitiveness of the economy in the five countries"

Statement No.	Content
23	Pharmaceutical substances produced by genetically engineered animals and plants (e. g. proteins, enzymes, hormones, antibodies) achieve a turnover share of at least 5 % of the pharmaceutical market in your country.
25	Enzyme systems are specifically developed to improve the environmental performance of conventional food-processing procedures.
43	Techniques based on modern biotechnology are practically used for on-line control of quality parameters in food processing (e. g. content of micronutrients or harmful substances).
48	A large variety of genetically engineered microorganisms which can be precisely controlled and regulated in their metabolic activities during food production processes has been developed.
52	The production of industrial enzymes in genetically engineered field crops is developed for large-scale application.
53	For the development of hybrids, genetic engineering is practically used in breeding of most field crops.
55	Genetically engineered plant varieties resistant to salinity and/or drought have improved agricultural productivity in arid environments significantly.
59	Plant cell cultures in large-scale bioreactors are widely used for the production of high value components (e. g. pharmaceuticals, fine-chemicals, proteins).
62	Rapid test systems based on modern biotechnology are widely used for pathogen identification in plant production and animal husbandry.

4.2 Differences between the expert groups

This Delphi survey involves the following specifically selected expert groups:

- Industry
- Researchers

- Farmers

- Consumers and users

- Biotechnology critics and experts for societal aspects

The U-Test of Mann-Whitney for non-parametric data has been used to filter out main differences concerning the personal attitude and the time of realization between the involved expert groups in the five countries (see Kähler 1995). The significance level of the U-test was set to 0.01. Overlooking the results of this analysis, all in all, a different response behaviour between the expert groups in the Central European and the Mediterranean countries can be registered (see table 4.11). The German and Dutch expert groups show rather polarized answering patterns, while the expert groups of the Mediterranean countries mostly indicate a relatively uniform response behaviour, both concerning personal attitude and time of realization.

Table 4.11: Differences between the expert groups concerning "personal attitude" and "time of realization" in the five countries

Country	Proportion of significant[1] different statements	
	Personal attitude	Time of realization
Germany	97 %	61 %
The Netherlands	45 %	6 %
Greece	28 %	46 %
Italy	23 %	7 %
Spain	18 %	0 %
[1] The distribution of the answers of the different expert groups has been analysed by using the U-Test of Mann-Whitney. The significance level was set to 0.01.		

The German panel is characterized by the most polarized response behaviour among the expert groups, compared to the other countries. Concerning the assessment of the "personal attitude" the extreme poles are represented by the experts from industry and research institutions on one hand and consumers and critics on the other hand. The response behaviour of farmers is placed between these two poles with clear tendencies towards the consumer/critics cluster. In general, industry and research experts tend to assess the stated development more positively, whereas most of the experts from the farmer, consumer and critics side are more sceptical or reject single developments. The highest differences between the expert groups in Germany are found in statements, for instance, dealing with the application of enzymes in food industry which are produced or optimized with the help of genetic engineering (statements 49, 50).

Besides, there are areas with similarities. For instance, all expert groups in Germany highly reject a possible job decrease in agriculture as well as negative health impacts (e. g. additional allergies) due to modern biotechnology (statements 20, 37, 40). On the other hand, they highly appreciate information of consumers concerning modern biotechnology in general or the health effects of this technology in particular (statements 1, 35). In addition, a great majority of all German experts agrees as well that modern biotechnology contributes to the improvement of the environmental situation or that different social groups participate on the assessment of genetically modified food (statements 30, 32, 5).

Concerning the time of realization, industry/research and critics represent the two extreme poles among the expert groups in Germany. Areas in which the future realization of modern biotechnology is heavily disputed among the different groups are the ecological effects of modern biotechnology on agricultural ecosystems (statements 13, 30), the integration of specific genetic engineering approaches in organic farming, and as the economic impacts of modern biotechnology in the Agro-Food sector e. g. reduction of production costs of agricultural bulk products, introduction of innovative processes and products in small and medium-sized companies (statements 33, 24, 15). While the majority of the critics assess, for instance, no negative effects on the maintenance of biodiversity in agriculturally influenced ecosystems due the use of modern biotechnology as unrealistic, a high proportion of the experts of the other two groups expect that such a development will be realized in future. In comparison to the critics, the consumers tend to estimate an earlier estimation of several statements. These mostly relate to scientific and technological developments of modern biotechnology.

In case of the Dutch expert panel, the original groups were modified to reach the minimum of questionnaires required to get a significant analysis. Hence, for the analysis on the group level only four groups are considered in the Netherlands: industry researchers from publicly funded research organizations, farmers and representatives of their organizations, including persons working in publicly funded agricultural organizations on the regional and local level and experts in the field of social and economic aspects ("biotechnology critics").

The personnel attitude of the groups towards the situation described in the statements shows a general pattern. Over the whole, two poles are represented by the experts from industry with the highest positive attitude towards the subjects described in the statements on one hand, and by the biotechnology critics on the other hand. The two other groups are in between, whereat the public researchers are generally close to the industry group, while the farmers tend towards the critics group.

Concerning the "time of realization", the differences between the groups are less remarkable. In general, the most significant differences can be found in statements, which are seen as not realizable at all, by a considerable part of the Dutch panel. In most cases, the farmers take a different position from the rest of Dutch expert groups. For instance, they expect later realization of EU regulations, no reduction of production costs of agricultural products (the rest expects this in the period six to ten years) or they expect that organic farmers are allowed to use specific genetic engineering approaches,while the other groups are more negative about this (statements 13, 18, 33).

In the Mediterranean countries a high proportion of statements are answered uniformly by the different expert groups. Nevertheless, differences between the respective expert groups in these three countries mostly arise concerning scientific and technological developments. For instance, consumersand critics reject that 90 % of enzymes used in the food industry are produced with genetically engineered organisms or the use of these kind of enzymes to improve the processing quality of food (statements 50, 46). In contrast, the experts from industry and research institutions appreciate these kind of developments.

Concerning the time of realization the differences between the expert groups are less remarkable in the Mediterranean countries, with the exception of Greece. There, the researchers and biocritics represent two positions, whereat the biocritics expect in general a later realization of the respective development in comparison to the researchers. For instance, this relates to the use of optimized enzymes in the food industry, or the use of genetic engineering in breeding of most field crops (statements 49, 53).

5. Future impacts of biotechnology in the Agro-Food sector

In this chapter the results of an in-depth analysis of selected key issues are presented. These key issues relate to the scientific and technical development of Agro-Food biotechnology as well as its future impacts on society, the economy, the environment and the health of consumers, users and employees. In order to facilitate the comparison between the results of these analyses and the results of the survey in the single countries, this chapter is structured according to the sequence of the domains in the questionnaire. Firstly, the future acceptance of modern biotechnology (5.1) as well as regulatory issues (5.2) are analysed. Afterwards, the impacts of Agro-Food biotechnology on the economy (5.3), the environment (5.4) and the health of consumers, users and employees (5.5) are discussed. This chapter concludes with a presentation of future scientific and technical trends in four selected areas of Agro-Food biotechnology and the most important prerequisites for these developments (5.6).

One partner of the project team was basically responsible for the analysis of each key issues, for which a specific analysis design adapted to the content and the nature of the issue was elaborated. In these analyses the statistical results of the Delphi survey were included (and often additionally analysed, using specific statistical methods which are mentioned in the single chapters) and compared with background information or the results of other research projects in order to discuss and interpret the results of this survey, taking into account the already available knowledge and information in this area.

5.1 Acceptance of biotechnology

5.1.1 Introduction

The issue to be analysed in this section, social acceptance of Agro-Food biotechnology, is not confined to a specific section of the questionnaire but is rather dispersed in several sections and categories. The parts of the questionnaire useful in the analysis of public acceptance are:

- Statements 1 to 7
- Answering category "personal attitude"

- Sub-category "social and ethical acceptance" in the category "most important influential factors in your country"

- Comments of experts to single statements

All the other issues, like e. g. scientific and technological developments, appear more or less related and intertwined with acceptance. In fact, the opportunities for economic development in the Agro-Food sector are clearly influenced by the degree of social acceptance, which in turn is influenced by the perception of the possible risks and benefits in the health and environmental domains. The regulatory framework is expected to be both influencing and influenced by the state of social acceptance of biotechnology. Finally, it should not be excluded in principle that the economic expectations related to the benefits of Agro-Food biotechnology might contribute to shape the social acceptance of biotechnology, although the results of the survey seem not to confirm this possibility.

In consideration of the many interrelations existing between acceptance and the other domains, the analysis will not be limited to the seven statements in the section "acceptance" (chapter 5.1.2), but will be also extended to the rest of the questionnaire (chapters 5.1.3 and 5.1.4). Chapter 5.1.4 will be devoted to the identification of major patterns and cultural trends in the public perception of Agro-Food biotechnology.

5.1.2 Analysis of the statements in the domain "acceptance"

This first part of the report will concern the "horizontal" analysis of the seven statements specifically devoted to the issue "acceptance" (see table 5.1) which represents the first section of the questionnaire.

Such statements may be grouped in three categories:

- Institutional initiatives aimed at informing and involving the public in the debate and decision-making (statements 1, 3, 5)

- Developments relevant to assessing consumers' demand for genetically engineered food products and market opportunities (statements 2, 6, 7)

- Possible negative impacts of widespread rejection of genetic engineering of animals on scientific research funding (statement 4)

Table 5.1:   Statements of the domain "acceptance"

Statement No.	Content
1	A governmental institution which provides information on modern biotechnology in the Agro-Food sector to all interested persons and institutions is established in your country.
2	Most consumers in your country have quickly got used to all kinds of food and beverages produced with the help of genetic engineering.
3	The widespread diffusion of information on genetic engineering from different sources (e. g. public and private institutions, several interest groups and associations, media) has increased the acceptance of food products made with the help of genetic engineering in your country.
4	Research on genetic engineering of farm animals is not funded by public institutions in your country due to ethical reasons (e. g. animal rights, animal welfare, preservation of the Creation).
5	An advisory committee with the objective to assess whether food products made with the help of genetic engineering meet the needs of consumers is jointly established by different research groups (e. g. consumers, industry, retailers) in your country.
6	Food companies and retailers proactively create specific labels for food and beverages produced without the help of genetic engineering.
7	A restaurant chain or catering service specialized in offering trendy meals with genetically engineered ingredients open branches in almost all large cities in your country.

Personal attitude

The analysis of the experts' personal attitudes for all the seven acceptance-related statements would not make much sense as two of them (statements 4 and 6) have a "negative" connotation (research on animals will not be funded, labelling of non-genetically engineered food). It seems to be more reasonable to perform such analyses only for the three statements concerning institutional initiatives (statements 1, 3, 5).

The attitudes of the experts towards governmental actions aimed at informing and involving the public seem to be mostly positive (figure 5.1). However, such attitudes are obviously influenced by the degree of public trust placed in the public authorities in these countries. The findings of the latest Eurobarometer on Biotechnology 46.1 show that such confidence in public authorities, compared to other information sources, is not very high on average in EU countries (see figure 5.2) (European Commission 1997, Zimmermann et al. 1994, Sparks et al. 1994).

Figure 5.1:   Personal attitudes towards institutional initiatives on public
                information and involvement in the five countries (state-
                ments 1, 3, 5)

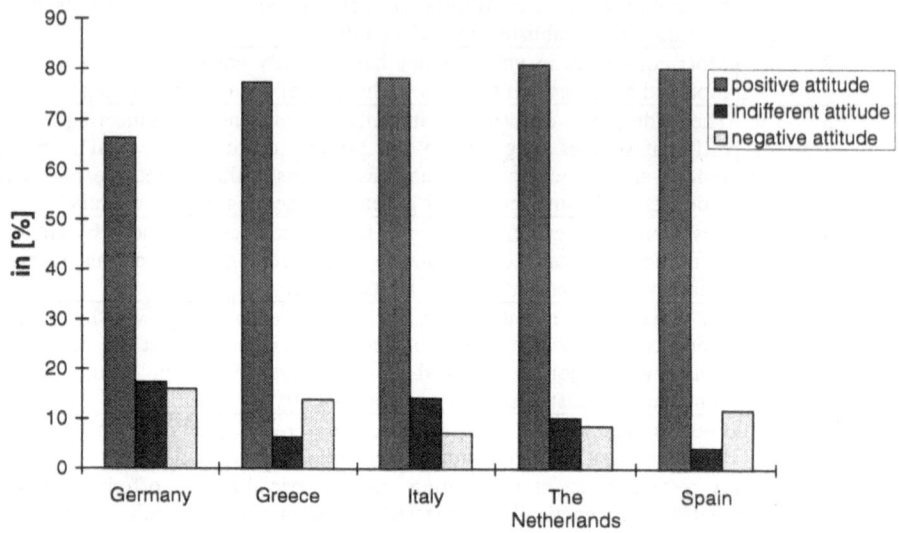

Figure 5.2:   Trust in different information sources on biotechnology

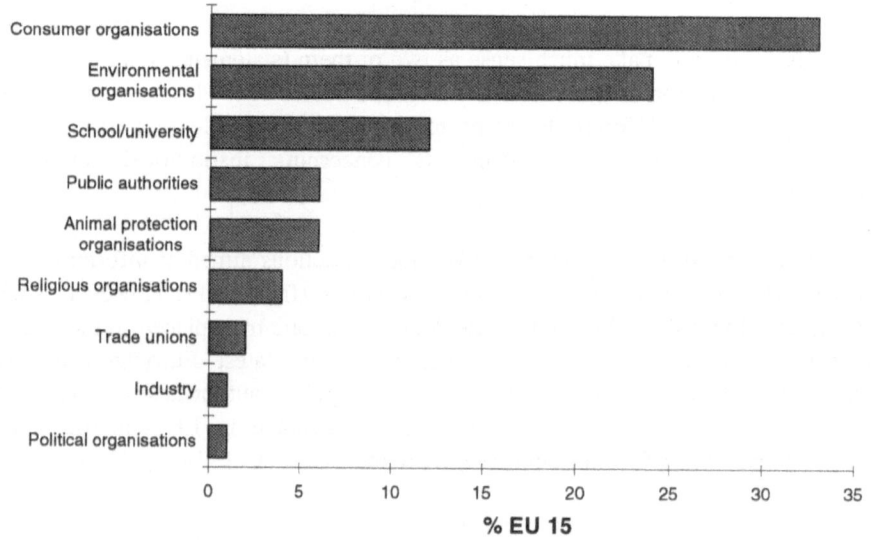

Source: European Commission 1997

Some differences can be observed between the countries. The Dutch respondents are the most favourable to an informed and participative process of development of Agro-Food biotechnology, while the German respondents are the most reserved in this sense.These results go in line with the findings of Eurobarometer 46.1, showing that the country with the lowest confidence in "public authorities" as sources of reliable information on biotechnology is Germany, together with Belgium.

It may therefore be interesting to identify which expert groups within the German panel are more reserved towards these developments (figure 5.3). These turn out to be critics, consumers and farmers, while the researchers and industry appear more favourable. The most relevant concerns here seem to be related to the correctness and unbiasedness of the information provided by the governmental institutions, as stressed by several comments of German experts (e. g. "The Federal government is not willing to give objective information").

Figure 5.3: Personal attitudes of German expert groups towards institutional initiatives on public information and involvement (statements 1, 3, 5)

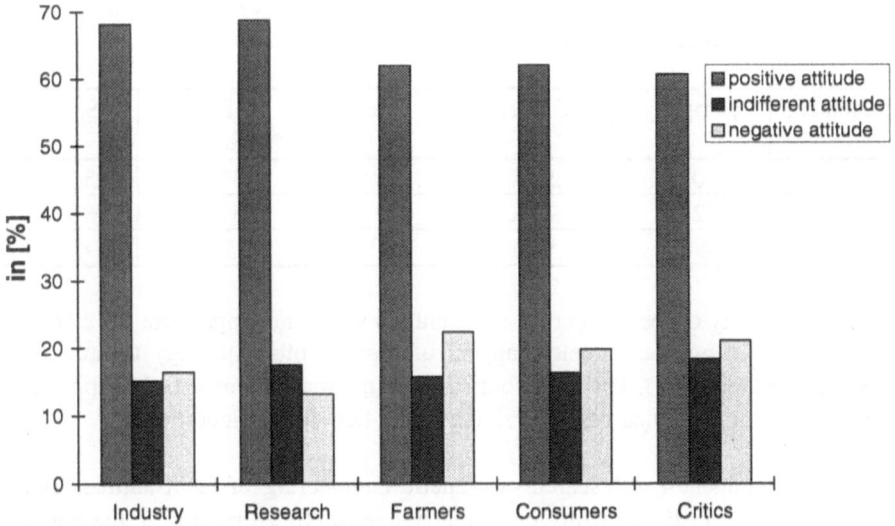

It is interesting to note that even industry and research experts in Germany show a less positive attitude towards institutional initiatives than the "critics" in the other four countries (share of positive attitudes towards developments 1, 3 and 5: Spanish "consumers and critics": 70.7 %, Dutch "critics": 73.0 %, Greek "consumers and critics": 74.3 %, Italian "consumers and critics": 74.0 %). The "nationality" factor seems once again prevailing.

As regards the three statements related to consumer demand and market developments, on the whole the experts' attitudes seem rather homogeneous without significant differences among the five countries.

A first conclusion which can be drawn from the analysis of table 5.2 is the relevance that the experts assign to the freedom of choice of the consumer, which should not result from a passive process of "habituation" (in all the countries, except for Spain, only a minority of the experts show a positive attitude towards statement 2 dealing with the so-called "habituation effect"), but rather from an informed and free choice by the consumers. Therefore, a majority of the experts in all countries, except for Italy, would appreciate the creation of labels for products produced without genetic engineering (statement 6) in order to make the consumer aware of the differences existing between the products. The reservation expressed by the Italian consumers are possibly related to fears of consumers being manipulated by unscrupulous producers, as occurs with so-called "natural" products which are sometimes purposely confused with organic products.

Table 5.2:   Positive attitudes towards some possible market developments of Agro-Food biotechnology

Statement	Germany	Greece	Italy	The Netherlands	Spain
2	20 %	21 %	28 %	37 %	56 %
6	62 %	90 %	49 %	57 %	67 %
7	8 %	9 %	9 %	5 %	4 %

The large majority of respondents in all countries would not appreciate the diffusion of restaurants using the "novelty" appeal of modern biotechnology to attract the consumers (statement 7). This also confirms that the majority of experts supports an informed consumer choice based on rational cost-benefit considerations.

Regarding statement 4 ("Research on genetic engineering of farm animals is not funded by public institutions in your country due to ethical reasons"), the national differences are very strong and even within the single countries the experts appear divided. However, it seems that the majority of experts in every country (except for Greece) are not supporting any limitation of research (see table 5.3), even in Germany and the Netherlands, where concerns for animal welfare are strong. All in all, it seems that the value attached to the creation of new scientific knowledge and/or to the possible benefits of R&D findings offsets the concern for animal suffering.

Table 5.3: Positive attitudes towards limitations of research on genetic engineering of farm animals

Attitude	Germany	Greece	Italy	The Netherlands	Spain
Positive	36 %	64 %	17 %	29 %	31 %
Indifferent	21 %	4 %	22 %	29 %	9 %
Negative	42 %	29 %	55 %	42 %	57 %

Time of realization

In each of the five countries a high proportion of the experts forsees realization of the statements 1 to 7 in the short term. Compared with the rest of the questionnaire, the statements included in the section "acceptance", as well as in "regulation" are among those with the shortest time of realization (see table 5.4).

Table 5.4: Time of realization in the single domains by country (relevance of sub-category "0-5 years")

Domain	Germany	Greece	Italy	The Netherlands	Spain
Acceptance	45.1 %	40.4 %	30.3 %	59.0 %	48.9 %
Regulation	40.2 %	45.2 %	46.7 %	78.3 %	63.5 %
Economy	10.8 %	20.4 %	17.9 %	18.5 %	26.1 %
Environment	22.0 %	24.2 %	35.7 %	30.2 %	35.3 %
Health	27.0 %	36.7 %	29.6 %	30.7 %	46.0 %
Scientific/ technical developments	18.2 %	41.7 %	33.8 %	32.3 %	47.1 %

Such results seem to indicate the existence of strong expectations concerning short-term public initiatives aimed at improving public information/communication on Agro-Food biotechnology, as well as ensuring a proper regulative framework. In this respect no significant difference is observed between the expert groups, though the underlying motivations of the single groups might be of course very different.

Influential factors

The analysis of the most influential factors in the domain "acceptance" leads to several important conclusions. The most striking one concerns the negligible importance attached to "ecology", which could be expected to be one of the most relevant factors shaping the social acceptance of Agro-Food biotechnology.

Moreover, the differences between the countries are not significant in this respect, especially between Central European and Mediterranean countries (with e. g. Spain scoring exactly as Germany). The most relevant factors, apart from the obvious prevalence of "social/ethical acceptance", are "information" (especially in Greece and in Spain), "policy" (especially in Germany and the Netherlands) and "markets" (see figure 5.4).

"Regulation" only gets a high score in Italy and "funding" is considered relevant by Greek and German experts. Technical factors (R&D infrastructure, technology transfer, skilled personnel) are more important in Mediterranean countries. From the perspective of potential European versus national policy initiatives focusing on the acceptance issue, it is interesting to note that the experts mean to consider the international dimension ("international collaboration", figure 5.4) as almost negligible. Obviously, the acceptance issue is perceived mainly as a country-specific topic, with no need for international, e. g. European, activities.

Figure 5.4:   Influential factors related to the "acceptance" domain
                (statements 1 to 7)

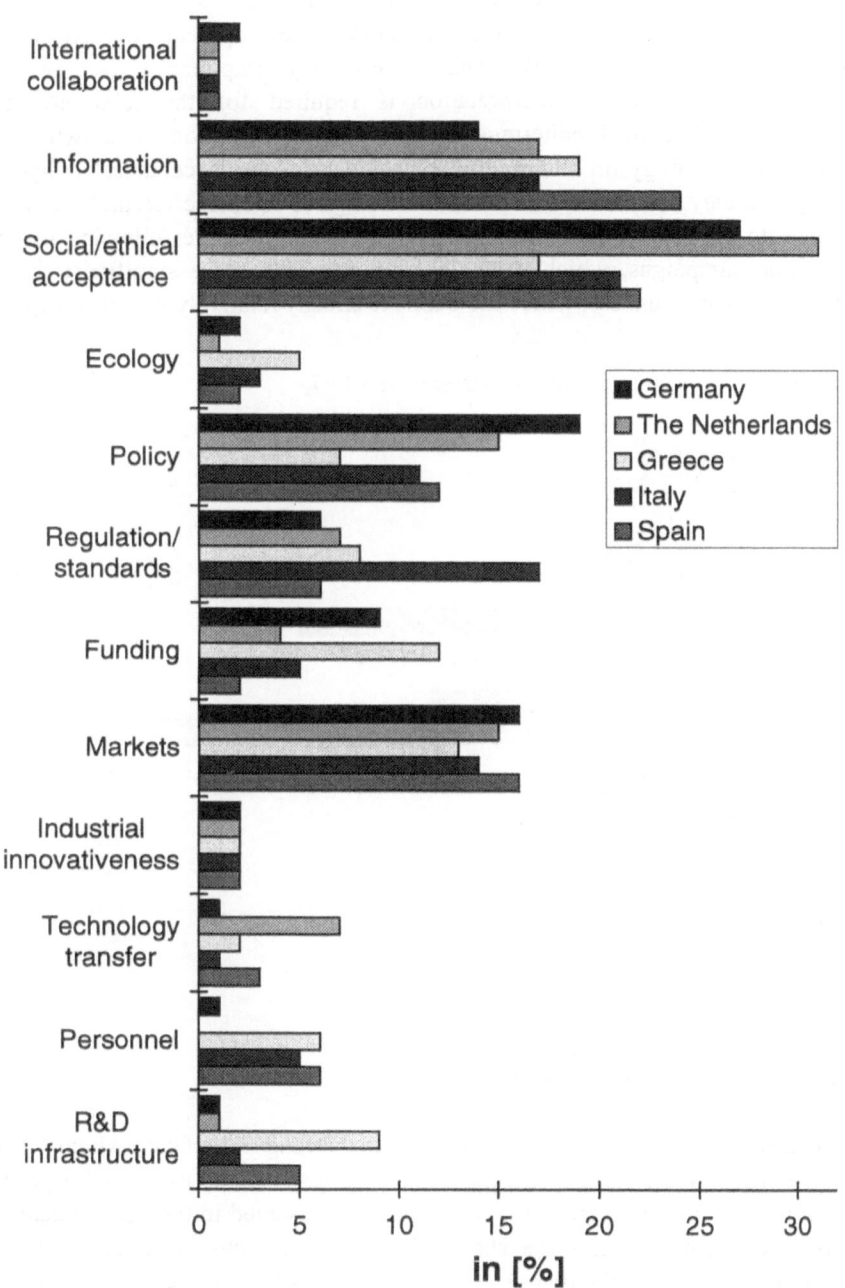

Importance

As concerns the category "importance", the most relevant for all countries seems to be "competitiveness of the economy", confirming that social acceptance is considered crucial for the future economic development of Agro-Food biotechnology (see figure 5.5). Again, very low importance is given to "environment". A careful interpretation is required for the relatively high importance attached by Mediterranean experts to the "creation of knowledge in science and technology" in relation to the statements of the "acceptance" domain. It is likely that they do not refer to possible findings of scientific research activities, but rather to the improvement in the level of public knowledge following massive information campaigns, which would be proportionally more significant than in Central European countries, where the public is already relatively well informed.

Figure 5.5:   "Importance" of statements 1 to 7

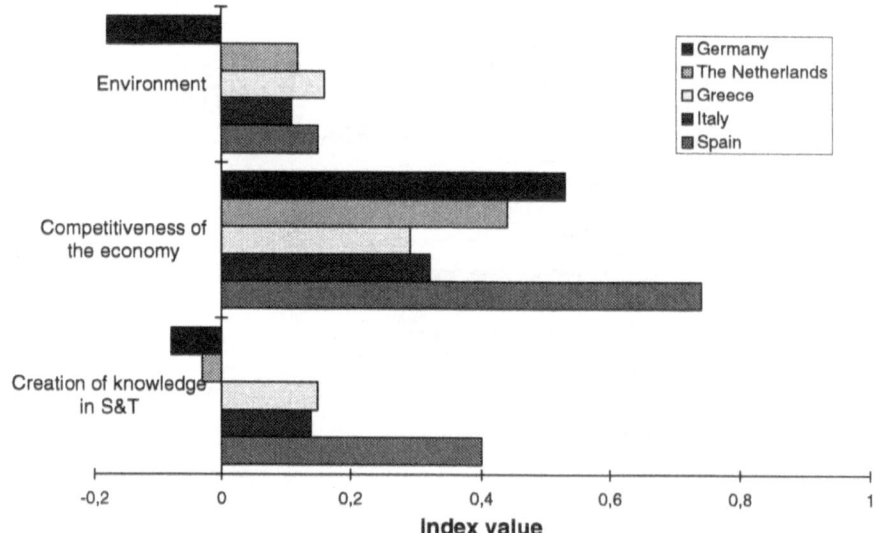

5.1.3   General overview of the results

This chapter considers the general findings concerning acceptance which emerge from the whole questionnaire. On the whole, the respondents expressed a positive attitude towards the majority of the developments presented in the questionnaire. In the two Northern countries the share of positive attitudes is smaller, while in Greece, Italy and Spain the degree of optimism seems higher (see figure 5.6). Such national population differences appear consistent with the results of EU-wide population opinion surveys, in particular the Eurobarometers on Biotechnology 1991, 1993 and 1996 (European Commission 1991, Marlier 1993, European Commission 1997).

Figure 5.6:   Personal attitudes of the respondents by country

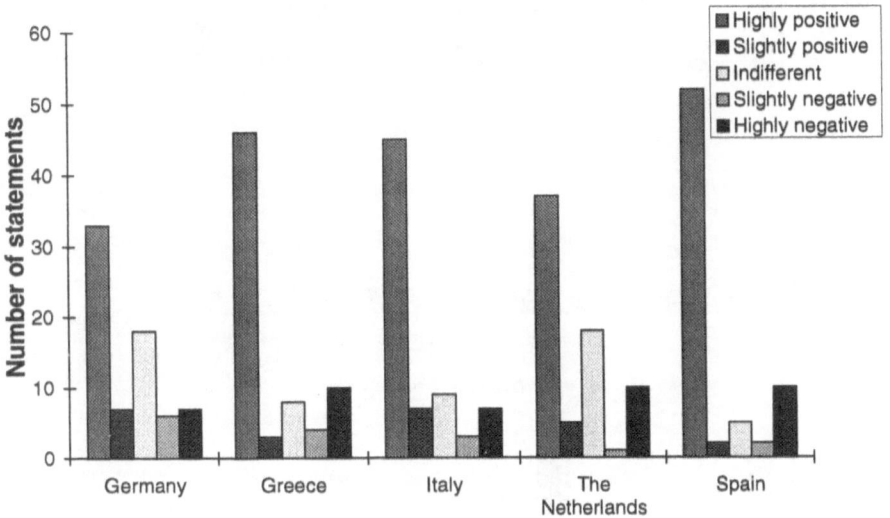

In order to obtain a more detailed picture of the attitudes of different social groups, it is interesting to compare this finding with a cross-country review of the attitudes of the respondents by expert groups. Also in this case a predominance of positive attitudes is found for every group, though the share of positive attitudes is higher for the two groups "industry" and "researcher" than for "consumers" and "critics".

Note to the following figure 5.7:

Consumers and critics originally formed two separate expert groups. However, due to the insufficient number of respondents in these groups, in three countries (Greece, Italy and Spain), it was necessary to aggregate the two groups to create a larger group, including those respondents who show a more critical attitude toward food biotechnology with respect to the general public. In the Netherlands the respondents in the group "consumers" did not answer. Other groups ("farmers" and "others") also did not reach the minimum size for significance in some countries.

Figure 5.7:   Personal attitudes of the respondents by expert group in the five countries

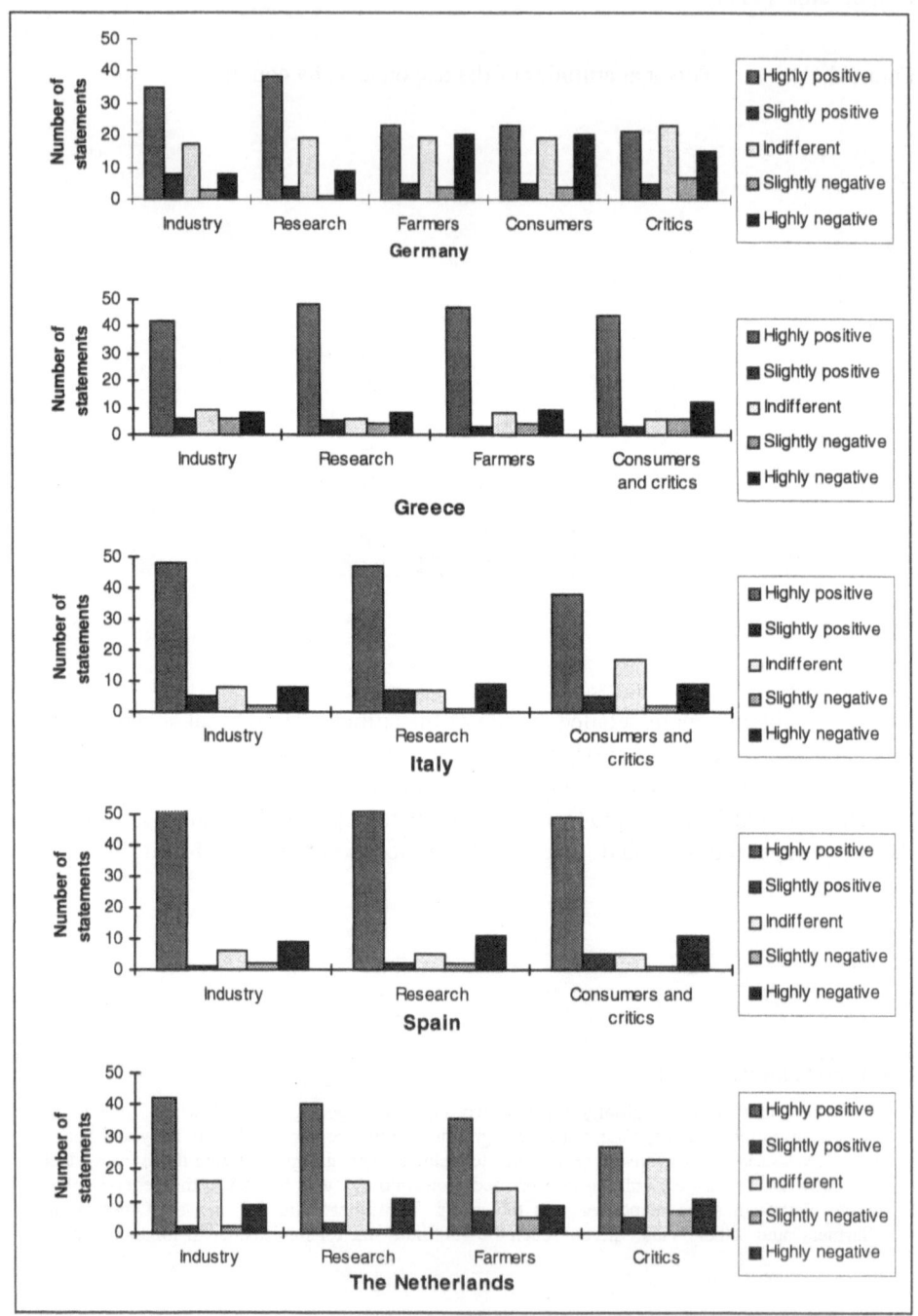

On the whole, such inter-group differences are smaller than one might expect, considering the controversial character of the public debate on food biotechnology within the EU (see figure 5.7), though in some countries the differences are much more significant, with the most remarkable differentiation reached in Germany, where the share of statements with positive attitudes within the groups "consumers" and "critics" is below 50 %.

In general, the national differences seem to be at least as significant as the differences between expert groups, reflecting the complex character of the debate on the application of Agro-Food biotechnology within the EU and the prevailing influence of socio-cultural factors.

Table 5.5:   Acceptance differences between countries

	Germany	Greece	Italy	The Netherlands	Spain
Mean "positive attitudes" (% values) (a)	0.49	0.62	0.62	0.53	0.69
Mean "negative attitudes" (% values) (b)	0.26	0.25	0.19	0.22	0.19
Index (a-b)	0.23	0.37	0.43	0.31	0.50

A comparison with the results of Eurobarometer surveys on biotechnology can be made by calculating a simple measure ("index") of acceptance for each country (see table 5.5).

Such findings, resulting in a ranking of countries according to their degree of acceptance of biotechnology, appear basically consistent with those of the latest Eurobarometer on Biotechnology (European Commission 1997) (see figures 5.8 and 5.9) with some minor differences. In particular, with respect to Eurobarometer the respective positions of Italy and Spain are exchanged and the rank of Greece appears higher. Of course this comparison has only an indicative value and should be carefully interpreted for several reasons. In the first instance, the questions are formulated differently in the Eurobarometer survey (where the interviewees are directly asked to evaluate the impact of the new technology ("will improve your way of life")) and in the Delphi survey, where such overall assessment is derived from the aggregation of the respondents' attitude towards single, both positive and negative developments.

In addition, the Eurobarometer surveys are conducted on samples of the general public, while the Delphi survey has been carried out on a selected sample of experts whose understanding of the technology is likely to differ significantly from that of the public at large. Such a difference may be particularly marked in those countries, where the debate on biotechnology has remained mostly confined to small circles. The latter consideration may also help to explain, for example, why the relative position of Greece emerging from the Delphi survey is somewhat different (more inclined to "optimism") from that indicated by the results of the Eurobarometer survey.

Figure 5.8:   Degree of acceptance of Agro-Food biotechnology in the five countries

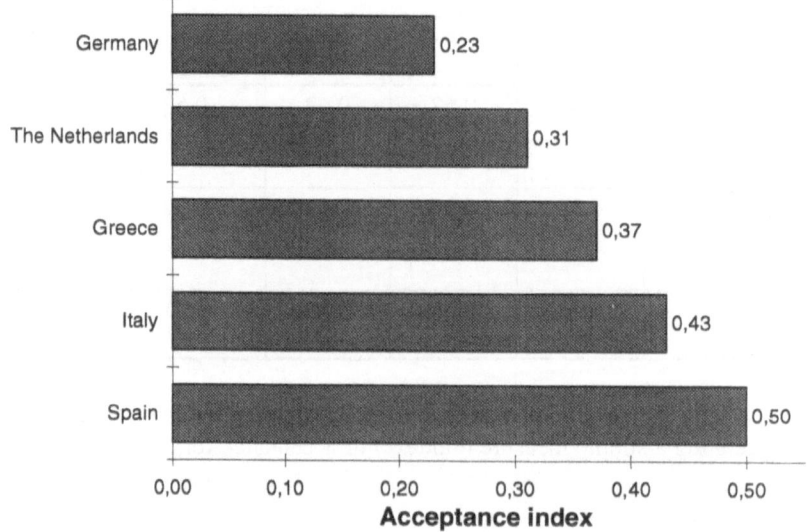

Figure 5.9:   Anticipated effects of biotechnology/genetic engineering
according to the Eurobarometer 1996

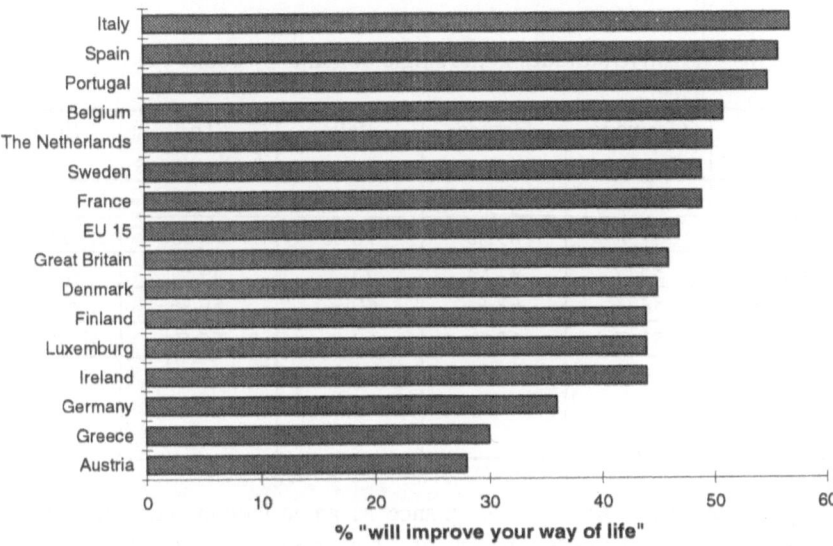

% "will improve your way of life"

Source: European Commission 1997

Other important findings are obtained by splitting the data concerning the positive attitudes by thematic area (see table 5.6). Though there are significant national differences, one common pattern clearly emerges, with the following features:

- The greatest consensus in four out of five countries concerns the developments included in "regulation", showing that the demand for public control is widespread. The only exception is Spain, which is consistent with the findings of Eurobarometers 1991 and 1993 on biotechnology showing that the demand for control on biotech activities in Spain is the weakest within the EU (European Commission 1991, Marlier 1993). This could be connected to a high perception of the risks involved in Agro-Food biotechnology for human health and the environment.

- The lowest share of positive attitudes is found in the economic area in every country: this is even more apparent for Germany and the Netherlands. It seems to indicate that the personal agreement with the economic impacts of biotechnology are lower than the agreement with the scientific and technological findings.

- The figures concerning the areas "health" and "environment" show that in every country the perceived risks are balanced by the perceived benefits.

- In the domain "scientific and technological developments" the national differences are very strong, ranging from 49 % positive attitude in Germany to 81 % in Spain.

Table 5.6:   Positive attitudes in the single domains by country

Domain	Germany	Greece	Italy	The Netherlands	Spain
Acceptance	46 %	59 %	48 %	53 %	57 %
Regulation	67 %	89 %	78 %	78 %	69 %
Economy	33 %	47 %	45 %	34 %	47 %
Environment	54 %	63 %	60 %	59 %	63 %
Health	51 %	62 %	63 %	55 %	64 %
Scientific/ technological developments	49 %	62 %	68 %	53 %	81 %

The importance of social/ethical acceptance as an influential factor is diversely assessed by the national panels. The role of social acceptance in the development of Agro-Food biotechnology appears proportionally much more crucial in the Northern countries than in the Mediterranean countries, confirming the pattern which emerged from the analysis of the experts' attitudes as shown in figure 5.10.

Figure 5.10:   Importance of "acceptance" as an influential factor

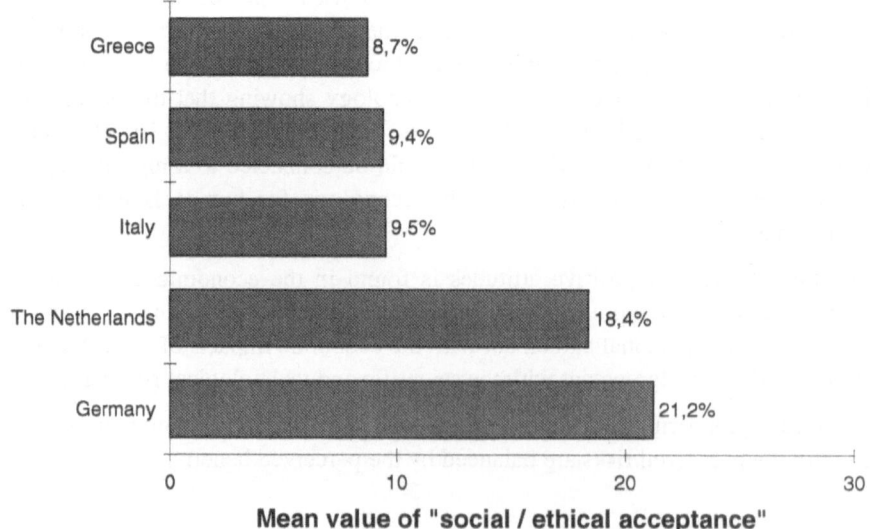

Mean value of "social / ethical acceptance"

The same ranking of the five countries can be found by considering simultaneously the other two factors "policy" and "market" (see figure 5.11). Actually, these three factors appear strictly associated at the national level. This could be interpreted in the sense that policy options and public initiative are seen as crucial in order to ensure a favourable social environment and to encourage the development of a sufficient level of market demand for the products of food biotechnology.

Figure 5.11:   Associated influential factors to "social/ethical" acceptance

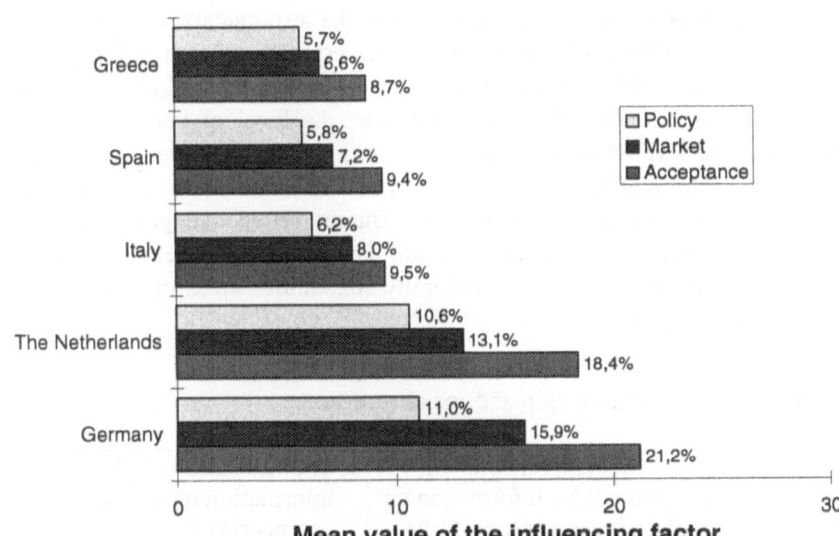

On the other hand, no positive correlation is found between the influential factors "social-ethical acceptance" and "information". As indicated by the comments of many experts to the statements 1 and 3, there is widespread concern about the correctness and unbiasedness of the information which would be provided to the public by public institutions. Some comments expressed by the experts might illustrate this estimation:

- "Neutral information is not to be expected because of the previous lobby-activity of the propagandists." (Germany)

- "The Federal government is not willing to give objective information." (Germany)

- "Impartiality of government institutions is not sufficiently guaranteed." (The Netherlands)

- "The public believes that governmental institutions have low objectivity." (The Netherlands)

- "In what way is this information subjective? Is it independent and neutral? Are the dangers mentioned as well?" (The Netherlands)

- "It is important that such information is correct." (Italy)

It is obvious that the development described in statement 1 (information provided by public institutions) receives much more support than statement 3 (information itself increases acceptance of biotechnology) (see table 5.7). This finding shows that, while public information is generally considered as positive, the concept of a direct link between public information and public acceptance of Agro-Food biotechnology is rejected by a significant share of the respondents, especially in the Northern countries. This could result from a widespread concern about possible manipulation of the public. These findings are supported by population surveys which indicate that no direct link exists between the level of knowledge of the population and the level of acceptance related to the use of genetic engineering in the Agro-Food sector (Heins 1992, Koschatzky and Massfeller 1994, Dixon 1995). In addition, recent research of Evans and Durant (1995) suggests that more knowledgeable members of society are more favourably disposed to science in general, but they are less supportive to morally contentious areas of research, such as genetic engineering.

Table 5.7:   Public information and social acceptance

Statement	1. A governmental institution which provides information on modern biotechnology in the Agro-Food sector to all interested persons and institutions is established in your country.			3. The widespread diffusion of information on genetic engineering from different sources has increased the acceptance of food products made with the help of genetic engineering.		
Attitude	positive	indifferent	negative	positive	indifferent	negative
Germany	84 %	15 %	1 %	39 %	23 %	39 %
Greece	98 %	0 %	0 %	42 %	14 %	39 %
Italy	91 %	7 %	2 %	59 %	24 %	16 %
The Netherlands	94 %	5 %	2 %	61 %	18 %	21 %
Spain	92 %	4 %	2 %	62 %	9 %	26 %

The argument that better information tends to improve the public acceptance of the products of biotechnology is also implicitly addressed in statement 27 ("The public debate about environmental risks of the deliberate release and marketing of genetically engineered microorganisms (e. g. uncontrollable spread, disturbance of natural balances) has lead to a broad rejection of such activities"). Such development is seen as likely in Germany, Italy, Spain and Greece (respectively

81 %, 77 %, 62 % and 24 % of respondents see it realized in five years), while in the Netherlands it is considered impossible ("no realization" by 64 % of experts). The emerging opinion might be interpreted in a sense that a public debate lacking a proper information background would lead to widespread rejection of Agro-Food biotechnology. In fact, the country with the highest share of "no realization" is the Netherlands, which is also the country with the highest level of "objective" knowledge of biotechnology according to Eurobarometer 1996 (European Commission 1997).

5.1.4   General trends and findings

This chapter focuses on the analysis of general features and trends in the public perception of biotechnology, in particular of Agro-Food biotechnology, as emerging from the pattern of the answers to the 71 statements of the questionnaire.

Identification of main areas of social rejection/acceptance

In a first step it is important to see which developments are significantly rejected by the national panels. Ranking the statements in decreasing order by "negative attitude", it is possible to see that the pattern of answers for the included countries is remarkably homogeneous (see table 5.8).

An analysis of tables 5.8 and 5.9 allows to draw the following conclusions to be drawn:

- The respondents' concerns are primarily related to the possible negative impacts of modern biotechnology on human health (statements 37, 40). Environmental concerns (statements 26, 34) appear slightly less important, though coming immediately after possible health impacts: these results go in line with recent research on consumer acceptance of food biotechnology in the Netherlands (Hamstra 1994, 1993, 1991).

- Worries about possible economic impacts are third in this "hit parade of public concerns" about Agro-Food biotechnology (table 5.9). These include concerns about possible job reductions (statements 20, 24) and concerns about possible marginalization of SMEs (statement 19), which is probably associated with a concentration process in the Agro-Food industry. A price increase for food products improved by the application of modern biotechnology is clearly rejected in every country.

- While in Germany and the Netherlands the most critical application area concerns the genetic engineering of farm animals (statements 63 to 69, see table 5.10) - which is underlined in other studies as well (e. g. Hampel et al. 1997, Hamstra 1993) -, in the three Mediterranean countries social acceptance is highly referred to specific framework conditions of Agro-Food biotechnology (statements 2, 3, 27).

Table 5.8: "Top 20" of the statements for "negative attitude" in the five countries (statement no.)

Rank	Germany	Greece	Italy	The Netherlands	Spain
1	40	40	26	37	40
2	37	26	37	26	27
3	26	37	40	40	19
4	24	20	34	19	26
5	20	34	20	20	37
6	34	19	24	24	13
7	19	24	19	15	34
8	27	15	15	34	24
9	65	65	4	47	20
10	15	64	7	27	15
11	2	67	27	63	17
12	66	2	65	65	4
13	7	7	2	66	47
14	63	47	47	4	7
15	69	70	6	64	3
16	4	69	64	69	14
17	67	52	14	7	6
18	45	45	70	67	61
19	33	3	67	14	2
20	64	50	66	21	12

Table 5.9: Most important statements for influential factor "social/ethical acceptance"

Rank	Germany	Greece	Italy	The Netherlands	Spain
1	66	7	7	63	7
2	63	27	4	4	10
3	33	2	3	7	2
4	7	47	27	65	3
5	65	64	47	33	27

Ranking the statements according to the share of "positive attitude", it is possible to assess the degree of social acceptance in the different application areas (table 5.10).

Table 5.10:   "Top 20" of the statements for "positive attitude" in the five
               countries (statement no.)

Rank	Germany	Greece	Italy	The Netherlands	Spain
1	62	1	58	62	62
2	58	12	41	57	42
3	41	9	25	1	31
4	43	8	42	25	43
5	42	41	32	42	58
6	57	35	62	9	32
7	32	58	43	12	41
8	35	42	31	43	57
9	36	11	1	32	55
10	1	10	59	41	48
11	25	32	36	31	56
12	31	5	51	58	9
13	55	6	71	5	71
14	71	55	44	36	23
15	29	43	55	35	18
16	30	25	9	59	52
17	5	62	57	30	49
18	22	71	35	10	59
19	11	44	5	71	1
20	12	57	12	55	29

The most supported applications can be grouped in two categories:

• Monitoring and control activities

This area includes two different segments: the first refers to monitoring and control
techniques based on modern biotechnology in food production and processing. This
relates in particular to the use of rapid test systems based on modern biotechnology
to identify pathogens in plant production and animal husbandry (statement 62), as
well as improvements in hygiene monitoring and the on-line control of quality
parameters in food processing with the help of modern biotechnology
(statements 42, 43).

The second area refers to monitoring and control of the impacts of Agro-Food
biotechnology. In this context, especially the investigation of possible long-term
health impacts of Agro-Food products is highly appreciated (statement 41). The
same applies - in most countries to a smaller extent - to the improvement of the

reliability of risk assessment of field trials with genetically engineered plants (statement 57).

- Positive contribution of biotechnological approaches to the environment

A very positive attitude is generally observed in all the countries toward those applications which produce some direct benefit for the environment. These include enzyme systems which are specifically developed to improve the environmental performance of conventional food-processing procedures, the reduction of emissions and waste from animal production, as well as the conversion of organic agricultural waste into marketable products with the help of modern biotechnology (statements 25, 31, 32).

A relatively high level of a personal positive attitude can also be observed for

- Fundamental scientific results such as statement 58 (The molecular basis of virus resistance mechanisms in economically important perennial plants like e. g. vine, olive and fruit trees is elucidated).

- Developments referring to institutional and private initiatives aimed at informing and/or involving the public in the debate and decision-making on the application of modern biotechnology in the Agro-Food sector (statements 1, 5, 12, 35).

- Potential positive health impacts of modern biotechnology, like food, in which the allergenic potential has been decreased by genetic engineering (statement 36).

- Genetically engineered plant varieties resistant to salinity and/or drought (statement 55).

- Introduction of genetically engineered vaccines in animal production (statement 71).

Instead, the attitudes expressed toward those developments directly related to consumption of genetically engineered food in general are less positive. In this sense, it is interesting to analyse the statements which have been given the lowest score in terms of "positive attitude" in all the countries involved (see table 5.11).

Table 5.11:   "Bottom 20" of the statements for "positive attitude" in the five
                countries (statement no.)

Rank	Germany	Greece	Italy	The Netherlands	Spain
52	21	50	64	64	3
53	64	3	6	21	2
54	2	69	66	27	60
55	63	70	67	14	61
56	47	14	65	69	4
57	69	64	14	63	14
58	66	47	2	66	45
59	14	2	27	65	17
60	65	45	4	67	47
61	67	67	47	47	13
62	15	65	45	45	19
63	45	19	19	7	20
64	7	7	7	15	26
65	19	37	20	24	27
66	20	15	15	40	15
67	24	20	40	34	7
68	26	24	34	26	37
69	37	26	37	19	40
70	34	40	24	20	24
71	40	34	26	37	34

Apart from the obvious rejection of statements such as 20, 24, 26, 34, 37, 40 (job
reduction, damages to human health and the environment), it is interesting to
observe the following statements as well among the less welcomed developments:

- Most consumers in your country have quickly got used to all kinds of food and
  beverages produced with the help of genetic engineering (statement 2).

- A restaurant chain or catering service specialized in offering trendy meals with
  genetically engineered ingredients open branches in almost all large cities in your
  country (statement 7).

- Food made with the help of genetic engineering achieves a turnover share of
  30 % or more of all the food consumed in your country (statement 14).

- In your country most of the beer is produced with genetically engineered yeast
  (statement 45).

All such statements directly refer to the market development of food biotechnology
and deal directly with the habituation of consumers to gene food products - an area

which is identified as being decisive for the future market development (see chapter 5.3.3). This clearly indicates that in general, the introduction of genetic engineering in food production and processing is not raising enthusiasm in any of the included countries.

Attitudes towards genetic modification as applied to different biological domains

Several opinion surveys carried out among the European population have shown that the attitudes of the public towards biotechnology and genetic engineering vary considerably, depending on the target of genetic modification. According to the findings of Eurobarometer 35.1 and 39.1 on biotechnology, research carried out on animals is generally less supported than that on plants and microorganisms, due to ethical considerations. In turn, genetic research on plants is less supported than that on microorganisms used in food production and processing (see figure 5.12).

Figure 5.12:   Acceptance of different applications of biotechnology and genetic engineering according to Eurobarometer surveys

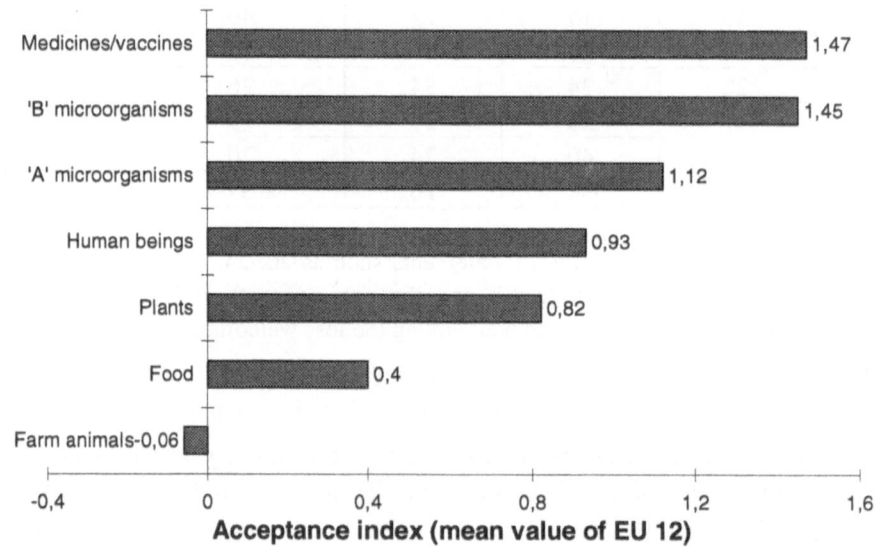

Source: Marlier 1993

This pattern of the Eurobarometer survey is basically confirmed by the findings of the Delphi survey. Also in the "expert" panel, the lowest support is given to the genetic modification of animals, and the support granted to genetic modification of plants appears slightly lower with respect to microorganisms/enzymes. In order to assess this point, it has been necessary to select the five most relevant statements for each of the three concerned areas (see table 5.12).

Table 5.12:   Personal attitudes by target of genetic modification (% value)

Statement No.	Germany		Greece		Italy		Spain		The Netherlands	
	pos.	neg.	pos.	neg.	pos.	neg.	pos.	neg.	pos.	neg.
Microorganisms/enzymes										
46	56	25	67	18	68	16	92	2	66	11
48	56	17	76	10	76	7	95	3	72	5
49	57	15	72	15	78	6	92	2	71	7
50	26	26	43	39	53	10	81	4	27	12
51	51	13	73	13	88	2	92	1	73	2
Plants										
52	52	20	48	42	74	10	93	2	36	21
53	41	20	67	20	72	9	85	2	54	18
54	43	21	69	20	61	16	91	3	63	18
55	80	8	90	5	86	8	95	2	82	10
60	61	10	50	18	74	12	55	14	56	13
Animals										
63	19	47	56	34	53	18	76	8	10	53
64	23	38	25	67	52	23	84	9	20	42
65	11	56	19	70	34	35	68	11	8	49
66	14	53	48	37	44	20	81	3	10	45
69	17	47	38	42	55	17	81	6	13	37

Also in this respect some national differences are observed:

- In Germany, contrary to what is observed for the other four countries, the genetic modification of microorganisms seems to be more controversial than that of plants.

- In Spain and Greece, the genetic modification of microorganisms seems to be more supported than that of plants.

- In Italy and the Netherlands, both positive and negative attitudes toward the genetic modification of plants are more frequently expressed than for microorganisms. Such polarization observed for plants possibly reflects the fact that the public debate focused on transgenic crops is currently very intense in those countries, while the genetic modification of microorganisms is more a "neutral" issue arousing more "indifferent" attitudes.

Genetic engineering applied to animals is confirmed as the most controversial issue. Concerns for animal welfare are clearly more widespread in the Netherlands and in Germany than in the Mediterranean countries, though in Greece some exception is observed in this respect. In the Netherlands 35 % and in Germany even 69 % of the group "consumers and critics" are favourable to refuse for public funds to research on genetic engineering of farm animals (statement 4). This share is lower among the consumers and critics in Spain (31 %) and Italy (21 %). The highest share is found in Greece, but the conclusion that the concern for animal welfare is highest in this country is not entirely consistent with other data, maybe the underlying motivations to the assessment of statement 4 are of a different kind (e. g. concerns for the level of public expenditure, religious motivations).

The high level of acceptance of transgenic animals used to produce pharmaceutical substances (statement 23) within the group "consumers and critics" confirms that the applications of genetic engineering in health care are generally better accepted than Agro-Food applications (see table 5.13).

Table 5.13:   Social acceptance of genetic engineering of animals for different purposes (proportion "positive attitude" of consumers and critics)

Statement No.	Content	Germany	Greece	Italy	The Netherlands	Spain
23	Market success of pharmaceuticals produced by genetically engineered animals and plants	39 %	74 %	52 %	68 %	88 %
66	Genetic engineering of early stages of embryo cells of farm animals	1 %	44 %	30 %	11 %	67 %

Specific trends in public perception

The latter finding is part of the more general conclusion that the acceptance of the application of genetic engineering to food production and processing depends on the nature and justification of the purposes of genetic modification in every single case. Such an "ethical dimension" of the social acceptance is a peculiarity of biotechnology which has been already indicated in previous studies (e. g. by Gofton et al. 1996, Grove-White et al. 1997). A clear indication in this sense is provided by comparing the attitudes expressed toward two specific developments which raise opposite emotional reactions among the experts:

- A restaurant chain or catering service specialized in offering trendy meals with genetically engineered ingredients open branches in almost all large cities in your country (statement 7).

- Genetically engineered plant varieties resistant to salinity and/or drought have improved agricultural productivity in arid environments significantly (statement 55).

In the first case, the application of genetic engineering to food production is made for futile reasons, not justified by any people's basic need. In every country, such adevelopment ranks very high in the "top 20" of "negative attitude" (see table 5.8) as well as in the "top 5" for relevance of "influential factors: social/ethical acceptance" (see table 5.9). The social and ethical acceptance of statement 7 is even seen as more problematical than that of statements such as e. g. 65 (cloning of farm animals). The comments of the respondents to statement 7 are also mostly critical, such as:

- "It would not be accepted now and is unlikely in the future." (Spain)
- "Where is the consumers' need for that?" (Germany)

Table 5.9 also suggests that the rejection of non-essential application of genetic engineering to food production and processing might be stronger in the Mediterranean countries. On the opposite pole, statement 55 (genetically engineered plant varieties resistant to salinity and/or drought) obtains the highest score for "positive attitude" among the developments associated to the genetic engineering of plants in every country (see table 5.10), probably because it is the statement most clearly justified on the basis of its usefulness ("It helps to fight against desertification and abandonment of unproductive land. Very positive ecological impact." - comment of a Spanish researcher).

On the whole, the advent of food biotechnology seems to be considered as an inevitable, rather than desirable, development. A sense of fatalism is suggested by the pattern of the answers to several statements: the application of genetic engineering is seen as inevitable even when it is not wanted by a significant share or even a majority of the public. The statement in which this opinion emerges most clearly is that concerning the cloning of farm animals (statement 65). In all countries negative attitudes are mostly prevailing over positive attitudes, except for Spain: if we exclude the latter country, in the other four countries a majority rejects this development. However, only a negligible minority of respondents in each country believes that this development will not be realized. Almost three quarters of the experts believe that it will be realized within ten years (see table 5.14).

Table 5.14:   Cloning of farm animals (statement 65)

Country	Personal attitude			Time of realization					
	posi-tive	indif-ferent	nega-tive	0-5	6-10	11-15	16-20	> 20	no reali-zation
Germany	11 %	32 %	56 %	21 %	40 %	21 %	4 %	4 %	4 %
Greece	19 %	9 %	70 %	42 %	29 %	16 %	8 %	2 %	1 %
Italy	34 %	26 %	35 %	42 %	38 %	5 %	1 %	1 %	5 %
The Netherlands	8 %	43 %	49 %	52 %	20 %	6 %	1 %	1 %	20 %
Spain	68 %	16 %	11 %	44 %	33 %	10 %	4 %	1 %	0 %

The same sense of inevitability emerges regarding statement 2 (consumers get quickly used to food and beverages produced with the help of genetic engineering). In this statement Spain again is an exception, for the majority of respondents is clearly supporting the application of genetic engineering in food production (see table 5.15).

Table 5.15:   "Habituation effect" of consumers to gene food (statement 2)

Country	Personal attitude			Time of realization					
	posi-tive	indif-ferent	nega-tive	0-5	6-10	11-15	16-20	> 20	no reali-zation
Germany	20 %	27 %	53 %	20 %	56 %	16 %	2 %	3 %	4 %
Greece	21 %	19 %	56 %	35 %	39 %	17 %	4 %	1 %	1 %
Italy	28 %	38 %	30 %	21 %	41 %	20 %	2 %	3 %	3 %
The Netherlands	37 %	43 %	20 %	45 %	46 %	3 %	0 %	1 %	4 %
Spain	56 %	23 %	19 5	28 %	53 %	15 %	1 %	1 %	1 %

The pattern of attitudes expressed toward those developments concerning the acceptance of genetic engineering in food production in general (statements 2, 7, 14) is very telling in this sense. Such statements, in fact, are among those with the lowest score for "positive attitude". This suggests that the introduction of genetically engineered foods does not raise any enthusiasm in the majority of respondents. It is rather seen as something that cannot be avoided, as shown by the forecasts expressed by the majority of experts who think that these developments will be realized.

The role of food traditions

A major factor among those contributing to shape the social environment for the development of Agro-Food biotechnology is the attachment of consumers to their traditional tastes and foods. The most relevant statement in order to analyse people's attachment to food traditions and the respective implications for the market perspectives of Agro-Food biotechnology is statement 47 (Modern biotechnology has failed in producing food and beverages satisfying traditional taste preferences of large consumer groups in your country).

In the first instance, the different patterns of answers to this statement observed at the national level indicate the existence of a polarization between the Northern and Mediterranean countries (see table 5.16). In particular, the share of respondents who see this development as not realistic (in other terms, who see food traditions as a non-influential factor) is much higher in Germany and the Netherlands compared to Italy, Greece and Spain. Two indications can be drawn from this finding:

- People's attachment to food traditions is considered much stronger in Mediterranean countries.

- Such attachment is seen as a possible constraint to the development of a market for the products of food biotechnology.

Table 5.16:   Modern biotechnology has failed in producing food and beverages satisfying traditional taste preferences of large consumer groups in your country (statement 47)

Time of realization	Germany	Greece	Italy	The Netherlands	Spain
0-5	14 %	39 %	14 %	6 %	28 %
6-10	32 %	42 %	33 %	8 %	26 %
11-15	10 %	4 %	17 %	0 %	7 %
16-20	2 %	1 %	0 %	1 %	1 %
> 20	2 %	4 %	0 %	0 %	1 %
No realization	24 %	2 %	8 %	75 %	4 %
Not able to answer	17 %	8 %	27 %	10 %	32 %

Another clue in this sense can be found by comparing the value of "social/ethical acceptance" as an influential factor for this specific statement with the mean value for all the statements in each country. Actually, in Southern countries the difference with respect to the national means is much more apparent: the value of this sub-category for statement 47 is 25 % for Greece (national mean: 8.7 %), 28 % for Italy (national mean: 9.5 %) and 21 % for Spain (national mean: 9.4 %), respectively. In Germany and the Netherlands it is 32 % (national mean: 21.2 %) and 18 % (national mean: 18.4 %), respectively. This suggests that, although in Northern European countries social acceptance is much more important for the development of Agro-Food biotechnology than in the Southern countries, the role of attachment to food traditions is not an important factor in shaping the public attitudes toward Agro-Food biotechnology. The motivations underlying the existing negative attitudes toward Agro-Food biotechnology in Northern countries are likely to be rather related to concerns about food safety or to ethical objections.

On the other hand, no remarkable differences are observed between the countries as concerns the time period, in which food traditions are expected to have an influence on the market development of Agro-Food biotechnology. Only a negligible minority of the experts in each country expects that such influence would last for more than ten years.

On the whole, the final success of food biotechnology in getting accepted by the consumers seems taken for granted by a majority of the experts in all countries. This also emerges from many comments stressing that price considerations would prevail over other factors: ("Not taste, but price and packaging decide on the acceptance.", "Not the taste decides but the prices and the social situation of the consumers."), or the power of marketing to influence the consumers ("Marketing can do almost anything.", "Taste is already manipulated.").

Another statement relevant in order to evaluate the role of food traditions is statement 45 ("In your country most of the beer is produced with genetically engineered yeast"). Here the situation observed for statement 47 is reversed, as the respondents in beer-drinking countries (Germany and the Netherlands) are clearly more sensitive on this issue. In fact, they tend to deny the realization of this development to a greater extent than respondents in wine-drinking countries (see table 5.17).

Moreover, in beer-drinking countries the influence of acceptance related to the use of genetic engineering in beer production is seen as more important than for statement 47 which refers to food in general (see table 5.17). This does not happen in wine-drinking countries without an established tradition in brewing (as epitomized by the comment of an Italian consumer: "I am a wine drinker, I do not like beer.").

It seems clear that in countries where beer is already seen as an industrial product without a national tradition, the acceptance of genetic modification in brewing is facilitated. On the contrary, in countries where beer is a traditional, still perceived as craft-made product, consumers' acceptance is more controversial. Conversely, in countries where food production is more industrialized, the acceptance of food biotechnology in general is less influenced by attachment to food traditions. This confirms that the role of attachment to food traditions is effectively important.

Table 5.17:   In your country most of the beer is produced with genetically engineered yeast (statement 45)

Time of realization	Germany	Greece	Italy	The Netherlands	Spain
0-5	15 %	47 %	35 %	26 %	36 %
6-10	39 %	37 %	33 %	32 %	43 %
11-15	18 %	8 %	17 %	8 %	6 %
16-20	3 %	2 %	0 %	4 %	4 %
> 20	6 %	0 %	0 %	2 %	0 %
No realization	13 %	1 %	0 %	19 %	1 %
Not able to answer	5 %	5 %	14 %	8 %	10 %

Table 5.18:   Importance of acceptance as an influential factor in relation to the use of genetic engineering

Statement No.	Content	Germany	Greece	Italy	The Netherlands	Spain
45	Beer production	36 %	16 %	11 %	32 %	22 %
47	Food production in general	32 %	25 %	28 %	18 %	21 %

It is also important to observe the different assessment of the category "personal attitude" for statement 45 and other statements with a similar technical content (use of genetically engineered microorganisms and enzymes in the food industry). Such statements are the following:

- Genetically engineered microorganisms and enzymes are widely used to improve the processing quality of food (statement 46).

- A large variety of genetically engineered microorganisms which can be precisely controlled and regulated in their metabolic activities during food production processes has been developed (statement 48).

- Enzymes optimized by protein engineering are practically used in specific sectors of the food industry (statement 49).

- Approximately 90 % of the enzymes used in the food-processing industry are produced by genetically engineered organisms (statement 50).

Table 5.19:   Personal attitudes concerning the use of genetically engineered microorganisms and enzymes in the food industry

Statement No.	Germany		Greece		Italy		The Netherlands		Spain	
	posi-tive	nega-tive	posi-tive	nega-tive	posi-tive	nega-tive	posi-tive	nega-tive	posi-tive	nega-tive
45	9	39	20	39	14	13	6	13	26	7
46	56	25	67	18	68	16	66	11	92	2
48	56	17	76	10	76	7	72	5	95	3
49	57	15	72	15	78	6	71	7	92	2
50	26	26	43	39	53	10	27	12	81	4

It is obvious that statement 45, which from a technical point of view does not differ significantly from the others, has a dramatically lower degree of acceptance (see table 5.19). This could appear as a logical inconsistency. However, statement 45 concerns not food in general but rather a specific, concrete product (beer), susceptible of raising also emotional reactions. Here again, the difference is even more apparent in Germany and the Netherlands. The comments to statement 45 are much more frequent and lively than for the other statements:

- "The economic advantages are far outweighed by the risk of erosion of the image of natural, healthy products." (Spain)

- "Because beer is drunk for reasons other than common sense, this may experience resistance." (The Netherlands)

- "Well-known brands will not do it because of the rejection of consumers." (The Netherlands)

- "Genetic engineering yes, but surely not with German beer." (Germany)

- "Beer is holy." (Germany)

- "Not in Bavaria!" (Germany)

Such comments clearly show that when it comes to a specific product with some cultural implication, instead of foodstuffs in general, consumers' acceptance of genetic engineering in food production may be more problematic. To mention just one well-known example, this concept has been confirmed in recent years by the considerable opposition experienced within the EU against the recombinant bovine somatotropine (rBST). In that instance the concerned product was milk, which is often considered as the symbol of "pure" food itself.

These considerations point to the conclusion that the question of safety cannot be considered as the only relevant factor in the social acceptance of food biotechnology, and probably not even the most important one, as cultural factors still play a major role in shaping consumers' attitudes and behaviour.

5.2   Legal framework conditions

5.2.1   Introduction

The core of this chapter is about the way European experts believe how modern biotechnology applied to the Agro-Food sector should/will be regulated and controlled, the role of the public in decision-making regarding the marketing of gene food products and beverages, and the impact such norms may have on the performance of industries working in the biotechnology sector.

A general trend in the EU seems to focus on the need to implement regulations and standards which would allow products produced/developed with the help of modern biotechnology to reach the same status of normality as conventional food and beverages[7]. Despite this, there is a fear that the current situation (the public's lack of information and widespread rejection of genetically engineered products) is not ideal for implementing regulations, especially labelling and public participation in decision-making, as they could lead to commercial disadvantages for such products.

The final objective of the analysis is therefore to define the general response of the sample as well as to highlight either similarities or differences between countries and different expert groups. Taking into account the amount of information referring to the five countries that participate in the survey, only relevant aspects will be described, rather than displaying a general description of the whole data as this seems to be more informative with regard to meeting this objective.

The data described correspond to the second round of the Delphi survey for the statements under the domain "regulation". The comments written by the experts

---

[7]   A new legal document on the issue has recently been approved by the European Parliament: the final version of the Directive regulating patents on biotechnological "inventions" (12.05.98). The document, based on the one which was already approved by the European Council of Ministers in 1995 and afterwards rejected by the European Parliament because of ethical reasons, excludes from patenting human clones, embryo manipulations and modifications of the genetic identity of human beings and animals. Once the Directive has obtained the approval of the Council of Ministers, whose negotiations will start in November 1998, the member States will have two years to incorporate the law in their national legislation. The Directive has the opposition of the Green parties, which believe that such regulation will foster "biopiracy" practices by Western firms, as they will exert a monopoly on patenting plants from Third World countries without their governments knowing it or being able to act against it.

who answered the questionnaire have also been used to further interpret the quantitative data.

The domain "regulation" is actually divided into four main issues:

- Controls and procedures for the approval of genetically engineered products
- Labelling of these products
- Participation of the public in that process
- Potential impact biosafety directives may have on the biotechnology industry

These specific issues are addressed in sub-chapters 5.2.4 and 5.2.5, while general aspects regarding regulation are dealt with in sections 5.2.2 and 5.2.3. Finally, this chapter closes with a summary (sub-chapter 5.2.6).

Table 5.20:   Statements of the questionnaire belonging to the domain
                "regulation"

No.	Content
8	The labelling of all food products made with the help of genetic engineering is compulsory and fully implemented in your country.
9	All foods and beverages which are produced with the help of genetic engineering are subject to an official case-by-case procedure for approval in your country.
10	For the marketing approval of Agro-Food products made with the help of genetic engineering, concerns on their social impacts and potential ethical conflicts are generally taken into account.
11	The responsible authorities in your country practically use specific technical equipment and standardized methods to identify food and beverages which were produced with the help of genetic engineering.
12	European and national authorities start initiatives which involve the public in the debate and decision-making on the application of modern biotechnology in the Agro-Food sector.
13	The uniform implementation of EU biosafety directives in all EU countries has led to a higher attraction of the EU for companies based outside the EU.

5.2.2   General overview of the results

Regulations on the marketing of biotechnological food products and beverages are expected to be realized soon, and this is seen as a positive step by most of the experts in the panels of the five countries involved. Regarding the self-assessed expertise, the chosen experts in Germany, Italy and the Netherlands also appear to have a better knowledge of this issue than other issues dealt with in the questionnaire. This relationship for the category "degree of knowledge" is not noticeable in Spain nor in Greece (see figure 5.13). Figure 5.13 shows the comparison between the results for all statements of the questionnaire (left column), and the results for the domain "regulation" (right column) mean of all statements.

Figure 5.13:   "Degree of knowledge" in the "regulation" domain by country

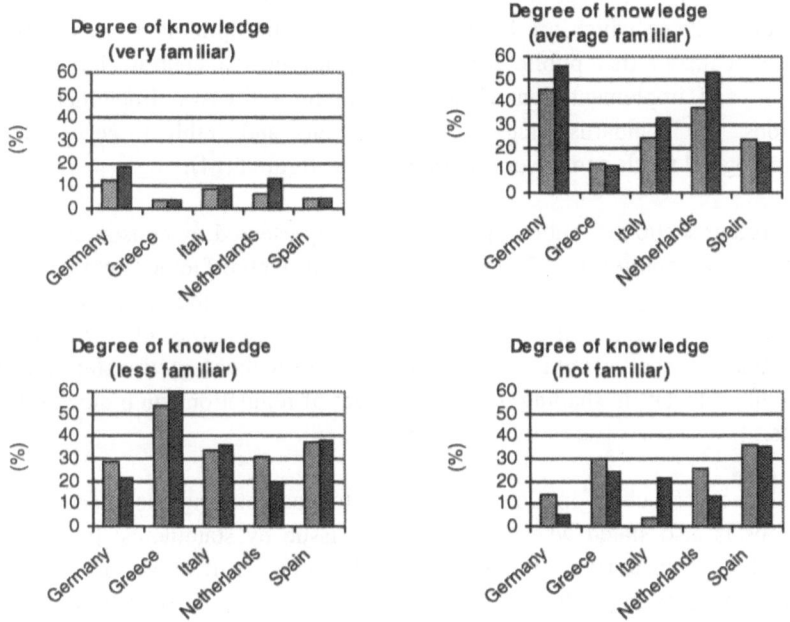

Most of the chosen experts in the five countries involved in the Delphi survey expect new regulations to emerge in the next ten years, although Dutch experts, followed by the Spanish, place the development of the legal framework on biotechnological food products sooner (more than half of their national samples think further regulations will come in the next five years).

Most experts in all countries have a positive attitude towards this situation, although there is a significant but small percentage of them who see the undertaking of legal actions in negative terms. Despite this, a proper legal framework would have a positive effect on economic development. In fact, respondents consider it especially important in order to foster the competitiveness of the economy, although this trend is not as strong among Italian experts as in the rest of the samples involved.

Having said that, some differences appear between countries. German and Dutch experts highlight the importance of new regulations for the competitiveness of the economy. The participants in the three Mediterranean countries, for their part, appear to believe that such regulations also foster the principles of sustainable development. Furthermore, the importance of the implementation of norms and standards for knowledge creation in science and technology is highlighted by Spain and Italy, whose expert panels seem to believe most strongly in the importance of new regulations on research.

In general terms, the factors which most influence regulatory issues can be interpreted as those which are highly related to public opinion. In this sense, the view can be taken that public participation represents a central element in the discussion and implementation of policies, in the successful implementation of regulations or standards, and in the social and ethical acceptance of biotechnological foods and beverages as well (see figure 5.14).

One important section of public opinion is also represented by consumers' opinion. And consumers' opinion is a central part of the influential factor "markets", which has been assessed by a relatively high percentage of the respondents as one of the "most influential factors " in designing and applying regulations. On the other hand, environmental concerns, industrial innovativeness or technology transfer appear not to be highly relevant in shaping the development of regulations on biotechnological products.

The relevance of public opinion with regard to guaranteeing the success of new regulations is also stated when looking at the issue by statements. Involving the public (i. e. consumers) in debate and decision-making on marketing biotechnological food products is the most urgent need, as it seen by the experts. Furthermore, it is expected to occur sooner than other actions dealt with in this domain.

As a matter of fact and in more general terms, regarding the category "time of realization" it is noticeable that the biggest percentage of the samples in the five countries involved believe that issues such as labelling, public participation in decision-making, or the general implementation of case-by-case procedures for market approval of genetically engineered food and beverages will occur in the next five years. It is also remarkable that a significant percentage of the German and Dutch experts think that such developments will not take place at all.

Labelling biotechnological food products is, then, an action very likely to happen in the following five years. In this sense, experts seem to believe that biotechnological foods and beverages cannot be regulated globally, but they need to be studied individually, and more than half of the expert samples think that a "case-by-case" procedure for their approval will be implemented in the next five years (this percentage is lower for Germany: 42.6 %).

Figure 5.14:   Influential factors of activities regarding regulations of biotechnological issues[8]

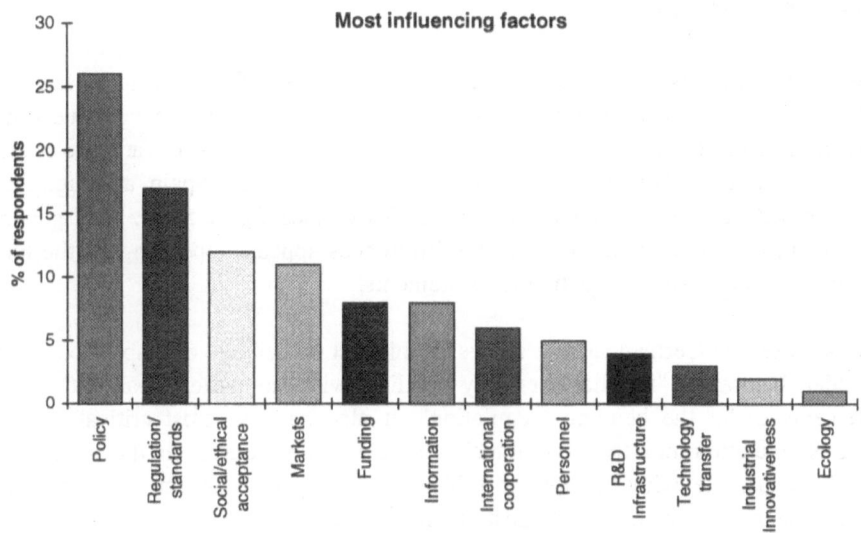

In short, more than half of the population believe that in the following five years a case-by-case procedure of approval of biotechnological food products will be developed; the public will participate in such approval (a situation that is seen happening very soon in the Netherlands and, as well, in Germany and Italy), and the final decision will be conditioned by the social and ethical implications of the product to be marketed. In all these processes, public authorities will be able to identify products made with the help of genetic engineering by means of using specific equipment and methods but the time span is longer, in which such technical advancement will take place.

Finally, experts do not seem to have a clear opinion on the impact that regulation may have on the performance of biotechnological industries. The lowest response rate of the area "regulation" and the highest percentage of experts who consider themselves not familiar with the issue belong to statement 13 ("The uniform implementation of EU biosafety directives in all EU countries has led to a higher attraction of the EU for companies based outside the EU"). This statement is also the one that would take longer to happen and besides, its realization is perceived quite negatively by the sample.

---

8   The "European mean" (mean of all the countries involved in the Delphi survey) is used here because, although the percentages vary between countries, all national samples point at the same "most influential factors", that is, they show the same trend for this category.

5.2.3 Expertise and attitudes towards norms and standards

Degree of knowledge

Most of the experts in Germany and the Netherlands affirm having at least an average familiarity with regulation issues regarding biotechnological products and show a positive attitude towards norms and standards. This positive attitude is also held by most parts of the national samples in Italy, Greece and Spain, although their self-assessed degree of knowledge is much lower (see figure 5.13). This general assessment has to be qualified, as some differences appear, depending on the kind of regulation addressed by each of the statements.

In this sense, the German sample shows the highest percentage of respondents who consider themselves "familiar" or "very familiar" with the matters dealt with in the statements under the heading "regulation". It also has the most critical attitude towards regulation initiatives on biotechnological food products. On the other hand, the Greek sample with the lowest self-assessed expertise of the five countries involved stands out with the strongest positive attitude.

The degree of familiarity increases in issues that do not include any technical concepts and therefore do not require any specialized knowledge on these matters or technological or scientific background to assess them. All participants in the sample define their formal knowledge or expertise in similar terms regarding statements 8, 9, 10 and 12. Such statements deal with labelling, procedures for approving new food goods, ethical and social concerns as well as public participation in decision-making, respectively.

Those matters are not specifically related to biotechnology, but affect all food products. In this sense, genetically engineered products could have been seen as just another case of food goods that may involve risks for the environment or human health. And so the assessment of their future developments may have been approached from a more general framework.

On the other hand, there is a decrease in the results for the category "degree of knowledge" in all countries for statement 11 and, especially, statement 13. In both cases the sub-category "not familiar" increases and, in fact, it is the highest of the whole domain "regulation" while the answers for "very familiar" are the lowest or among the lowest (in Germany and Spain). Statement 11 deals with technologies and methods specifically aimed at identifying biotechnological products. Statement 13, on the other hand, focuses on the attraction that the EU may have for foreign companies once biosafety directives are applied.

Regarding the level of knowledge, two groups of countries can be distinguished (see table 5.21):

- Germany and the Netherlands: more than two thirds of the expert panel see themselves as "familiar" or "very familiar", so experts' self-assessed expertise is higher than in the rest of the countries included.

- Italy, Spain and Greece (in this order): more than half of the expert panel assess their degree of formal knowledge on those matters as "less familiar" or "not familiar".

German experts also show the most constant knowledge through all the statements, this means that experts consider themselves very familiar or not familiar with the whole subject and their degree of knowledge does not change significantly depending on the specific issue addressed. This feature also occurs in the Netherlands, while the results for the category "degree of knowledge" in the Mediterranean countries are more variable.

Table 5.21:   Degree of knowledge by country regarding "regulation" issues

Country	Very familiar	Familiar	Less familiar	Not familiar
Germany	18 %	56 %	21 %	5 %
The Netherlands	13 %	53 %	21 %	13 %
Italy	10 %	33 %	36 %	21 %
Spain	5 %	22 %	38 %	35 %
Greece	4 %	12 %	60 %	24 %

As has already been mentioned, there is a general decrease in the level of knowledge in all countries for statement 13 ("The uniform implementation of EU biosafety directives in all EU countries has led to a higher attraction of the EU for companies based outside the EU"). The level of knowledge also decreases for statement 11 in all countries, but this decrease is much more acute in Italy, Greece and, especially, Spain.

In summary, the self-assessment of the degree of formal knowledge varies sharply between the countries; considering the largest part of the expert panel (more than 50 %) and excluding statement 13 the following distribution emerges:

- Germany: high (familiar and very familiar)
- The Netherlands: high (familiar and very familiar), except for statement 11.
- Italy: average (familiar and less familiar)
- Spain: low (less familiar and not familiar)
- Greece: low (less familiar and not familiar)

Personal attitude

The category "personal attitude" appears to be quite uniform through all the issues considered by the domain "regulation", with statement 13, again, the "exceptional" case here. Leaving this statement aside, most of the samples in the five countries involved think that further regulation on biotechnology food products is positive (the range going between 67 % of positive answers in Germany and 90 % in Greece). Furthermore, there is not a strong negative attitude on the issue, although opinions contrary to regulations are quite significant in Germany (13 %) and Spain (21 %) (see figure 5.15).

Figure 5.15:   Personal attitude of regulation issues by country

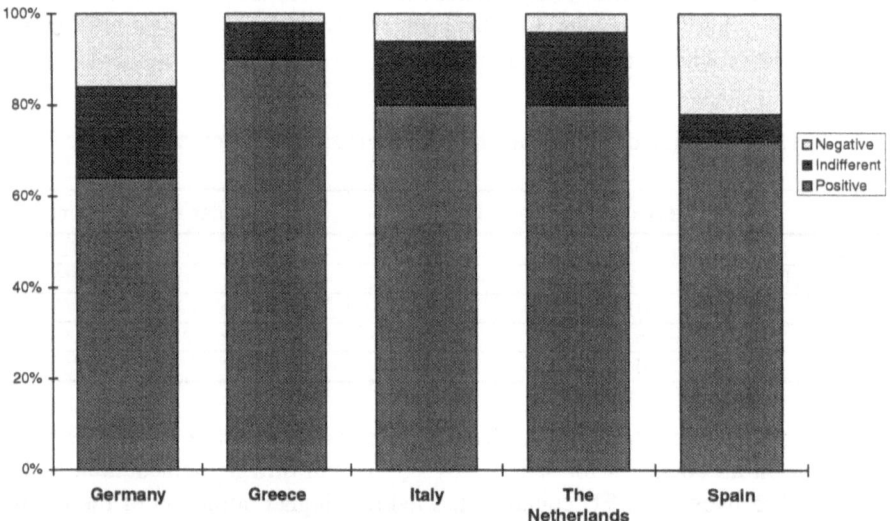

As shown in table 5.22, Greek experts appear to be the most optimistic towards the general positive influence of norms and standards (90 % share this opinion). German experts, on the other hand, appear to be the most critical and Spanish experts also show a higher percentage of negative attitudes than the other three countries. It is relevant to point out here that the relatively low percentage of positive attitudes in Spain cannot be considered as a general trend (which is quite positive), but it is strongly conditioned by the negative opinion in statement 13 (77 % of respondents consider its contents in negative terms).

According to the results of the Delphi survey, Germany is also the country where the largest percentage of the experts does not feel very strongly about regulation issues: 18 % of the sample is indifferent towards the potential effects that new norms may have on modern biotechnology applications (the highest percentage for this sub-category in comparison to the other four countries involved). On the other hand, Spanish and Greek samples, whose self-assessed expertise is lower than in the rest of countries, tend to define their attitudes more clearly or, as in the case of Spanish experts, not to answer the question (3 % of the Spanish experts did not answer the category "personal attitude", the highest percentage for all countries).

Table 5.22:   Positive attitude of "regulation" by country

Country	Positive
Greece	90 %
Italy	78 %
The Netherlands	78 %
Spain	70 %
Germany	67 %

5.2.4   Controls on marketing of gene food products

In recent years there has been an intensive discussion concerning the procedure to regulate the market approval of gene food products in the EU. In this context especially the labelling practice has been in the centre of this debate since consumer organizations of all member countries of the EU claimed more or less a compulsory labelling of these products to give consumers the freedom of choice to decide whether they want to buy gene food products or conventional food. In contrast, especially industrial companies and their respective associations argued in favour of a more relaxed labelling procedure of gene food products.

After several years of intensive debate the "Novel Food Regulation" of the EU (No. 258/97 EU) was put into force on May 15th, 1997. This regulation is valid for food which contains genetically modified organisms (GMOs) or consists of GMOs. In addition, food produced from GMOs but not containing them is covered in this regulation as well (see table 5.23). According to the "Novel Food Regulation" the characteristics of the "Novel Foods" have to be stated which explain the missing equivalence with conventional food. In this case, the gene technology procedure which was used to achieve these characteristics has to be published. In addition, existing GMOs in the food product have to be named as well (Amtsblatt EG 1997).

Since the genetically engineered, herbicide-resistant soybean of Monsanto as well as the genetically modified, insect- and herbicide-resistant maize of Novartis ("B.t. maize") got market approval in single member countries of the EU (United Kingdom, France) before the passing of the "Novel Food Regulation", a specific regulation (No. 1813/97 EU) was put into force in order to regulate the labelling of these products, basically in the same way as in the "Novel Food Regulation" (see table 5.23). Products based on genetic engineering which are not food or gene food products, for which their producers have applied market approval before May 15th, 1997 (with the exception of Monsanto's soybean and the Bt-maize of Novartis) have to be labelled according to the 90/220/EC directive (in some cases also according to a new version of 1997), in which labelling is practised less stringently than in the "Novel Food Regulation" (see table 5.23).

Table 5.23: Overview of the current regulation of labelling of gene food products in the EU

Name/ Number of regulation	No. 1813/97 EU	No. 1813/97 EU	No. 258/97 EU ("Novel Food Regulation")	90/220/EC in the version of 97/35/EU	90/220/EC (directive on the delibe- rate release of GMOs)
Type of food	Food produced from genetically modified maize, which got mar- ket approval based on the decision of the EU Comm. 97/98/EU	Food produced from genetically modified soy bean, which got market approval based on the decision of the EU Comm. 96/281/EU	Food which contains GMOs, consist of GMOs or food produced from GMOs but do not contain GMOs itself	Products which con- tain GMOs or consist of GMOs but are no food	Products which con- tain GMOs or consist of GMOs
Additional require- ments			Market approval has been applied for after May 15th, 1997	Market approval has been applied for after June 28th, 1997	- Market approval has been applied for before June 28th, 1997 and products are not food - Products are food but market approval has been applied be- fore May 15th, 1997
Type of labelling	Note that the production pro- cedure based on genetic enginee- ring and/or exi- sting GMOs	Note that the production pro- cedure based on genetic enginee- ring and/or exi- sting GMOs	Note that the production pro- cedure based on genetic enginee- ring and/or exi- sting GMOs	Note that (potentiall y) existing GMOs	Note that genetic mo- dification only due to safety reasons

Source: Dederer 1998a

Up to date, the "Novel Food Regulation" has not been implemented in the EU member countries. Details concerning the text of the labels of gene food products as well as scientific procedures which are required to identify these products are not clarified at present, because detailed rules on how to implement the "Novel Food Regulation" have not been elaborated so far (Dederer 1998a). But the EU Commission has worked out basic guidelines concerning the labelling of gene food products which comprise - among others - the following aspects (Dederer 1998b):

- Gene food products should be labelled along the entire food chain ("from the stable to the table").

- Consumers should be adviced of the character of the genetic modification of the food product as neutrally as possible. The label should facilitate the buying decision of the consumer without stigmatizing modern biotechology or raising doubts concerning the safety of the products.

- Labelling of gene food products should be based on scientific rules and methods which allow the application of gene technology to be proved.

In the Delphi survey there are three statements regarding the way in which gene foods and beverages are approved for marketing. They refer, more specifically, to the use of special equipment by national authorities to identify genetically engineered food and beverages (statement 11), to the implementation of a case-by-case procedure for the market approval of gene food products (statement 9), thereby taking into account ethical and social concerns (statement 10).

Personal attitude

Experts in the panels of the five countries involved in the Delphi survey regard control and monitoring systems to identify gene food products before being marketed quite positively. The same applies to the establishment of specific procedures to approve their sale, different to the ones applied to other foods produced by conventional methods.

This positive attitude is more noticeable in the case of the Spanish and Dutch experts, together with the Greek one, which is, as has already been said, the sample that views these developments most "optimistically". The Italian and, above all, German sample appear to be more critical: 21.3 % of the German experts see these new regulations and control procedures in a negative light.

This criticism of German experts seems to be due to "technical" reasons as could be inferred by their comments. In the first place, they state the difficulty or even the impossibility of predicting the social effects that marketing gene food products may have, and thus the difficulty of taking them into account while making the decision to market them. Also side-effects of such products are still unknown. Another quite recurrent argument among German experts is that compulsory controls on biotechnological food products may cause high costs.

In their opinion, such increases in the production and marketing costs cannot be fully justified, because the identification of some genetically engineered components, such as additives, is technically impossible. Dutch experts, in their case, also point out these technical difficulties since, for instance, genetically engineered soy bean is verifiable, but genetically engineered soy bean oil is not.

Time of realization

In the time schedule foreseen by the experts for the application of controls on marketing gene food products, the establishment of a case-by-case procedure will come first, followed by the consideration of ethical concerns and finally, the use of specific equipment and methodologies by national authorities to identify gene food products will be the last development to occur.

Figure 5.16:   Distribution of the category "time of realization" by country for statements 9, 10 and 11

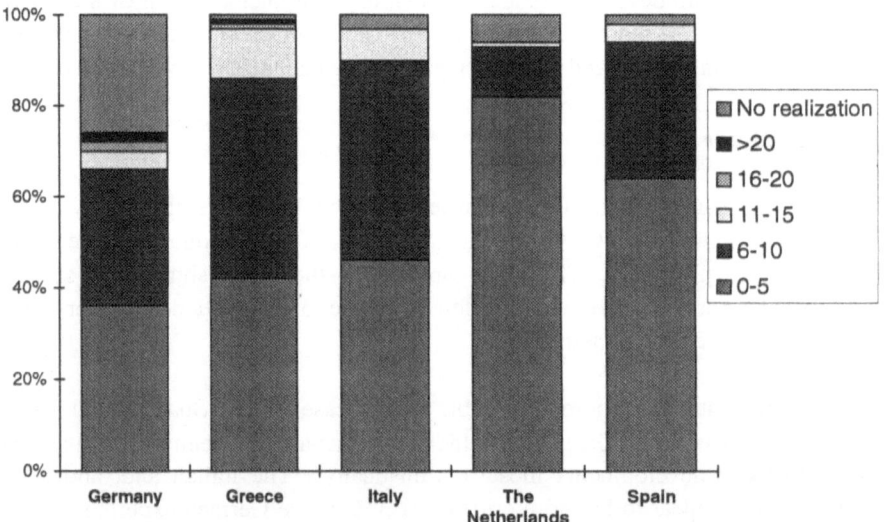

Regarding the time of realization, the Dutch sample, followed by the Spanish one, expects these regulations to be implemented quite soon, while German and Greek experts do not think these changes will happen so quickly (see figure 5.16). In fact, the Netherlands and Germany already have legal regulations which apply in some cases: the Dutch Novel Food Resolution and the food regulations in Germany.

Paradoxically, the two countries with the highest degree of self-assessed expertise in the whole sample, and the ones that seem more advanced in the development of regulations, are the ones whose experts believe in higher percentages that those regulations will not come at all or, better said, will not be fully implemented.

More than one third of the German sample affirm that a case-by-case approval process and the considering of social and ethical concerns during this process will not be implemented (statements 9 and 10). Besides this, 20 % of German experts and 13 % of Dutch experts believe that national authorities will never use specific techniques to identify genetically engineered food (see figure 5.17).

Figure 5.17:   Relevance of "no realization" for statements belonging to the "regulation" domain

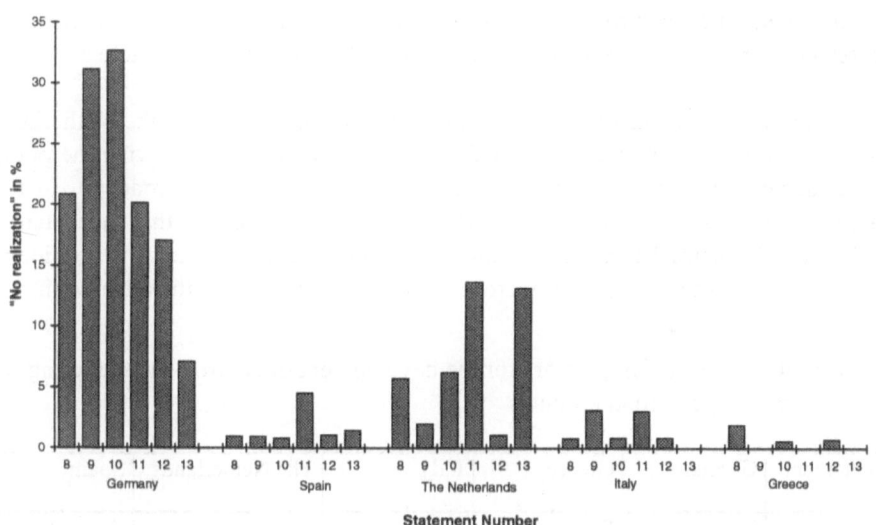

On the other hand, the Mediterranean countries present very low percentages of "no realization", especially Greece where more than 99 % of the population think such regulations on the marketing of biotechnological food products will be implemented, sooner or later (see figure 5.17). This does not seem to be the case of experts in the panels of Spain and the Netherlands, though, for whom regulations which are not implemented in the next 15 years, will not be implemented at all.

Influential factors

In the category "influential factors", "policy", "regulations/standards" and "acceptance" are perceived as the most crucial factors for the issues dealt with in the "regulation" domain. However, some differences appear between countries.

Broadly speaking, the German sample sees "policy" as the most influential factor for the development of new controls on commercial gene food products (27 % of the participants). This trend varies a little for statement 11. In that case, 32 % of the German sample assess that "funding" is the most influential factor. In contrast to this, the other national samples point out the importance of the factor "R&D infrastructure" in developing specific equipment and methods to identify genetically engineered food products.

In fact, the Mediterranean countries highlight, in all statements, the technical ability to implement such controls and give a high relevance to "R&D infrastructure", "personnel" and, in the case of Italy and Spain, to "information" (see table 5.24).

Public acceptance seems to be a very relevant factor in the case of the Netherlands and Spain, especially regarding the realization of statement 10 (involving the public in the debate and decision-making while approving gene food products). In that case, more than 30 % of the participants in both countries believe that "acceptance" is the most influential factor (see table 5.24). In Germany "markets" is considered among the three most influential factors, which is not the case in the other countries.

Table 5.24:   Influential factors for the development of controls on marketing of gene food products

Statement No.	Germany	Greece	Italy	The Netherlands	Spain
9	Policy Regulation Funding	Regulation Policy R&D infrast.	Regulation Policy Personnel	Policy Regulation Acceptance	Regulation Policy Acceptance
10	Policy Markets Acceptance	Policy Regulation Acceptance	Regulation Acceptance Information	Acceptance Policy Regulation	Acceptance Policy Information
11	Funding Policy Personnel	R&D infrast. Personnel Funding	Regulation Personnel R&D infrast.	R&D infrast. Regulation Policy	Regulation Personnel R&D infrast.

Importance of the statements

According to the view of the experts, new regulations on the marketing of gene food products will have a positive effect on the competitiveness of the economy, especially the one implementing a "case-by-case procedure" to approve the sale of those products. Despite this, and despite the fact that most of the five national samples involved in the Delphi survey think that those new regulations contained in statements 9, 10 and 11, will be developed in the next ten years at the latest, there is a relatively high percentage of experts who think that a "case-by-case procedure" to approve the sale of biotech products will not be carried out (see figure 5.17).

All the national samples show the same trend, the Spanish one being the most optimistic about the beneficial influence that these new controls on marketing gene food products may have on the economy, and Italy the most sceptical about it (see figure 5.18). In fact, the Italian sample is the one which gives least importance to the issue "regulation" in order to improve the national economy, while its experts consider that further regulations will, above all, reinforce environmental protection.

Controversely, in the view of the Spanish experts' sample, those regulations aiming to establish a standardized approval process are the most important ones for the competitiveness of the national economy. German experts, for their part, hold a completely different opinion: such processes will not improve the competitiveness of the German economy significantly compared with the positive economic effects that actions such as compulsory labelling, the involvement of the public in decision-making, or the uniform implementation of the EU biosafety directives will have (statements 8, 12 and 13).

Besides this, the German sample does not seem to associate these control procedures with a greater protection of the environment or the fostering of sustainable economic practices. The other four countries, and especially those of the Mediterranean area, seem more likely to affirm that further control will lead to sustainability (see figure 5.19).

Finally, such regulations will not have a great impact on encouraging knowledge creation in science and technology, an opinion which is stronger in the cases of Germany and the Netherlands than in the Mediterranean countries (see figure 5.19). There is a small qualification to be made here. The fact that national authorities practically use specific equipment and methods to identify gene food products (statement 11) will have a greater effect on stimulating science and technology developments than other regulations: this interrelation is seen especially in the Netherlands, Italy and Spain.

Figure 5.18:   Importance of "regulation" statements for "competitiveness of national economy"

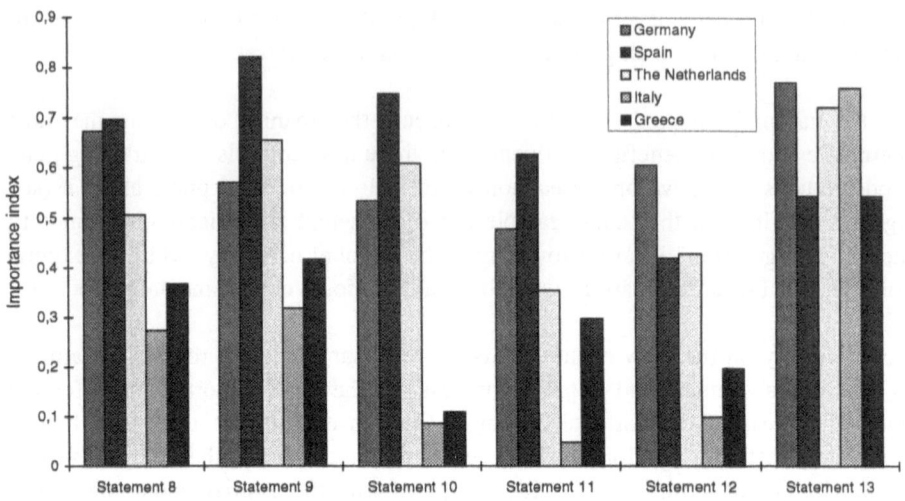

Figure 5.19:   Importance of "regulation" statements for "protection of the environment/sustainable development"

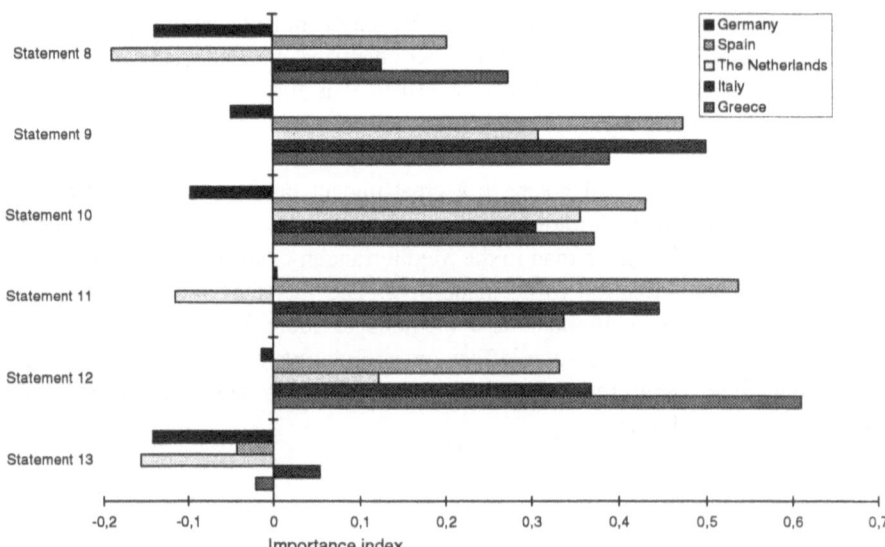

Figure 5.20:   Importance of "regulation" statements for "knowledge creation
                 in science and technology"

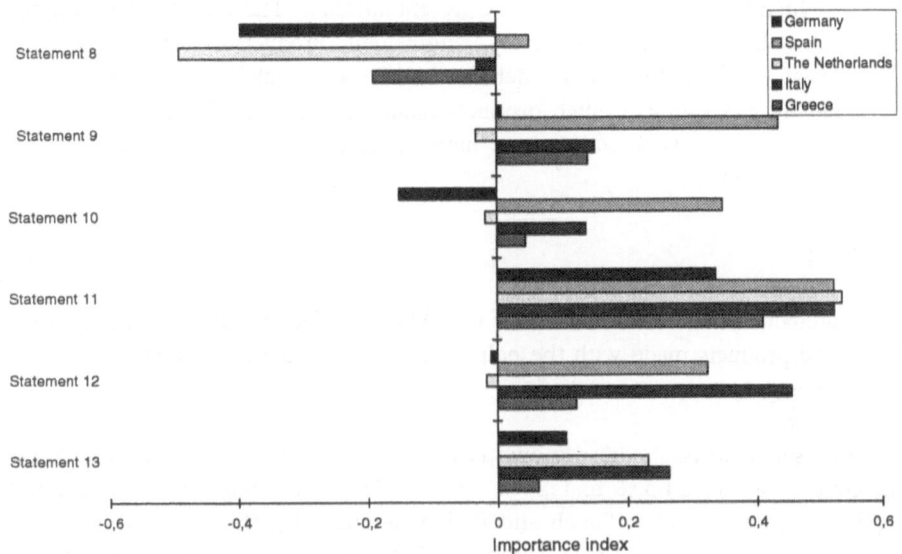

Summary

To sum up, according to the opinions expressed by the experts in the panel,
Germany seems to be the country whose experts seem most sceptical and least
positive about the quick implementation of procedures to control and approve
biotechnological foods and beverages. They also believe that those new regulations
will not be very relevant for economic competitiveness, although they will be of
benefit for the national economy to a greater degree than they will improve
scientific and technological knowledge or the fostering of sustainable economic
practices.

In the Netherlands those regulations are expected to be fully implemented in the
shortest time span and this is seen very positively by their experts. Despite this, a
relatively high percentage of the sample believe that a case-by-case procedure to
approve the sale of genetically engineered food products, and taking into account
ethical and social concerns in the approval process, will not be developed.

There are no significant differences among Mediterranean countries, despite the fact
that Italy appears to be the most critical towards the implementation of a case-by-
case procedure to approve the sale of gene foods and, as was the case in Germany as
well, Italian experts see difficulties in considering social and ethical concerns.

5.2.5   Labelling, public participation and economic impacts

The rest of statements of the area "regulation" deal with three different issues, towards which European experts show different attitudes. These three issues are the compulsory labelling of genetically engineered products (statement 8), the involvement of the public in the debate and decision-making in the process of approving the marketing of such products (statement 12) and finally, the impacts that European biosafety directives may have in attracting firms located outside the EU (statement 13).

Labelling

Only statement 8 addresses the issue of labelling gene food products: "The labelling of all food products made with the help of genetic engineering is compulsory and fully implemented in your country".

Labelling seems to be the best-known issue among the experts in Germany, Spain and Greece (84 %, 24.5 % and 24 % of "very familiar" and "average familiar" respondents, respectively). Dutch and Italian samples also show a very high self-assessed degree of knowledge in labelling (82 % and 52 % in each case) although the percentage of "very familiar" and "average familiar" is not the highest of the domain "regulation". This relatively high self-assessed expertise may be explained because there was a long discussion concerning the labelling procedure connected to the "Novel Food Regulation" and now labelling is already regulated at an EU level although, as many experts comment, is not implemented yet in the member countries.

Table 5.25:   "Personal attitude" for compulsory labelling of gene food products
              (statement 8)

Country	Positive attitude	Negative attitude
Greece	96 %	1 %
The Netherlands	80 %	4 %
Italy	77 %	12 %
Spain	76 %	13 %
Germany	74 %	11 %

Again, Greece presents the highest percentage of positive attitude towards compulsory labelling of gene food products (see table 5.25), an action which is also seen as one of the most positive regulations by the German sample, despite its low "ranking" in comparison with the other countries. Although 77 % of the Italian experts appreciate compulsory labelling of gene food products, this percentage represents the lowest value of "positive attitudes" related to statements of the "regulation" domain in this country. Despite this, it is one of the most immediate to happen as foreseen by the Italian experts, as well as for the rest of the involved countries (see figure 5.21).

Figure 5.21:   Short-term realization (0-5 years) for compulsory labelling of gene food products

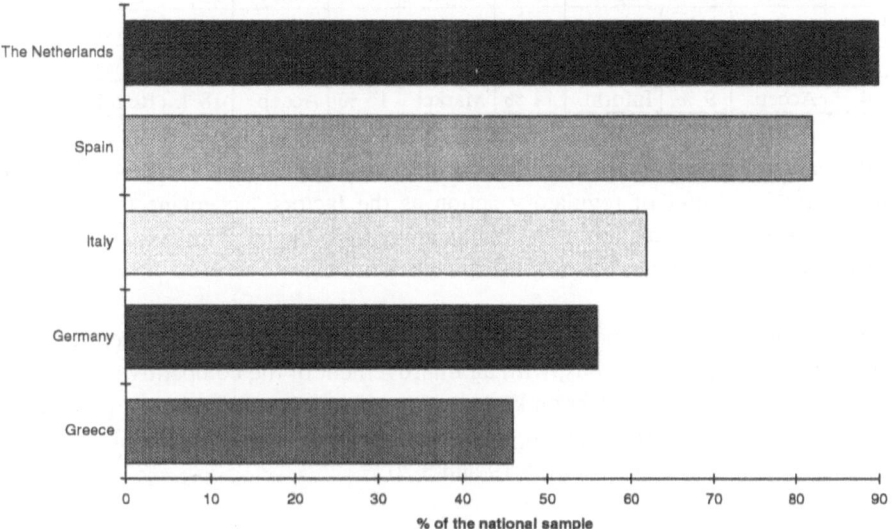

In this sense, compulsory labelling is seen as the first regulation on gene food products to be implemented in Germany, which already has a decree on the issue (1997) and Spain, although this development will imply problems to further development of gene food products. Some experts from these two countries, together with the Italians, consider that compulsory labelling of gene food products may be contrary to public acceptance as, given the actual situation of consumers' misunderstanding, a label will be seen as a "warning" and so might lead to a discrimination against labelled products.

Over 20 % of the samples in Germany, the Netherlands and Spain believe that the full implementation of compulsory labelling is mainly conditioned by "markets" - mainly by consumers' demand for information about the characteristics of the food products - besides the classical factors "regulation" and "policy". "Public acceptance" is also considered influential to a lower extent (see table 5.26). Although the percentage of Greek and Italian experts that regard "markets" as a very influential factor is lower than in the rest of countries (18 % and 16 %, respectively), they also place it among the four most influencing ones (see table 5.26).

Table 5.26: "Top 4" influential factors related to compulsory labelling of gene food products by country (statement 8)

Rank	Germany		Greece		Italy		The Netherlands		Spain	
1	Policy	30 %	Regul.	29 %	Regul.	32 %	Policy	23 %	Accep.	24 %
2	Market	24 %	Policy	23 %	Policy	17 %	Regul.	22 %	Market	23 %
3	Regul.	16 %	Market	18 %	Accep.	17 %	Market	21 %	Inform.	19 %
4	Accep.	9 %	Inform.	11 %	Market	16 %	Accep.	18 %	Regul.	19 %

In the case of Spain, compulsory labelling seems to be more a matter of public concern than a matter of regulatory action as the factors "acceptance", "markets" and the release of information to the public are seen as the most influential, ahead of "regulations" and "policies".

Germany and the Netherlands strongly relate labelling, and so public identification of genetically engineered goods, with an improvement of the competitiveness of the economy, with almost no effect on knowledge creation or environmental protection. Spanish and Greek experts, though, think that labelling is also important to protect the environment (see figures 5.18, 5.19 and 5.20).

Public participation in decision-making

The involvement of the public in the debate and decision-making on the application of modern biotechnology to the Agro-Food sector is addressed by statement 12. The Spanish experts deviate a little from the trend of relatively high percentages of "positive attitude" in the rest of the countries involved in the Delphi survey, as almost 18 % of the Spanish sample think public participation in decision-making is negative, besides being an innovation which will not happen in the shortest time span.

Such an opinion could be related to the contents of a comment made by one of the Spanish panel members, who suggests that it would be better to involve experts (before the lay public) as decision-makers are not very knowledgeable on biotechnological issues.

In the experts' view, public participation depends basically on the display of policies that promote it, as "policy" is the most influential factor in the opinion of all countries, except in the Netherlands, where it comes closely after "information". In this sense, making information available and working to increase public acceptance are the other two most influential factors in all countries.

Furthermore, few experts relate involving the public in decision-making with having an educated population, which does not seem to be the case nowadays according to their comments. In this sense, comments by some Greek and German experts show a fear that such involvement could become a handicap to research, development and investment and, so, it may be obstructive for economic advancement in the biotechnological sector.

Besides this, Dutch experts are the ones who see this initiative in the least problematic light and, in fact, some experiments have already been started there, where the government has fostered some exchanges with an important consumer organization, although the results of the contacts are not being taken sufficiently into consideration, as one of the experts affirms.

Impact of EU biosafety directives

The potential impacts that EU biosafety directives may have on the performance of companies seem to be very difficult to assess by experts. In all countries involved in the Delphi survey, the lowest figures for the category "degree of knowledge" are found for the issue "The uniform implementation of the EU biosafety directives in all EU countries has led to a higher attraction of the EU for companies based outside the EU" (statement 13).

Table 5.27: Assessment of attractiveness of EU after uniform implementation of biosafety directives (statement 13)[9]

Category	Mean (all statements of "regulation")	State-ment 13	Mean (all statements of "regulation")	State-ment 13	Mean (all statements of "regulation")	State-ment 13
Degree of knowledge	Very familiar		Not familiar		No response	
(% of the total)	12 %	6 %	15 %	29 %	-	-
Personal attitude	Positive		Indifferent		No response	
(% of the total)	74 %	49 %	14 %	30 %	1 %	4 %
Time of realization	0-5 years		No realization		No response	
(% of the total)	50 %	28 %	12 %	6 %	3 %	8 %

---

9 The "European mean" (mean of all countries involved in the Delphi survey) has been used here, as all the samples show the same trend regarding the "deviation" in the answering pattern for statement 13.

In fact, the response behaviour of the samples changes dramatically in this statement, by comparison with the rest of the statements belonging to the area "regulation" (see table 5.27). This is why the interpretation of the results has to be very cautious as it seems that the contents of the statement has not been fully understood by all experts. Few comments written by some respondents can be interpreted in that way.

A movement from companies into the EU is the situation which will take the longest to develop:almost half of the samples in the Netherlands, Spain and Greece place it in the coming six to ten years. It is seen as a feasible development and, deviating from the rest of the countries and from their own behaviour while addressing regulation issues, German experts consider it as the most likely situation to develop: only 7 % answer "no realization", the lowest percentage in this section. Finally, "policy" is the most influential factor and, in addition, "international collaboration" has a great influence on this issue as well.

5.2.6   Summary

- The experts in the five countries involved in the Delphi survey assess they have a fair knowledge of the legal framework of biotechnological food processing and marketing, compared with their self-assessed expertise for the other issues addressed by the questionnaire.

- Two groups of countries can be defined, depending on the level of knowledge estimated by more than half of their experts: the dominant part is familiar or very familiar in Germany and the Netherlands, whereas most experts in the Mediterranean countries consider themselves less familiar or not familiar at all.

- Regulations on gene food products already exist in Germany and the Netherlands, although experts in those countries do not think they are fully implemented. Norms are not that advanced in the Mediterranean countries, although they will be developed soon (in the next ten years at the latest), especially those regarding labelling and the implementation of specific processes to approve marketing of gene food products.

- Despite the fact that these norms seem imminent, or are already in progress, there is a significant percentage of experts who think such regulations will never come into force. This opinion is stronger in Germany and the Netherlands than in the other countries, i. e. the higher the self-assessed degree of knowledge, the greater the percentage of experts who think that strict regulations will not happen.

- In general terms, experts in all national samples share a positive attitude towards further regulation on gene food and beverages, German experts being the most sceptical about its beneficial results. In fact, the comments written by German experts highlight the technical difficulties and the high costs of implementing such regulations.

- In the experts' view, regulation favours, above all, economic innovation and, unlike the German and Dutch experts' perception, improvements in environmental protection are also foreseen in the three Mediterranean countries.

- Regarding this, the Mediterranean countries seem to look at the application of modern biotechnology as a way to improve the environmental performance of food production and processing, and not only a possibility to improve economic efficiency.

- Public opinion appears to be the key issue in promoting the formulation of new regulations on biotechnological products and in guaranteeing their successful implementation. The most often mentioned influential factors (policies, regulation/standards, acceptance, markets) are all related to the different ways in which the public can be understood: in the role of voters, consumers, pressure groups, etc.

- In this sense, public involvement in decision-making is one of the most imminent regulations to come about, besides being the one which is seen most positively. This trend, though, does not fully apply for Spain, whose sample places such realization a little further back in time and does not share a very positive opinion of it, compared to the results for other statements.

- On the other hand, all national samples seem to be quite indifferent to the potential impacts of new regulations on improving the attractiveness of the EU as a site for biotechnological companies, again the Spanish sample deviating a little from this trend, as it holds a relatively strong negative attitude on this issue.

- Germany and the Netherlands seem to share a similar framework for the development of new controls on biotechnological commercial food products, a framework where policies and regulations are always the most powerful influential factors.

- The Mediterranean countries, on the other hand, highlight the technical ability to implement such controls and give a relevant influential role to the development of "R&D infrastructures", the availability of skilled professionals ("personnel") and, in the cases of Italy and Spain, to the release of reliable "information" on gene food products.

5.3   Economic implications

5.3.1   Introduction

In the opinion of the European Commission but also of national policy-makers, biotechnology has emerged as one of the most promising and crucial technologies for economic development in the next century (CEC 1993). Biotechnology has important impacts as it bears the potential to create new products and processes, increase productivity in existing industries and stimulate demand for highly skilled work forces and lead to new jobs. Periodically, national governments and international organizations like the OECD and EuropaBio monitor the economic impact of biotechnology and try to measure its social and economic spin-off.

The expected socio-economic impact of biotechnology is for all national and supranational governments a very important reason for stimulating this technology. So far, mainly the pharmaceutical sector has been a focus of biotechnology applications. However, also in the agricultural sector, since the basic technologies for transgene plants were developed in the 1980s, herbicide resistance, disease resistance and pest resistance in a wide variety of crops species have come to the market. In the food and drink sector the use of biotechnology has been much slower, compared to the pharmaceutical sector. One important impact in the food and drinks sector has been in the enzyme industry, with the improvement of the performance of production processes for food ingredients (EuropaBio 1997). As can be understood from this, the economic effects of modern biotechnology in the Agro-Food sector in terms of new products, new jobs or bigger market shares have been very moderate so far. What will be the future impact of this technology?

In the Delphi survey the economic aspects of biotechnology have been addressed in three ways. In the first place, eleven statements in the survey specifically focus on the economic aspects of biotechnology. Second, the members of the Delphi panels were asked to assess the most important conditions for the realization of the development/situation described in the statement, including economy-related influential factors. Finally, the panels were asked to estimate the importance of the described situation in the statement for the competitiveness of the economy.

5.3.2   Relevance of the economic issues

The eleven of the 71 Delphi statements (statements 14 to 24) which were specifically focused on issues related to the domain "economy" (table 5.28) address the following issues:

- Costs, prices and market shares of genetically modified Agro-Food products (statements 14, 15, 18, 19, 21, 22, 23),

- Influence of biotechnology on small and medium-sized enterprises (SME(s)) (statements 16, 19),

- Shifts in employment within industry and agriculture (statements 17, 20, 24).

Table 5.28:   Statements of the domain "economy"

Statement No.	Content
14	Food made with the help of genetic engineering achieve a turnover share of 30 % or more of all food consumed in your country.
15	The prices of food products in specific market segments, of which the quality is significantly improved by the application of modern biotechnology are, at least, 30 % higher than that of corresponding conventional products.
16	The widespread use of modern biotechnology enables small and medium-sized companies of the Agro-Food sector in your country to introduce a large variety of innovative processes and products
17	The practical use of modern biotechnology in the food industry creates a small number of new jobs (e. g. in specialized service companies and the supplying industry).
18	The widespread use of modern biotechnology in agriculture reduces the production costs for agricultural bulk products (e. g. cereals, milk) by approximately 30 %.
19	Small farmers in your country are not able to afford new genetically engineered plants and animals.
20	The widespread use of modern biotechnology in animal and plant production leads to an approximate 30 % decrease in traditional jobs in agriculture.
21	Due to the widespread use of modern biotechnology in EU fish breeding and aquaculture, the imports of fish and fish products from outside the EU are significantly reduced.
22	Due to the application of modern biotechnology, farmers produce renewable resources (e. g. biofuel, starch, fatty acids) which are used outside the food sector on 20 % or more of the arable land in your country.
23	Pharmaceutical substances produced by genetically engineered animals and plants (e. g. proteins, enzymes, hormones, antibodies) achieve a turnover share of at least 5 % of the pharmaceutical market in your country.
24	Due to the hesitant application of genetic engineering (compared to other EU member states) the Agro-Food industry cuts 10 % or more of the jobs in this sector in your country.

Degree of knowledge

There are remarkable differences in the degree of knowledge between the involved countries concerning the statements belonging to the "economy" domain. In Germany, for nine of the eleven statements, the majority of respondents indicates being "very" or "average" familiar with the subject of the statement. In the Netherlands, more than half of the respondents indicate to be "very" or "average" familiar only for three of the eleven statements. Even lower is the average degree of knowledge of the panel in Spain, Italy and Greece: for none of the statements in the economy domain does a majority indicate being "very" or "average" familiar. All five panels show the least affiliation with the economic aspects of biotechnology on fish breeding (statement 21).

Personal attitudes and expected time of realization

Considering the personal attitude of the panel members to the economy statements, a comparable picture arises from all five member states. The panels in all countries take a negative attitude towards statements 15, 19, 20 and 24 dealing with high prices for gene food products and a potential loss of jobs due to the application of modern biotechnology. Also similar for all the countries are the positive stances taken for statements 16, 17, 18, 21, 22 and 23. These statements describe additional innovation and employment possibilities, a reduction of the production costs as well as new market opportunities outside the food markets due to the use of modern biotechnology.

For statement 14, in most countries, the most common attitude is indifference. Only in Greece is a small majority in favour of a high turnover share of food made with the help of genetic engineering.

No significant differences could be observed among the statements of the five countries for the expected time of realization. For most statements the most likely time of realization was estimated to be between six to ten years. With respect to the expected time of realization, the Dutch panel has a rather sceptical point of view: this accounts in particular for the statements of the "economy" domain. The majority of the Dutch panel assess statement 15, 19 and 20 as being completely unrealistic at all. This estimation relates in particular to a premium price strategy for gene food and to a potential job decrease in agriculture. In Germany a disbelief in realization is registered only for statement 15 (premium price). In all the other countries for none of the statements a relatively large portion indicates "no realization".

Influential factors

"Market", "technology transfer" and "industrial innovation" are indicated by the five national panels as the most important influential factors for the realization of the eleven economy-related statements. From these three, "market" is selected as the most important influential factor, especially in Germany and the Netherlands. In Germany it is mentioned as the most important factor for eight statements, and in the Netherlands for six statements. The results of the Delphi survey in Germany are

221

comparable to the findings of two surveys among biotechnology companies, in which especially the conditions on the relevant markets, financing difficulties as well as the general acceptance of biotechnology were mentioned as the most important constraints for the marketing activities of the companies (Reiss et al. 1995, Streck and Pieper 1997). For statement 15 (high prices of speciality food products), "market" is indicated as the most important influential factor in all countries. Only for statement 24 on the hesitant application of genetic engineering which leads to cuts of 10 % or more of the jobs in the Agro-Food sector, "market" is indicated as the most important influential factor in none of the countries. For Spain, Italy and Greece "technology transfer" is regarded as the most important influential factor in the economy statements. It scores as the highest factor in Spain five times, in Italy five times and in Greece six times.

Surveys among biotechnology companies partly support the findings of the Delphi survey. Executives of European biotechnology companies identified the availability of public and private capital, adequate patent protection, efficient product approval systems as well as qualified personal as the most important factors decisive to locate R&D investments in the biotech industry (Ernst and Young 1996). An additional survey among biotechnology companies with experience of operating in both Europe and the USA, shows that in overall terms, the most important external factors which influence their decisions to invest in and use biotechnology (per se) are market conditions (including consumer acceptance), the effectiveness of the intellectual property protection, and the impact of the regulatory framework (EuropaBio 1997). The last two factors are included in the Delphi survey in the influential factor "regulation". Industrial innovativeness (which is comparable with "entrepreneurship" in the EuropaBio list) and technology transfer (which is included in "science base") are not perceived as bottle necks by the interviewed companies. Reasons for the different outcomes of the two surveys could be that in the Delphi survey SMEs play a more important role (they are not represented in the set of companies interviewed for the EuropaBioreport) and that the Delphi panels also include consumers, critics, farmers and scientists. In addition, the Delphi survey is focused on the Agro-Food sector, whereas the EuropaBio-study is related to the biotech industry in general.

Importance of the economy-related influential factors

In order to assess the importance of the three economy-related influential factors, the frequency with which these influential factors are mentioned as most important in one of the 71 statements is analysed, showing specific patterns for each country (table 5.29). Germany and the Netherlands give "markets" the highest score and Spain and Italy "technology transfer". Greece takes a separate position between these two poles.

Table 5.29:   Most important influential factor for "economy" statements in the
                five countries

Country	Number of statements with the highest score for influential factor		
	Market	Industrial innovation	Technology transfer
Germany	14	0	0
The Netherlands	10	7	0
Greece	7	0	0
Spain	2	7	16
Italy	2	0	18

Importance of the statements for the competitiveness of national economies

The panels have been asked to indicate the importance of all statements to the
national economy and the competitiveness of the industry of their member state. To
the response "very important", a value of +1 was addressed, to "average important"
a value of zero and to "not important at all" a value of -1. Consequently, a mean
average per country was calculated. Over all 71 statements, this average is positive
for each of the countries, i. e. the experts regard biotechnology-related scientific and
technical developments as well as the framework conditions, under which this
technology is applied as crucial factors for the future economic competitiveness of
the Agro-Food sector in the included countries. Table 5.30 indicates the different
averages for "importance for national economy". These figures do not correspond
with the general pattern found in the Delphi; especially the strongly positive
German position is not in accordance with the criticism modern biotechnology
meets within this country.

Table 5.30:   Average indicator over all 71 statements "importance for economy"

Germany	The Netherlands	Greece	Italy	Spain
+ 0.62	+ 0.49	+ 0.33	+ 0.54	+ 0.76

5.3.3   Market aspects

In this paragraph the market opportunities deriving from the application of modern
biotechnology in the Agro-Food sector are analysed, using a specific statistical
methodology. This analysis is based on the assumption that the personal attitude of
the single experts to the different developments described in the questionnaire will
influence their individual behaviour. For this reason the analysis is based on the
answering category "personal attitude" taking into account the other categories in a
later stage.

Methodology

In a first step, the individual answers of the experts in the five countries in the category "personal attitude" are used as the statistical basis for a factor analysis. Factor analysis is a statistical technique commonly applied to identify a relatively small number of factors, which can be used to represent relationships among sets of many interrelated variables. The main target is to find typical patterns in the response behaviour of the experts to market-relevant aspects.

At the beginning a set of relevant statements was selected which deals with market aspects related to the application of modern biotechnology in the food and non-food area. The statements considered in the factor analysis are shown in the tables 5.31 to 5.34. The answers of all participating experts in the five countries for the answering category "personal attitude" were converted into a modified three-point Likert type scale including "positive", "indifferent" and "negative" personal attitude. As only around one quarter of the respondents answered each of the selected statements without assessing themselves as being "not familiar" with a certain development or using "not able to express", every missing value was replaced by the value of the median.

Principal components' analysis was performed with varimax rotation. After running the factor analysis, four factors with eigenvalues greater than 1.0 were produced. The tables 5.31 to 5.34 present the results of the factor analysis including the amount of variance explained by each factor on the varimax rotated components. In the factor analysis carried out in the study factor loadings of at least 0.50 are included (Thorndike 1978).

Factor 1 (see table 5.31) is the most important factor, accounting for 33.0 % of the variance of the total factor solution. In this factor, nine statements are included, attaining loadings from 0.55 to 0.73. The statements of factor 1 deal with the habituation of consumers to gene food or genetically engineered products. The second factor accounts for 9.9 % of the variance of the total factor solution. In total, seven statements load on this factor, attaining loadings from 0.53 to 0.76. This factor represents market niches related to modern biotechnology both in the food and the non-food area. Statement 39 dealing with fruit and beverages of high nutritional value which are produced with the help of genetic engineering has a factor loading of more than 0.50 in relation to factor 1 and factor 2 (see table 5.31, 5.32). Factor 3 and factor 4 explain 6.1 % and 5.1 % of the variance of the total solution respectively. The statements loading on these two factors attain rather high loadings, ranging from 0.67 to 0.83 (see tables 5.33, 5.34). The three statements loading on factor 3 deal with regulation aspects of gene food products, like the labelling or market approval procedures of these products. Factor 4 only consists of statement 15, which deals with a premium price strategy of food products of superior quality which is improved by the application of modern biotechnology (see table 5.34).

Table 5.31:   Results of the factor analysis: Factor 1: Habituation of consumers to gene food

Statement No.	Content	Factor Loading
2	Most consumers in your country have quickly got used to all kinds of food and beverages produced with the help of genetic engineering.	0.676
7	A restaurant chain or catering service specialized in offering trendy meals with genetically engineered ingredients open branches in almost all large cities in your country.	0.591
14	Food made with the help of genetic engineering achieve a turnover share of 30 % or more of all food consumed in your country.	0.699
38	Food enriched with specific microorganisms having positive health effects achieve approximately 25 % turnover share in their product group (e. g. yoghurt).	0.558
39	A large variety of food and beverages of superior nutritional value (e. g. vitamin, protein and fibre content; fatty acid ratios) supporting dietary health requirements is produced with the help of genetic engineering.	0.556
45	In your country most of the beer is produced with genetically engineered yeast.	0.699
49	Enzymes optimized by protein engineering are practically used in specific sectors of the food industry (e. g. starch processing, bakeries, breweries, cheese/dairy production).	0.638
50	Approximately 90 % of the enzymes used in the food-processing industry are produced by genetically engineered organisms.	0.734
52	The production of industrial enzymes in genetically engineered field crops is developed for large-scale application.	0.547
Factor 1: Total Variance: 33.0 %		

Table 5.32:   Results of the factor analysis: Factor 2: Market niches related to modern biotechnology

Statement No.	Content	Factor Loading
22	Due to the application of modern biotechnology, farmers produce renewable resources (e. g. biofuel, starch, fatty acids) which are used outside the food sector on 20 % or more of the arable land in your country.	0.622
23	Pharmaceutical substances produced by genetically engineered animals and plants (e. g. proteins, enzymes, hormones, antibodies) achieve a turnover share of at least 5 % of the	0.526

Statement No.	Content	Factor Loading
	pharmaceutical market in your country.	
31	Modern biotechnology is widely used to reduce emissions and waste from animal production (e. g. less manure due to enzymes as feed additives, improved anaerobic waste treatment, biofilter).	0.731
32	Modern biotechnology significantly contributes to the transformation of 40 % or more of the organic agricultural waste into marketable products (e. g. energy, secondary raw materials).	0.761
36	Allergy-sufferers are offered special food, of which the allergenic potential has been reduced by genetic engineering.	0.61
39	A large variety of food and beverages of superior nutritional value (e. g. vitamin, protein and fibre content; fatty acid ratios) supporting dietary health requirements is produced with the help of genetic engineering.	0.529
59	Plant cell cultures in large-scale bioreactors are widely used for the production of high value components (e. g. pharmaceuticals, fine chemicals, proteins).	0.595
Factor 2: Total Variance: 9.9 %		

Table 5.33: Results of the factor analysis: Factor 3: Regulation of gene food products

Statement No.	Content	Factor Loading
8	The labelling of all food products made with the help of genetic engineering is compulsory and fully implemented in your country.	0.673
9	All food and beverages which are produced with the help of genetic engineering are subject to an official case-by-case procedure for approval in your country.	0.787
10	For the marketing approval of Agro-Food products made with the help of genetic engineering, concerns on their social impacts and potential ethical conflicts are generally taken into account.	0.784
Factor 3: Total Variance: 6.1 %		

Table 5.34: Results of the factor analysis: Factor 4: Price strategy

Statement No.	Content	Factor Loading
15	The prices of food products in specific market segments, of which the quality is significantly improved by the application of modern biotechnology are at least 30 % higher than that of corresponding conventional products.	0.826
Factor 4: Total Variance: 5.5 %		

Composition of different attitude groups of experts

Since one of the targets of factor analysis is the reduction of a great number of variables to a smaller amount of factors, it is desirable to calculate factor scores for each respondent. Based on the loadings of the 19 statements to the four identified factors as well as the individual response behaviour of the included experts in the category "personal attitude", four factor scores were calculated for each of the 775 considered experts. Afterwards, the individual expert was sorted to the "attitude groups", of which he had the highest value in his calculated four factors scores. This procedure results in eight "attitude groups" ranging from (+1, -1) to (+4, -4) including respondents with a similar personal attitude related to a specific factor, and a ninth group (the so-called "neutrals"), who had individual factors scores between -0.5 and +0.5 for each factor which do not allow the definite attachment of this expert to one of the eight attitude groups. A characterization on the importance of the selected attitude groups is shown in table 5.35.

Table 5.35:   Importance of selected "attitude groups" of experts

Name of selected "attitude group"	Characteristic response behaviour	Number of experts	Percentage
Gene food cons	Indifferent/negative attitude to habituation of consumers to gene food	106	13.7 %
Gene food pros	Positive attitude to habituation of consumers to gene food	141	18.2 %
Speciality opponents	Negative attitude to market niches related to modern biotechnology	62	8.0 %
Speciality proponents	Positive attitude to market niches related to modern biotechnology	38	4.9 %
Deregulators	Negative attitude to strict regulation of gene food	115	14.8 %
Proregulators	Positive attitude to strict regulation of gene food	83	10.7 %
High price sceptics	Negative attitude to premium price strategy of gene food products	115	14.8 %
High price enthusiasts	Positive/indifferent attitude to premium price strategy of gene food products	109	14.1 %
Neutrals	No strong attitude to one of the four identified factors	6	0.8 %
Total		775	100 %

At first glance, it can be registered that there is a rather homogeneous distribution of experts between the different attitude groups. Almost one third of the experts express strong emotions related to the habituation of consumers to gene food, whereby positive feelings concerning this aspect ("gene food pros") outweigh a rejection of this development ("gene food cons"). A premium price strategy related to gene food products arouses specific emotions in almost 29 % of the experts, with an almost similar weight of supporters ("high price enthusiasts") and experts who reject such a development ("high price sceptics"). Around a quarter of the experts has specific feelings related to the regulation procedure of gene food ("deregulators", "proregulators"), whereas strong personal attitudes towards market niche products related to modern biotechnology are found in around 13 % of the experts ("speciality opponents", "speciality proponents") (see table 5.35). Only six experts are attached to the attitude group of the so-called "neutrals", who show no strong personal attitude to one of the four identified factors. Due to the small relevance, this group will not be taken into account in subsequent analyses.

In table 5.36 and 5.37 the distribution of the attitude groups by country or expert group according to the original coding of the experts by the project teams are shown. When interpreting the distribution of the experts, the general structure of the expert group analysed with the help of the factor analysis has to be taken into account, i. e. the structure of different attitude groups should be compared with the structure of the total sample. The gene food cons are overrepresented in Germany, whereas this group has a lower relevance especially in the Netherlands, Italy and Spain compared to the average distribution of the expert sample (see table 5.36). As expected, especially critics, consumers and farmers have an overproportional importance among the gene food cons, whereas experts from industry and research institutions are underrepresented in this group (see table 5.37). The group of the gene food pros who are in favour of the habituation of consumers to gene food products shows a mirror-imaged composition, thereby being more influenced by experts from industry and research institutions and an underproportional weight of the other three analyzed expert groups (see table 5.37). The gene food pros are highly overrepresented in Spain and have only a relatively low importance in Germany (see table 5.36).

The relatively small attitude groups of the speciality components and speciality proponents show relatively small deviations in the distribution by country, with the exception of a relatively high relevance of speciality opponents in Greece (see table 5.36). In contrast, there are strong variations in the distribution of these two attitude groups by expert groups, resulting in a strong concentration of biotechnology critics in speciality opponents, whereas experts from industry and research institutions are rather lowly represented in this group (see table 5.37). An interesting composition can be found in the speciality proponent group, where a relatively high weight of experts from research institutions and biotechnology critics can be registered, whereas experts from industry and farmers' organizations are underrepresented in this group (see table 5.37). The relatively high weight of the biotechnology critics in this group might be an indication that at least some of the critics might accept the use of modern biotechnology in specific market niches of the Agro-Food sector, especially in the non-food area. On the other hand, the relatively low relevance of experts from industry and farmers' organizations in this group may indicate that these two groups do not fully oversee or are rather sceptical about the future possibilities which might result from new scientific and technological developments related to modern biotechnology in specific non-food application areas.

The attitude groups of the deregulators, who are in favour of a kind of laissez-faire regulation of gene food products are highly concentrated in Germany, whereas they are of minor relevance in the Netherlands, Greece and Spain (see table 5.36). This type of regulation has strong supporters among industry experts and is rejected by most of the experts from consumer organizations and biotechnology critics (see table 5.37). As expected, proregulators show a mirror-imaged composition by expert groups. Proregulators have a rather high relevance in the Netherlands and Greece, whereas they are underrepresented in Germany (see table 5.36).

Table 5.36:   Distribution of "attitude groups" by country

Attitude group	No. of experts	Germany	Spain	Greece	Italy	The Netherlands
Gene food cons	106	76.4 %	0.9 %	12.3 %	1.9 %	8.5 %
Gene food pros	141	25.5 %	25.5 %	17.0 %	9.9 %	22.0 %
Speciality opponents	62	56.5 %	3.2 %	22.6 %	3.2 %	14.5 %
Speciality proponents	38	60.5 %	2.6 %	13.2 %	0.0 %	23.7 %
Deregulators	115	80.9 %	1.7 %	0.9 %	9.6 %	7.0 %
Proregulators	83	34.9 %	3.6 %	22.9 %	6.0 %	32.5 %
High price sceptics	115	51.3 %	9.6 %	16.5 %	9.6 %	13.0 %
High price enthusiasts	109	62.4 %	4.6 %	7.3 %	9.2 %	16.5 %
Total	775	55.2 %	7.9 %	13.3 %	7.1 %	16.5 %

Concerning the distribution by country, there are no significant deviations by the attitude groups in high price sceptics and high price enthusiasts. High price sceptics are overrepresented in researchers and farmers and have a rather low relevance in biotechnology critics and other experts (see table 5.37). On the other hand, high price enthusiasts have a rather high relevance in experts from industry, consumer organizations and other institutions, whereas they are underrepresented in experts from research institutions and biotechnology critics (see table 5.37).

Table 5.37:   Distribution of "attitude groups" by expert groups

Attitude group	No. of experts	Industry	Research institutions	Farmers	Consumers	Critics	Other experts
Gene food cons	106	9.4 %	19.8 %	15.1 %	17.0 %	27.4 %	11.3 %
Gene food pros	141	25.5 %	39.0 %	6.4 %	7.1 %	10.6 %	11.4 %
Speciality opponents	62	4.8 %	16.1 %	12.9 %	9.7 %	46.8 %	9.7 %
Speciality proponents	38	10.5 %	36.8 %	2.6 %	10.5 %	26.3 %	13.2 %
Deregulators	115	27.8 %	32.2 %	10.4 %	3.5 %	8.7 %	17.4 %
Proregulators	83	13.3 %	26.5 %	9.6 %	14.5 %	22.9 %	13.3 %
High price sceptics	115	20.9 %	33.0 %	15.7 %	11.3 %	12.2 %	7.0 %
High price enthusiasts	109	24.8 %	16.5 %	10.1 %	14.7 %	12.8 %	21.1 %
Total	775	19.0 %	28.3 %	10.7 %	10.7 %	18.2 %	13.2 %

Response behaviour of the different attitude groups

There are clear differences in the response behaviour of the different attitude groups related to the habituation of consumers to gene food products. While gene food pros and gene food cons express per definition a pronounced personal attitude towards this development, only speciality opponents have a rather negative personal attitude concerning this issue in addition.

Table 5.38:   Assessment of factor 1: Habituation of consumers to gene food

Attitude group	Personal attitude			Time of realization			
	positive	indifferent	negative	0-5	6-10	11-15	no real.
Gene food cons	7.5 %	25.5 %	66.2 %	17.1 %	36.3 %	21.1 %	11.2 %
Gene food pros	72.2 %	25.2 %	1.9 %	41.0 %	37.7 %	12.7 %	2.3 %
Speciality opponents	9.1 %	27.8 %	62.8 %	24.0 %	34.9 %	17.2 %	10.7 %
Speciality proponents	35.1 %	37.7 %	26.6 %	22.2 %	45.6 %	16.1 %	6.3 %
Deregulators	45.9 %	43.3 %	9.6 %	23.7 %	43.5 %	20.0 %	4.6 %
Proregulators	36.5 %	49.3 %	11.2 %	27.4 %	42.7 %	15.4 %	4.5 %
High price sceptics	40.7 %	28.0 %	28.5 %	25.2 %	42.1 %	16.0 %	8.0 %
High price enthusiasts	34.5 %	44.7 %	19.8 %	24.6 %	41.6 %	18.8 %	5.3 %

In contrast, the other groups express an indifferent or slightly positive personal attitude (see table 5.38). In spite of their personal attitude, most members of the gene food cons and speciality opponents expect a medium-term realization, although around 11 % of them see such a development as unrealistic (see table 5.38). In contrast, the majority of the gene food pros expect a relatively short-term realization. While the gene food pros mention a rather broad set of influential factors, including technology-influenced factors (like R&D infrastructure, technology transfer and industrial innovativeness) as well as demand-pull factors like social and ethical acceptance or the market conditions, the gene food cons limit themselves to few influential factors, such as mainly social and ethical acceptance and market conditions. The assessment of the other groups is placed between these two poles.

In addition, the following aspects can be registered if the response behaviour related to specific statements with significant loadings on factor 1 is analyzed in detail:

- Considerable market shares of gene food products mainly depend on the social and ethical acceptance of these products and the conditions on the relevant markets. This estimation comes up independently from the individual personal attitude of the experts. In all attitude groups a 30 % market share of genetically engineered food products is seen as a realistic option within six to ten years from now. A Delphi survey in the UK indicates that the British panel has a comparable time frame in mind, because around 40 % of them expect that gene food will achieve a market share of 20 % or more within five to ten years, whereas 33 % think that this will happen five years later (Loveridge et al. 1995).

- In case of probiotic food a rather positive personal attitude is mentioned from the experts compared to gene food products in general. From all attitude groups a short-term realization is expected, whereby the conditions on the relevant markets are identified as main influential factor. Acceptance barriers for probiotic food seem to be lower than for gene food products in general. In addition, information, especially concerning the health effect of probiotic food, is regarded as a supplementary key factor for the market success of these products. The estimations of the experts asked in the Delphi survey seem to turn into realization in the coming years. One example in this direction is the market success of probiotic yoghurt in Germany. In 1995 this type of yoghurt was launched on the German market by Nestlé ("LC1") and Südmilch ("Vifit"). In the following year probiotic yoghurt had already achieved a market share of 7 % in Germany (Lebensmittelzeitung 1996). An impressive growth of more than 100 % for this product segment has been registered during 1997, resulting in a total turnover of more than 300 mio. DM (Heimig 1997, Lebensmittelzeitung 1998). This equals around 13 % of the yoghurt market in Germany. The market growth of probiotic yoghurt has been massively supported by intensive advertising and information activities of the dairy industry, amounting to around 40 mio. DM in the first half of 1997 (Heimig 1997).

- The production of beer with the help of genetically engineered yeast is regarded indifferently or critically from all attitude groups, i. e. there are only few protagonists for such a development. Among all attitude groups a relatively strong uncertainty concerning the time of realization of such a development can be registered with considerable doubts (especially among gene food cons and speciality opponents) about the realism of this vision. In addition to market conditions and social and ethical acceptance, the regulation procedures and applied standards are regarded as key factors for the use of genetically engineered yeast in beer production.

- In case of the application of genetic engineering in the production of enzymes used in the food industry, an almost similar constellation concerning personal attitude and time of realization can be registered as for factor 1 in total. Interestingly, another set of influential factors is decisive for this estimation compared to the use of genetic engineering in food processing in general. While market conditions and social and ethical acceptance lose relative importance as long as enzymes are concerned, factors related to the scientific and technical innovation process like R&D infrastructure, technology transfer and industrial innovativeness are regarded as more relevant in this field. The market expectations of the experts expressed in the Delphi survey correspond with the results in other studies. At present, only few commercially available enzymes which are used in the food industry are produced with the help of genetically modified organisms (Novo Nordisk 1995, 1996). Nevertheless, Hüsing et al. (1997) come to the conclusion that in the medium term most of the enzymes applied in the food and drinks sector will be produced with the help of recombinant organisms.

With the exception of speciality opponents, all other attitude groups assess market niches related to modern biotechnology very positively. There are slight differences between the different attitude groups concerning the time of realization, resulting in a short to medium-term realization expected by the gene food pros and speciality proponents, whereas gene food cons and speciality opponents expect a medium-term realization and express in a few cases some doubts concerning the realism of the visions summarized in factor 2 (see table 5.39). While gene food pros and speciality opponents see a rather high relevance of the technology-oriented influential factors (like R&D infrastructure and technology transfer), gene food cons and speciality opponents emphasize the importance of the conditions on the relevant markets as well as the social acceptance of the use of modern biotechnology for the future market success of niche products based on modern biotechnology.

Table 5.39:   Assessment of factor 2: Market niches related to modern
                biotechnology

Attitude group	Personal attitude			Time of realization			
	positive	indifferent	negative	0-5	6-10	11-15	no real.
Gene food cons	63.1 %	19.4 %	15.6 %	12.6 %	40.4 %	27.5 %	6.3 %
Gene food pros	96.5 %	2.7 %	0.3 %	32.3 %	48.1 %	14.7 %	0.3 %
Speciality opponents	22.0 %	33.1 %	44.2 %	16.3 %	37.9 %	24.7 %	7.4 %
Speciality proponents	94.8 %	2.0 %	2.0 %	16.1 %	57.4 %	19.3 %	2.4 %
Deregulators	87.2 %	9.3 %	2.5 %	20.2 %	46.8 %	23.5 %	1.0 %
Proregulators	88.8 %	9.4 %	0.3 %	21.4 %	50.2 %	19.8 %	1.6 %
High price sceptics	85.2 %	8.5 %	4.4 %	18.0 %	52.5 %	22.0 %	1.3 %
High price enthusiasts	80.4 %	13.1 %	5.7 %	20.0 %	45.8 %	22.3 %	2.2 %

Going more into detail, a differentiated assessment of the statements with high
loadings on factor 2 can be registered:

- The increased production of renewable resources for non-food purposes due to
  the application of modern biotechnology is generally appreciated by all attitude
  groups, with the exception of speciality opponents. All attitude groups expect a
  medium- to long-term realization, mainly influenced by the conditions on the
  targeted non-food markets, as well as political activities. In addition, almost all
  experts see certain problems in the innovation process and funding of the
  required R&D, product development and marketing activities of the renewable
  resources. In contrast to most other products in the Agro-Food sector, renewable
  resources produced with the help of modern biotechnology have hardly any
  acceptance barriers. The latter finding corresponds with the results of several
  population surveys and discourse projects in Germany (Menrad 1995). Although
  a significant increase in the turnover of non-food products produced with the
  help of transgenic plants is expected from 15 mio. US$ in 1996 to 320 mio. US$
  in 2005 (Demicheli and Laget 1996), the expectations of the experts expressed in
  the Delphi survey seem to be rather optimistic. A technology assessment study in
  Germany came to the conclusion that around 400,000 ha are required to meet the
  domestic demand of renewable resources for chemical und technical purposes
  (Wintzer et al. 1993), in addition to the 486,000 ha which have been used for the
  production of raw materials for technical purposes in 1996 in Germany (BML
  1997a). This equals in total less than 10 % of the arable land in Germany. In the
  energy sector, the realization of the potential of biomass produced in specific
  short rotation plantations mainly depends on the shaping of the set-aside
  programmes of the European Union (Pontenagel 1995). Due to the increased
  productivity of food production since 1995, it seems to be rather unlikely that in
  the coming ten years substantial additional activities will be put into force by the

EU Commission in order to stimulate the additional production of renewable resources, not least because of additional budget requirements.

- The enhanced production of pharmaceutical substances with the help of genetically engineered animals and plants follows the typical attitude pattern of the entire factor. Almost independently from their personal attitude, all groups expect that such products will achieve a 5 % market share of the pharmaceutical market in the included countries within the following ten years. In this context the realism of this time estimation has to be questionable, considering the rather long development times of pharmaceuticals (in the range of ten years) and taking into account that there are at present only few transgenically produced recombinant human proteins like Anti-Thrombin III, Alpha-I-anti-trypsin and Alpha-glycosidase in clinical trials (Rudolph 1997). The experts mention a complex set of influential factors, mainly relating to the technology-influenced innovation process. In addition, regulation and standards achieve a certain relevance which can be regarded as typical for the market approval procedure of pharmaceuticals. Compared to other non-food application areas included in factor 2, the production of pharmaceuticals with the help of transgenic plants and animals faces relatively high acceptance barriers, mostly probably due to the required genetic modification, especially of animals.

- The reduction of emissions and waste from animal production and the transformation of organic waste into marketable products with the help of modern biotechnology are assessed extremely positive by all attitude groups. All groups are relatively uncertain concerning the time of realization of these developments and mention a complex set of influential factors. The response behaviour of the experts in the Delphi survey indicates that the contribution of modern biotechnology to improve the environmental situation in animal production and animal husbandry would offer an excellent opportunity to promote the application of this technology in agriculture, thereby gaining a relatively high social acceptance. The realization of these developments in the future mainly depends on common initiatives from science, industry and politics, thereby taking into account the needs of other groups as well. Some examples of the application of modern biotechnological approaches to reduce emissions from animal production are the use of phytase in animal feed which is partly produced with the help of genetically modified organisms (Hüsing et al. 1997). At present, mainly the Netherlands represent one of the most important markets for this enzyme due to the legislation in this area. Although the producers of phytase argue with the environmental benefits of their product, it is not totally clear at present to which extent the phosphor emission from animal production can be reduced in practice (Hüsing et al. 1997).

- The supply of food with positive health effects is assessed slightly positively by the identified attitude groups. In general, a mid-term realization is expected, with certain reservations of some experts concerning the realism of this development.

The pattern of influential factors related to health-supporting food based on modern biotechnology goes in the direction of the application of modern biotechnology in the food area in general, in which market aspects, the social and ethical acceptance of biotechnology, factors influencing the technology-oriented innovation process, as well as information activities, are seen as decisive for the future development.

A clear regulation of the market approval of gene food products is positively assessed by all identified attitude groups with the exception of deregulators, who favour a kind of laissez-faire regulation and intend to liberalize the existing standards and regulations in this area. The majority of the experts expects a rather short-term realization of enhanced labelling activities of gene food products and an official case-by-case procedure for market approval, thereby taking into account the social impacts and potential ethical conflicts of these products. Among the attitude groups, two types can be registered concerning the proportion of "no realization" (see table 5.40). A rather small proportion for this sub-category emerges for proregulators who expect that the mentioned strict regulation procedure of gene food products will be realized in the following years. A similar response behaviour is found for the attitude groups of high price sceptics, speciality proponents and gene food pros (see table 5.40). The reasons and motives for this response behaviour most probably differ between the pro regulators and high price sceptics on the one hand, and the speciality proponents and gene food pros on the other. While the first two groups want to limit potential risks or negative impacts of modern biotechnology especially to the consumer, the latter two groups intend to apply modern biotechnology approaches and to sell the respective products. In order to achieve the economic benefits related to this biotechnology, speciality proponents and gene food pros obviously seem to accept that they are obliged to consider the valid regulations and legal framework conditions in this field.

The second type among the attitude groups is represented by the deregulators, of which almost one third regards a strict regulation of gene food products as unrealistic (see table 5.40). Relatively high proportions of "no realization" can also be registered in the attitude groups of high price enthusiasts, gene food cons and speciality opponents as well. While high price enthusiasts might have the same motives as deregulators for their response behaviour, i. e. these two groups intend to liberalize the existing standards and regulations, the other two groups might have other reasons for their response behaviour. In principle, these two groups appreciate clear regulations for gene food products, but at the same time express some reservations concerning their realization.

Table 5.40:   Assessment of factor 3: Regulation of gene food products

Attitude group	Personal attitude			Time of realization			
	positive	indifferent	negative	0-5	6-10	11-15	no real.
Gene food cons	93.1 %	3.8 %	3.1 %	40.2 %	26.3 %	4.4 %	26.6 %
Gene food pros	79.3 %	13.5 %	6.5 %	73.4 %	17.6 %	0.2 %	6.8 %
Speciality opponents	87.2 %	6.7 %	4.6 %	47.9 %	18.2 %	4.2 %	26.0 %
Speciality proponents	85.8 %	11.5 %	2.7 %	67.5 %	15.8 %	1.8 %	14.0 %
Deregulators	20.7 %	29.2 %	49.9 %	42.7 %	19.2 %	3.7 %	32.1 %
Proregulators	98.0 %	0.0 %	0.0 %	75.9 %	16.2 %	2.0 %	4.0 %
High price sceptics	85.1 %	8.2 %	5.5 %	58.9 %	27.3 %	1.2 %	11.1 %
High price enthusiasts	76.2 %	15.7 %	8.0 %	48.3 %	22.6 %	3.4 %	21.7 %

There are no fundamental differences between the eight attitude groups in the relevance of the different influential factors related to factor 3. In general, political activities as well as the national and international laws and regulations are regarded as decisive for the regulation of the market approval of gene food products. Besides market aspects, social and ethical acceptance of modern biotechnology as well as information activities are mentioned in this context.

There are moderate differences in the personal attitude related to a premium price strategy of gene food. While only high price enthusiasts per definition appreciate such a strategy to a certain extent, especially high price sceptics, proregulators, gene food pros and deregulators reject such a development (see table 5.42). Each attitude group assesses premium prices for a gene food product as a rather unrealistic scenario, ranging from around 45 % in this sub-category by high price enthusiasts to 53 % by deregulators (see table 5.41).

Table 5.41:   Assessment of factor 4: Price strategy

Attitude group	Personal attitude			Time of realization			
	positive	indifferent	negative	0-5	6-10	11-15	no real.
Gene food cons	2.9 %	41.2 %	53.9 %	3.0 %	18.0 %	6.0 %	51.0 %
Gene food pros	0.0 %	14.2 %	83.3 %	19.5 %	27.1 %	0.8 %	41.5 %
Speciality opponents	14.8 %	36.1 %	47.5 %	16.9 %	16.9 %	11.9 %	40.7 %
Speciality proponents	2.9 %	42.9 %	54.3 %	5.7 %	34.3 %	5.7 %	40.0 %
Deregulators	0.0 %	27.7 %	72.3 %	8.8 %	18.6 %	9.7 %	53.1 %
Proregulators	0.0 %	27.6 %	68.4 %	9.2 %	26.3 %	7.9 %	43.4 %
High price sceptics	0.0 %	0.0 %	92.7 %	9.4 %	24.5 %	6.6 %	46.2 %
High price enthusiasts	41.3 %	58.7 %	0.0 %	13.1 %	33.6 %	5.6 %	34.6 %

In addition, the broad variation of answers between the different time categories mentioned in the questionnaire indicates the uncertainty of the experts concerning this issue. Some experts see an opportunity for premium prices in specific market niches like dietary products, products in the fitness and wellness area, or specific functional foods. One example in this context represents probiotic yoghurts which are offered at a 30 % higher price than conventional yoghurt in Germany (Lebensmittelzeitung 1998a). Mainly the market conditions as well as the social and ethical acceptance of modern biotechnology are mentioned by all attitude groups as the most important influential factors for the pricing policy of gene food.

5.3.4 Price and cost effects of genetically engineered products

What is the opinion of the panels about the direct economic benefits of genetically engineered products: Will they be more expensive (specialties) or cheaper (bulk products)?

A market share of genetically engineered food products of over 30 % of all food consumed in the country (statement 14) does not seem to be a controversial issue. Indifference is the most common attitude towards this big market share for all countries but Greece, where a positive attitude is the most common response. Also in all countries such a market share is assessed as a realistic option, within five to 15 years from now. Market and acceptance are the most important influential factors, agreement on this can be found among the panels. The Italian panel shows a high level of uncertainty on this point: there is a strong difference in the assessment of "industry" and "researchers". A Delphi survey in the UK indicates that the British panel has a comparable time frame in mind, because around 40 % of them expect that gene food will achieve a market share of 20 % or more within five to ten years, whereas 33 % think that this will happen five years later (Loveridge et al. 1995).

All panels express a negative personal attitude towards the 30 % higher prices of food products with a significantly improved quality in specific market segments, compared to corresponding convential products (statement 15). The Dutch and German panels strongly express their doubts about the realization of such development: "will not be realized" is the most frequently given response to the question about the "time of realization". In the other countries, these speciality products with 30 % higher prices are expected in six to ten years. Again, for all countries, market and acceptance are indicated as the most important influential factors.

The 30 % decrease in production costs for bulk products as the result of the use of genetic engineering (statement 18) is welcomed and expected by all panels. A positive personal attitude is indicated by the majorities in all countries and six to ten years are expected as a realistic time frame for such development. However, differences can be observed in the assessment of the importance of influential factors. For Germany and the Netherlands, market and acceptance are again indicated as of most importance, while in Italy and Spain technology transfer gets the highest score. Funding and R&D infrastructure are indicated to be the most important influential factors in Greece. Existing results of different studies indicate that some specific biotechnology application might decrease the production costs in animal husbandry and plant production to a certain extent (see Henze et al. 1995, Pezzatti et al. 1996), depending on the particular situation of the single farms as well as the production technique applied.

Taken together, the Delphi results show that lower production costs in agriculture and a market share of genetically engineered food of over 30 % can be expected, due to the introduction of modern biotechnology. In this context a basic discussion on agricultural production systems also arises. Many respondents comment on statement 18 by addressing the socio-economic and environmental impacts constraints of large-scale bulk production. As well, a number of respondents indicates that a price decrease is likely, but that 30 % is probably too high.

5.3.5   Employment in the Agro-Food sector

The employment effects of biotechnology is one of the main arguments for governments to stimulate this technology. Discussion in the recent years stated that biotechnology is a pervasive technology, like the information and communication technology, and would have the same impact on employment in industry and the service sector. However, employment assessments - of which only a few have been made - show no impressive figures about such direct effects of biotechnology. Biotechnology has not resulted in a totally new, not previously existing industry like the information technology industry.

A lot of new research-based biotechnology companies or dedicated biotechnology companies have been set up in the Agro-Food sector as well. These small companies have mainly been - or will be - absorbed by existing food (multinational) companies. So, the profile of the European biotechnology food industry is more or less the same as the profile of the "traditional European food industry" which has incorporated biotechniques. Also, the structural development of the food industry is not so much influenced by the technology as such. More influential are the process of globalization, increase of scale, the broadening of the knowledge base of product development and sector-specific developments such as the concentration of retailers, the growing influence of the market, facilitated for instance by the introduction of Effective Consumer Response systems, etc. However, in the knowledge-based economy of today, knowledge of biotechnology is an important asset (Enzing 1997).

Studies show that the most important growth in the past was in R&D jobs (due to the publicly funded R&D programmes) and some in the production jobs (Enzing 1991). Recently, job creation is expected to be directly linked to the service sector. One reason for this is that the fruits of biotechnology in the Agro-Food sector have not yet reach the ripening stage. The other is that biotechnology will, for a considerable percentage of its applications, mainly be a substitution of already existing products (Becher and Schuppenhauer 1996). Substitution is the most dominant employment effect, which means no extra jobs, but new skills requirements. The optimistic scenario (scenario B) in the EuropaBio-report (1997) with a very stimulating environment for biotechnology, came to very high employment figures. However, as this still is a positively constructed future, in which also all related industries are included in the biotechnology, industry, these figures do not give a realistic impression about employment effects of biotechnology.

Job creation in specialized companies and suppliers

However, it is generally expected that the European biotech industry will continue to grow. Most new jobs are expected to be created through the establishment of new companies to exploit the potential of biotechnology. According to the annual surveys of Ernst & Young, Europe counted in 1996 fully 716 dedicated biotechnology companies with a total number of employees of 27,500; 13,700 of them carrying out research functions. Compared to 1995 there was a 23 % increase in the number of companies and a 60 % increase in the number of employees (Ernst & Young 1996). The much sharper increase in the number of employees compared to the number of companies indicates that there seems to be also a considerable growth in the size of the already existing companies.

In all five countries the creation of a small number of new jobs in specialized service companies and supplying industry (statement 17) is assessed as being realistic, with a short-term time horizon for the Netherlands and six to ten years for the other countries. Comments emphasize that the character of such employment will be specialized and not in great numbers. In the most influential factors there are some differences among the countries, as can be observed from table 5.42.

Table 5.42: Most important influential factors for the realization of (limited) job growth in specialized companies and supplying industry

Germany	Greece	Italy	The Netherlands	Spain
Market	Personnel	Personnel	Personnel	Personnel
Acceptance	Infrastructure	Technology Transfer	Industrial Innovation	Industrial Innovation

Germany holds a unique position with first ranking for the market-pull factors. In the other countries conditional factors are mentioned. Both the German and the Dutch panel have a neutral to indifferent attitude towards this statement.

Loss of jobs in traditional agriculture and in agribusiness

Statements 20 and 24 describe both a decrease in employment due to the introduction of modern biotechnology in animal and plant production (statement 20), or the slow implementation of biotechnological applications in the Agro-Food industry (statement 24).

In both cases, in all five countries, the majority of respondents indicates a negative personal attitude towards the loss of employment. Remarkable are the differences in the assessment of time of realization. In all countries but the Netherlands respondents assess the most likely time of realization for both statements as within six to ten years. The majority of the Dutch respondents expects that the situations described in statement 20 and 24 will never be realized at all.

In general, in all countries (with Spain least pronounced), policy is seen as the most important influential factor for these developments. For statement 20, market conditions are seen as very relevant as well. For statement 24, acceptance (Netherlands, Germany), industrial innovation (Spain), market (Greece) and regulation/standardization (Italy) are the second most important influential factor.

The employment issue is a very hot issue in Italy, where the "industry group" of the Italian Delphi panel tends to deny the risk of a loss of jobs in agriculture (31 % opt for "no realization"), with "industrial innovativeness" as important influential factor, while the group "consumers and critics" believe that this is an inevitable development (0 % for "no realization), with "market" as important influential factor. Nevertheless, this specific response could, according to the Italian project team, also be interpreted as a general pessimism of the "consumers and critics" group concerning employment trends. But also in the Netherlands and Germany, comments are strongly pronounced and very controversial. On the decrease of traditional jobs in agriculture, comments range form "False statement!" to "This statement will certainly be the case". Some respondents see the loss of jobs in traditional agriculture as a direct result of biotechnology, while others stress that structural changes in agricultural production (larger scale, less employment) are necessary anyway.

The employment issue is also addressed in statement 37, although in a more qualitative way. The statement says that the increased use of modern biotechnology in food production and processing results in additional allergies in people actively involved in these processes. The statement is assessed as unrealistic in the Netherlands (61 % of the panel says: "will not be realized"). In the other countries the respondents expect such effects within ten years. In some cases, this response seems to be initiated by criticism of the biotechnology industry in general. A German response states: "Hardly avoidable, as money is more important than health".

5.3.6   Small and medium-sized enterprises (SMEs)

The stimulation of new entrepreneurship based on biotechnological innovations has been a central focus point in European and national biotechnology policies. Some clear examples of very successful SMEs based on specific biotechnological knowledge are present in Europe and the USA. In one of the annual assessments of the European biotech industry, Ernst and Young (1996) specifically focus on these high tech biotech companies and their fate during several years. As has also been mentioned above, a great deal of new employment in biotechnology is linked to these "dedicated biotech firms". However, on the contrary it is also understood that SMEs in general experience difficulties in getting a good entrance to new biotechnological developments in the public infrastructure or in private companies. The policy relevance of this is quite clear and in 1997 the Dutch government started a specific project to gain more insight, in which sectors SMEs are most ready to adopt new biotechnological techniques; specific activities for technology transfer will follow. In the Delphi survey two statements addressed the SME issue.

Innovative biotech SMEs

In all five countries, a majority of the Delphi panels is in favour of the situation that the widespread use of modern biotechnology enables small and medium-sized companies to introduce a large variety of innovative processes and products (statement 16). The score for positive personal attitude is over 70 % in each country. The statement is also assessed as realistic. Again, a high level of correspondence among the countries can be observed: six to ten years is seen as the most probable time-horizon. However, the dual character of the biotech arena in Germany also expresses itself in this statement: 12 % of the German respondents (most from the farmers, the consumers and critics groups) believe that biotechnology will not enable SMEs to introduce a large variety of innovative processes and products. In the other countries this level lies below 5 %.

The assessment of the importance of various influential factors on the possibilities biotechnology offers to SMEs, diverges among the countries. Infrastructures and technology transfer are seen as the most important factors to stimulate the application of Agro-Food biotechnology in the Netherlands, Italy and Spain. Funding and infrastructures are mentioned as most important in Greece. Market and funding are mentioned as most important in Germany.

In the comments, respondents write that funding and infrastructures are important, because they feel that biotechnology is capital-intensive and this makes it difficult to adopt these new technologies, especially for SMEs. Therefore, governmental funding and transparent R&D infrastructures are needed to enable SMEs to get access to new technologies. This argumentation goes along with the findings in other projects dealing with the specific problems of biotechnology SMEs (Reiss and Koschatzky 1997, Reiss et al. 1995). A number of respondents, especially from Germany, indicate that large companies are able to use genetic engineering to a greater extent and will have much more benefits from this technology than SMEs. It is often mentioned that SMEs will be taken over or face competition from bigger companies.

Can small farmers afford genetically engineered crops and animals?

Statement 19 suggests that small farmers will not be able to buy the expensive genetically engineered crops and animals. Such a situation is assessed negatively by the majority of the respondents in all countries (over 70 % of the respondents indicate a negative personal attitude towards the statement). The realization of the statement is also strongly questioned by the Dutch respondents: 65 % of the panel expects that the situation will not be realized. Therefore, it can be concluded that the Dutch experts expect that farmers will be able to buy the genetically engineered crops and animals in the near future. In the other countries, over 60 % of the panels believes that the situation described in statement 19 will occur within ten years.

In Germany and the Netherlands it is indicated that this development is already taking place under existing market conditions. The specific effects biotechnology could have are not considered as positive, but as less relevant for these countries. Market is recognized as the most important influential factor in this respect in the Netherlands and Germany, while funding is seen as an important factor for Spain and Greece as well as technology transfer for Italy. Funding is seen as important in Spain and Italy. In both countries comments are added to underline the importance of these issues: "Subventions and aid should be provided to boost this sector!" (Spanish comment), "Small farmers are the priority!" and "The key problem is represented by regulations on patenting." (Italian comments).

5.3.7 Branch-specific impacts of biotechnology

The future impact of biotechnology in the Agro-Food sector - the central issue in the Delphi survey - relates mostly to new products and processes in agro-business and the food and drinks industry. The Delphi survey also includes some specific branches of the Agro-Food sector, each with its own specific conditions for integrating biotechnology. First, this is the cluster of retailer, restaurants and catering services, all located at the beginning of the food chains. Second, there are the ecological farmers who might welcome biotechnological pest control for its "natural" origin, instead of chemical pesticides, just like some vegetarians prefer genetically modified chymosine in their cheese instead of the chymosine from the calf. Finally, there is fish on the Delphi dish.

Retail, restaurants and catering

Since the beginning of the nineties, the supply-driven agricultural production system very drastically changed into a market driven food production system. The consumer and his attitude towards buying and consuming food, mediated by the retailer, became an important source for innovations in the Agro-Food sector. In theory, for each product a separate food chain can be distinguished, each with specific aspects. In a growing number of fresh fruit and vegetable chains, the retailer is the central conductor who organizes the whole production processes, including the growing - mostly on contract basis - of fruit and vegetables in all parts of the world. However, food companies with strong brand names such as in the dairy or in the beer industry, have strong positions in the process of dictating the conditions in the food supply chains, which they are part of. Bijman and Enzing (1995) concluded that biotechnology contributes to the growing influence of retailers on activities upstream in the food supply chain.

The marketing activities of retailers, catering services/restaurants and food companies with respect to genetically engineered products is the central issue in two statements. The development of food companies and retailers proactively creating specific labels for food and beverages produced without the help of genetic engineering (statement 6) is assessed positively by the respondents in all countries. In Greece, 90 % of the respondents indicates having a positive attitude towards this, while in the other countries this level lies around 60 %. Concerning the time of realization, the majority (around 60 % for all countries) expects the introduction of specific "no biotechnology" labels within five years. Also equal over the countries is the assessment of the influential factors: "market" and "acceptance" are the most important ones. Comments indicate the appearance of two groups of consumers: consumers who prefer ecologically grown food without any use of genetic engineering, and consumers who prefer food produced with modern techniques. Specific labelling is expected to be viable because of these specific market demands. However, some respondents show great scepticsism on this matter. "Sounds like science fiction" is an outspoken Spanish comment.

In some member countries of the EU, especially in Germany and Austria, specific efforts are undertaken to define and label non-genetically engineered foods. In Austria in December 1997 a coalition called AGL, incorporating supermarkets, food manufacturers, organic farmers and environmental groups, produced criteria for controlling the production of GMO-free food and started to label the respective food (Abbott and Roeper 1998). In Germany, environmental groups like Greenpeace or

Friends of the Earth (BUND) are in favour of such a label or have elaborated specific criteria in this context (Friends of the Earth 1997). In July 1997 an initiative for a GMO-free label has been started by the German consumer association (AgV) and the Green Party in the European Parliament (dpa 1997). In addition, there are initiatives in different federal states of Germany to establish such a label. The coalition "Gentechnikfrei aus Bayern" has collected enough signatures among the population in Bavaria to start a referendum (Abbott and Roeper 1998). Similiar activities are being carried out in Lower Saxony ("Gentechnikfrei aus Niedersachsen") and Bremen (Mühlenberg 1998).

A restaurant chain or catering service specialized in offering trendy meals with genetically engineered ingredients which opens branches in almost all large cities of the country (statement 7), receives different responses from the different countries. The most common personal attitude in the Netherlands and Spain is indifference; in the other countries the attitude is negative. In the Netherlands 69 % of the respondents do not believe the statement will come true, with as most influential factors "market" and "acceptance". These are also the two most important influential factors in the other countries. However, for each of the other countries, the opening of these restaurant chains or catering services that use genetically engineered ingredients is expected in six to ten years. Comments express that the link between genetically modified organisms and trendy meals is not important. If biotechnology is accepted on a bigger scale, catering services will use it in their products. This is the general tendency found in the Delphi results.

Ecological farming

Using biotechnology in organic farming is an issue in two statements. Statement 28 deals with the use of genetically engineered microorganisms and plants in biological pest control, and statement 33 refers to the use of specific genetic engineering approaches (biocides and animal vaccines) by organic farmers. The use of genetically engineered microorganisms and plants producing biocides in biological pest control is assessed as positive by all panels. It is also evaluated as a realistic option, with an expected time of realization within ten years. Only in the influential factors is there some difference to be found among the countries. Acceptance is seen as most important in Germany, acceptance and R&D infrastructure are most important in the Netherlands. R&D infrastructure is seen as most important in Greece and Spain. Finally, in Italy regulation and acceptance are the most important influential factors.

Integrating genetic engineering approaches in organic farming is an issue of which the panels indicate only a limited level of knowledge. In all countries the majority indicates having "less" to "no knowledge" about the issue. Of those addressing the statement, in all countries the bigger fraction indicates a positive personal attitude towards the integration of genetic engineering in ecological farming. Only in

Germany is this majority small. The 42 % experts with "positive" attitude is counterbalanced by a 39 % proportion of "negative" attitude.

In Spain, Italy and Greece, the biggest fraction of the panel believes that genetic engineering will take its place in organic farming in six to ten years. Although this is also the "time of realization" answer in the Netherlands and Germany, in these countries there is also a considerable group of around 25 % who believe that genetic engineering will not get a place in ecological agriculture at all. In Germany, the Netherlands and Spain, acceptance is seen as the most important influential factor for this. In Italy regulation and standards are considered most important, while in Greece, funding and R&D infrastructure are indicated as such.

The use of genetic engineering for the reduction of the use of chemicals in agriculture is less controversial than the use of genetic engineering in organic agriculture in general. The principal point of view becomes clear in a number of comments: "In the biological agriculture this (genetic engineering) is being rejected for fundamental reasons". Also, it is brought forward in one of the comments that in the Netherlands the organization of ecological farmers (SKAL) rejects the use of genetic engineering for fundamental reasons. A German respondent illustrated his (negative) attitude by asking: "When will the pope get married?".

Biotechnology, fish farming and fishery

The economic impacts of the use of biotechnology in fish farming and fishery is the central issue in statement 21, which says: "Due to the widespread use of modern biotechnology in EU fish breeding and aquaculture, the imports of fish and fish products from outside the EU are significantly reduced".

The remarkable thing about this statement is the low level of knowledge that the panels claim to have on this subject. The fractions indicating "less" or "no knowledge" range from 74 % of the respondents in Germany up to as much as 95 % in the Netherlands. To make it even more difficult, in the Netherlands and Germany, the majority indicates an indifferent personal attitude towards the subject. In the other countries, a more positive attitude can be found. In Spain, Italy, Germany and Greece the biggest fraction of the panel estimates that the import of fish will indeed be reduced as a result of the use of biotechnology. The German experts from research institutes and farmers stress the relevance of policy for the future development on the markets of fish products, whereas industry experts give an underproportional weight to this factor.

Market is indicated as the most important influential factor in most of the countries. In Italy, a lot of weight is given to market and technology transfer as well. Only in Spain R&D are infrastructures and technology transfer seen as more significant than market.

5.4   Environmental impacts

In recent years the environmental impacts and risks of modern biotechnology have been intensively discussed in the member countries of the EU. They have been one of the areas of major concerns related to Agro-Food biotechnology. This has resulted in several regulations and other legal activities on national and international level, aiming at preventing or at least reducing the effects of modern biotechnology on the environment. Despite these initiatives and R&D activities in this area there are still a lot of open questions not only related to scientific and technical aspects, but as well concerning the consequences which should have been derived from a specific finding.

The issue to be analysed in this section, environmental impacts of Agro-Food biotechnology, is not confined to a specific part of the questionnaire but is rather dispersed in several sections and categories. The parts of the questionnaire useful in the analysis of environmental impacts of Agro-Food biotechnology are:

- Statements 25 to 34, as well as specific statements in the domain "scientific and technological development"

- Sub-category "ecology" in the category "most important influential factors in your country"

- Sub-category "protection of the environment/sustainable development" in the category "importance of the statement to"

- Comments of experts to single statements

Within the Delphi survey the following areas are considered in the questionnaire:

- Biotechnological approaches to reduce environmental problems in the Agro-Food sector

- Scientific developments in plant production which have indirect impacts on the environment

- Potential negative environmental impacts of modern biotechnology

- Specific topics like herbicide-resistant plants, impact of modern biotechnology on biodiversity, and the introduction of genetic engineering approaches in organic farming

In addition, the relevant factors for the future environmental impacts of Agro-Food biotechnology as well as the importance of these aspects will be briefly analyzed in this chapter.

5.4.1 Biotechnological approaches to reduce environmental problems in animal husbandry, plant production and food industry

In a first step, the response behaviour of the experts in the five countries involved will be analysed with respect to biotechnological approaches which might contribute towards reducing the environmental burden or problems in animal husbandry, plant production and food processing. In table 5.43 the statements which are included in the Delphi questionnaire with relevance for this area are shown.

Table 5.43: Biotechnological approaches to reduce environmental problems in the Agro-Food sector

Statement No.	Content
25	Enzyme systems are specifically developed to improve the environmental performance of conventional food-processing procedures.
28	Genetically engineered microorganisms and plants producing biocides are widely used for biological pest control.
31	Modern biotechnology is widely used to reduce emissions and waste from animal production (e. g. less manure due to enzymes as feed additives, improved anaerobic waste treatment, biofilter).
32	Modern biotechnology significantly contributes to the transformation of 40 % or more of the organic agricultural waste into marketable products (e. g. energy, secondary raw materials).

Environmentally friendly enzyme systems in the food industry

As the roots of modern biotechnology can be found in the manufacturing of food and beverages, biotechnical approaches are broadly and traditionally used in this industrial branch. In table 5.44 some examples are pointed out which currently contribute to a reduction of environmental load by the use of specific enzymes in the food industry (Hüsing et al. 1998).

An increased use of biotechnological and enzymatic procedures in the food industry is mainly restricted by country-specific legal regulations, hand craft type processing procedures, negative impacts of these approaches on food quality (e. g. change in taste), traditionally low R&D investments and profits of the food industry, problems in the management and integration of these new approaches in traditional food-processing procedures, and a rather low acceptance of consumers towards enzymes produced with the help of genetically modified organisms (Schell and Mohr 1995, Hüsing et al. 1997, Videbaek 1997).

Table 5.44:   Modes of action of enzyme use with positive environmental effects
in the food industry

Mode of action	Example	Effect on environmental load
Reduction of viscosity leads to favourable rheological properties, reduced water-binding capacity	Use of enzymes for the reduction of viscosity in waffle dough; use of enzymes for fruit liquefaction; enzymatic digestion of cloudy substances in juices, wine and beer	Reduced energy consumption e. g. in pumping/stirring, baking, pressing; improved exploitation of raw materials by increased yield, waste reduction; filters can be used longer
Shorter fermentation and ripening times	Shorter ripening time for cheese by adding lipases and proteases	Energy-saving due to requirement for fewer cold rooms
Substitution of chemical processes	Starch and protein hydrolysis by acids is replaced by enzymatic hydrolysis	Substitution of acid by aqueous solutions
Substitution of resolution processes by selective reactions	Enzymatic degumming of vegetable oils enzymatic modification of fats for the production of mono- and diglycerides	Downstream processing is no longer needed
Substitution of chemical synthesis by biosynthesis	Biosynthesis of colours, flavours, fragrances and vitamins	Production under milder reaction conditions, fewer side products

Source: Hüsing et al. 1998

A broad majority of the experts in all five countries included highly appreciates the development of environmentally friendly enzyme systems for food processing (see table 5.45). Such enzyme systems are regarded to be realizable in the short to medium term (see table 5.46) mainly influenced by typical "technology push" factors like R&D infrastructure, technology transfer mechanisms, the innovativeness of the food industry as well as funding possibilities. In this sense the experts involved in the Delphi survey support the argumentation deriving from the studies mentioned above that the further penetration of these new approaches in the food industry is depending on a complex set of different activities, which often represent a severe constraint, especially for small and medium-sized companies.

Biocide-producing plants and microorganisms in biological pest control

There has been much public debate in Germany and other European countries about the production of biocides, in particular insecticidal toxins from *Bacillus thurengiensis (B. t.)*, in transgenic plants. This issue was put on the agenda mainly by organic farmers, who have been applying solutions of B. t. toxins successfully for years as a means of biological pest control, and who fear the widespread emergence of resistant insect strains due to the cultivation of such transgenic plants. The experts in all included countries consider the environmental situation as an important influential factor with respect to statement 28, too. Whether these experts intend to express concerns about possible negative influences or hopes for a pesticide-free, environmentally friendly crop protection remains open. Consumers probably worry about the effects of active toxins present in plants - when applied in solutions - B. t. toxins are inactive until they reach the digestive tract of insects, and they can be washed off before being taken up by consumers. These points might explain the relatively low value this statement achieves in the assessment of personal attitudes (see table 5.45).

Strictly speaking, the development of biocide-producing plants and microorganisms does not require polygenic approaches, though a combinatorial expression of several toxins increases efficiency and reduces the risk of pathogen resistance. Biocide-producing plants with single gene insertions are available on the market in e. g. USA and France. The question is therefore mainly when they will be widely used, and whether they will be considered in biological pest control. The expressions "biological" and "organic" are clearly defined and protected by European law and the agreements of biological farming organizations and biocide-producing plants would currently not qualify for these labels. A fundamental change would have to occur in the minds of German organic farmers, biotechnology critics and consumers which constitute an influential lobby in this area to include transgenic organisms in their concepts. Therefore, estimates for the time of realization are consequently later from German and Dutch experts compared to the Mediterranean ones (see table 5.46).

Biotechnological approaches to reduce environmental problems in animal husbandry

In the Delphi survey two different biotechnological approaches are included in order to reduce environmental problems in animal husbandry. The first approach can be considered as a kind of production integrated technique since it aims at reducing emissions and waste by shaping the process of animal production (statement 31). In this context the use of enzymes as feed additives is one possibility. Recently, enzymes are mainly used as feed additives in poultry and piglet feed, a diffusion to pork feed is intended (Hüsing et al. 1997). The main reasons for the use of enzymes as feed additives are improvements in the performance of animal production or cost-saving strategies (e. g. lower feeding costs due to a higher digestibility of the feed or use of cheaper raw materials), but in most of the cases they are not targeted to reduce the environmental impacts of animal husbandry.

The second biotechnological approach included in the Delphi questionnaire can be regarded as a kind of end-of-pipe technology, since it aims at transforming considerable proportions of organic agricultural waste into marketable products. The production of methane from agricultural waste or other organic raw materials like straw or wood represents one possible technique in this context. Currently, this technique is of minor importance in Europe, with the exception of Denmark, where several plants exist in which biogas is produced from agricultural organic waste (BML 1995). In Germany there are 440 biogas plants on farms and 15 central plants in 1997 which are able to produce the electricity needed in around 3,000 households (BMELF-Information 1998). Taking into account the current energy prices, the production of methane is not economic for an individual farmer, mostly due to the high investment costs for the biogas plant and the limited possibilities to use the produced methane during summer (BML 1995).

Both approaches are highly appreciated by a broad majority of experts in all five included countries (see table 5.45). Nevertheless, most of the experts expect a medium- to long-term realization, in particular for the transformation of organic agricultural waste into marketable products (see table 5.46). This time estimation is substantiated by a complex set of influential factors which are mentioned by the experts. These factors relate in particular to the required R&D infrastructure, the management of the technology transfer from research institutions to industry, the innovation management in relevant industrial companies as well as the financing of the entire process. In Central Europe social and ethical acceptance of these processes is additionally regarded as crucial for their future penetration.

In contrast to the other countries the Dutch experts expect a rather short-term introduction of biotechnological approaches to reduce emissions from animal husbandry (see table 5.46). This rather specific response behaviour is mainly due to the fact that since 1990 the amount of phosphorpentoxide per hectar is limited in the Netherlands - a country with a very intensive animal production, in which the disposal of agricultural organic waste represents a severe problem. Therefore, the Netherlands are one of the major markets for phytase at present - an enzyme which should contribute towards reducing the phosphor emissions from agricultural waste, according to the information of the producer companies (Hüsing et al. 1997).

Summing up the answers of the experts, it is possible to say that the use of modern biotechnology is accepted when it contributes to the improvement of the environmental situation in the Agro-Food sector. This relates to the following cases:

- Use of specific enzymes in conventional food-processing procedures
- Reduction of environmental impact in animal production
- Management of organic waste in animal husbandry thereby producing secondary raw materials
- Biological pest control
- Reduction of herbicide use

Table 5.45:   Personal attitude towards biotechnological approaches to reduce
environmental problems in the Agro-Food sector

	Development of enzymes to improve environmental performance of conventional food-processing procedures (statement 25)		
	Positive	Indifferent	Negative
Germany	84 %	11 %	4 %
Greece	89 %	2 %	9 %
Italy	97 %	3 %	0 %
The Netherlands	94 %	3 %	3 %
Spain	88 %	7 %	1 %
	Biocide-producing, genetically engineered microorganisms and plants are used for biological pest control (statement 28)		
	Positive	Indifferent	Negative
Germany	65 %	13 %	20 %
Greece	85 %	4 %	11 %
Italy	81 %	7 %	8 %
The Netherlands	73 %	14 %	13 %
Spain	89 %	2 %	5 %
	Biotechnological approaches to reduce emissions and waste in animal production (statement 31)		
	Positive	Indifferent	Negative
Germany	82 %	9 %	7 %
Greece	77 %	7 %	13 %
Italy	94 %	4 %	3 %
The Netherlands	91 %	3 %	6 %
Spain	98 %	1 %	0 %
	Transformation of organic agricultural waste into marketable products (statement 32)		
	Positive	Indifferent	Negative
Germany	87 %	8 %	3 %
Greece	92 %	4 %	3 %
Italy	95 %	4 %	1 %
The Netherlands	92 %	5 %	2 %
Spain	96 %	2 %	2 %

Table 5.46:   Time of realization of biotechnological approaches to reduce
                environmental problems in the Agro-Food sector

	Development of enzymes to improve environmental performance of conventional food-processing procedures (statement 25)				
	Up to 5 years	6 to 10 years	11 to 15 years	More than 16 years	No realization
Germany	48 %	43 %	5 %	0 %	1 %
Greece	53 %	40 %	6 %	1 %	0 %
Italy	41 %	49 %	7 %	0 %	1 %
The Netherlands	86 %	11 %	2 %	0 %	0 %
Spain	48 %	39 %	7 %	1 %	0 %
	Biocide-producing, genetically engineered microorganisms and plants are used for biological pest control (statement 28)				
	Up to 5 years	6 to 10 years	11 to 15 years	More than 16 years	No realization
Germany	18 %	58 %	17 %	3 %	2 %
Greece	39 %	47 %	12 %	2 %	0 %
Italy	65 %	30 %	2 %	0 %	1 %
The Netherlands	17 %	54 %	18 %	3 %	7 %
Spain	39 %	52 %	8 %	0 %	0 %
	Biotechnological approaches to reduce emissions and waste in animal production (statement 31)				
	Up to 5 years	6 to 10 years	11 to 15 years	More than 16 years	No realization
Germany	16 %	49 %	26 %	5 %	2 %
Greece	22 %	56 %	14 %	6 %	1 %
Italy	38 %	47 %	7 %	0 %	0 %
The Netherlands	72 %	20 %	5 %	0 %	1 %
Spain	37 %	50 %	10 %	0 %	0 %
	Transformation of organic agricultural waste into marketable products (statement 32)				
	Up to 5 years	6 to 10 years	11 to 15 years	More than 16 years	No realization
Germany	3 %	26 %	46 %	17 %	4 %
Greece	22 %	53 %	19 %	5 %	0 %
Italy	20 %	49 %	11 %	1 %	7 %
The Netherlands	12 %	55 %	22 %	4 %	4 %
Spain	26 %	60 %	12 %	1 %	0 %

5.4.2 Scientific developments in plant production

In a second step the response behaviour of the experts will be analysed related to specific scientific developments in plant production, which might have an indirect effect on the environment. In table 5.47 the statements which are included in the Delphi questionnaire with relevance for this area are shown.

Table 5.47:   Specific scientific developments in plant production with indirect effects on the environment

Statement No.	Content
54	Genetic engineering approaches are developed which allow the alteration of polygenic traits in most economically important plant species.
55	Genetically engineered plant varieties resistant to salinity and/or drought have significantly improved agricultural productivity in arid environments.
56	To prevent resistance against biocides in pathogens, new genetic engineering approaches for plant defence mechanisms have been developed (e. g. combination of different resistance genes, increase in pathogen tolerance).
58	The molecular basis of virus resistance mechanisms in economically important perennial plants like e. g. vine, olive and fruit trees is elucidated.

New approaches for genetic engineering of polygenic traits

German experts feel more negative about the development of approaches which allow the alteration of polygenic traits in plants than their European colleagues (see table 5.48). This result is surprising and seems inconsistent, given that other projects depending on exactly these techniques received much higher appreciation. Maybe the potential of a powerful technique as such arouses more fears of abuses than enthusiasm about the opportunities, though these are clearly recognized, too. German experts also share the opinion of their European colleagues about the high impact such advances could have on the creation of scientific and technological knowledge and on the competitiveness of the national economies.

Technological advances in the genetic engineering of plants are in a steady flow and it is difficult to define a stage by which statement 54 would count as realized. The molecular basis of many economically important polygenic traits is in the process of being elucidated, but the stage of scientific progress varies, depending on species and fields. The combined expression of genes in transgenic plants is practised already, but limited to a few genes with current techniques. It is difficult to estimate how long it will take for revolutionary breakthroughs, such as the use of laser microsurgery for genetic engineering, to facilitate this difficult task. Anything from the average time of realization of around five years in Greece to the German estimate of more than 13 years can be regarded as a possible development (see table 5.49), depending on how far these new developments are expected to go.

Polygenic strategies for engineering salinity/drought tolerance in plants

By comparison to other traits, genetic engineering of stress tolerance in plants has proven particularly difficult. So far, non-transgenic methods such as the selection of tissue cultures with the desired properties and the subsequent regeneration of complete plants seems the faster way of obtaining tolerant crops (Brandt 1995).

The most common environmental stress factor limiting plant growth is water deficiency - not only during drought, but also as a consequence of elevated temperature and high salinity. Due to the polygenic nature of water tolerance, research on its molecular and biochemical basis is only beginning to increase our understanding of the mechanisms involved. Modelling of pathways in transgenic plants have opened several promising strategies for engineering tolerance (Bohnert and Jensen 1996). The transfer of individual genes in these cases resulted in marginal increases in water stress tolerance - enough to reveal a function - but too little to produce marketable strains. It seems clear that the engineering of efficient tolerance will require the transfer of multiple genes or even several pathways. High stress resistance inevitably leads to reductions in productivity (Bohnert and Jensen 1996). The aim should therefore be to increase tolerance relatively moderately. However, it will have to be determined case-by-case to what degree this will affect yield.

In the Delphi survey the successful use of transgenic water-stress-resistant plant varieties is generally welcomed (see table 5.48) and the expected impact on knowledge creation is among the highest of all statements. The benefits for the national economy depend on the climate and the crops preferably grown: in the Netherlands, a country with high rainfall and water resources, the use of such plants would be relatively limited. It is surprising in this respect that this development receives relatively low values for personal attitude by the Greek experts. One could argue that Greek experts, being familiar with the special conditions of arid environments, are more critical due to a better understanding of the possible negative ecological implications, e. g. the exploitation of water resources or the spread of transgenic plants into natural habitats. However, the ecological situation was not pointed out as a major influential factor by the Greeks or by the experts of any other country.

The realization of this development seems to be well under way. Several approaches are being pursued and the first promising results have been obtained under laboratory conditions. The biggest open question is how long it will take to achieve a compromise between water-stress tolerance and productivity which is economically viable in arid areas. This would be an important requirement for their marketability and the realization of the second part of statement 55. The very low Greek and Spanish time estimations do not seem to be impossible, but the average of almost ten years in the other countries is probably more likely (see table 5.49).

Co-expression approaches in crop protection

One purpose of the development of transgenic plants is to improve crop protection. Aiming for cost-effective, flexible and efficient weed control, various crops resistant to non-selective herbicides have been developed. While the ecological impacts of non-selective herbicides (which could theoretically be used at lower doses than mixtures of selective herbicides) are debated, the use of genetically engineered plants for pest control most probably will offer clear potential benefits for the environment, since transgenic disease-resistant plants constitute an alternative to chemical crop protection and could significantly reduce the amount of pesticides used.

Genetic engineering of resistance to pathogens is challenged by two problems, namely, the relatively narrow spectrum of many defence substances and the loss of efficiency due to the selection of resistant pathogens. For example, insects can develop resistance against the toxin from *Bacillus thurengiensis* (B. t.) within a few generations, whether applied in a solution or produced by transgenic plants (Brandt 1995). A combinatorial deployment of different transgenic strategies might offer a way to achieve broad and durable resistance. Indeed, co-expression of the plant defence genes chitinase and glucanase, each of which provides only limited protection individually, has recently proved very effective in providing resistance to fungal pathogens in several organisms (Shah 1997). Similarly, the protection of plants against insects by the introduction of the B. t. toxin can be increased by simultaneously expressing other genes, e. g. a trypsin inhibitor from melon. The same principles apply for B. t. toxins produced in transgenic bacteria which constitute an alternative way of controlling insect pests (Brandt 1995). A growing number of examples for co-expressing two genes demonstrate the potential of combinatorial approaches. The understanding of plant-pathogen interactions at the molecular level has increased dramatically over the past few years, greatly adding to the repertoire of potent fungicidal and anti-microbial substances available for genetic engineering.

The problem of preventing pathogen resistance, touched on in the discussion of statements 28 (see chapter 5.4.1) and 56, is a clear case for polygenic and combinatorial approaches. Advances in this field are generally welcomed, as this development would reduce the negative influences transgenic biocide-producing plants might have on organic farming, in particular in Germany. Experts from all included countries agree that the development of resistant-proof plant defence mechanisms would be very beneficial for the economy, resulting in the highest overall rating.

The combinatorial expression of two or three defence genes has proven effective in increasing the efficiency (broader spectrum, less resistance) of biocide-producing transgenic plants already. Though researchers have only started to exploit the enormous potential of these approaches and some time will be needed until such plant varieties are widely used for pest control, the statement could be regarded as being realized already. The experts in all countries who choose a very short-term realization period have been probably better informed than the majority of experts with medium- or long-term expectations (see table 5.49).

Virus resistance mechanisms in perennial plants

In the case of virus resistance the sequence of scientific discovery and technological progress was the reverse of combinatorial approaches: while the first transgenic approaches were developed as early as 1985 and virus resistance based on various strategies has been engineered successfully in a number of plants subsequently, it is only in the last few years that our understanding of viral pathogenesis and disease is slowly catching up (Beachy 1997). For coat protein mediated resistance, the first and one of the widely used strategies, the viral target molecules and the exact mechanism of its action remain widely unknown. It has been suggested that coat proteins can confer resistance via a variety of mechanisms. Work on the three-dimensional structure of coat proteins is in progress to uncover the protein's function (Beachy 1997). Despite major gaps in the understanding of the molecular basis, progress in genetic engineering has resulted in more efficient and broader virus resistance. In one case, resistance to three different strains of viruses was achieved by co-expressing the three corresponding nucleoproteins (Prins et al. 1995).

Most basic research in the field of virus resistance and pathogenesis is carried out in tobacco plants. It is usually quite straightforward to transfer the results obtained in this model organism to other annual and biennial plants. However, genetic engineering of perennial plants, in particular trees, has attracted relatively little scientific attention. Along with huge genomes and large sizes of trees, the main obstacles are slow growth and long generation times. It is not surprising, therefore, that one of the few trees with engineered virus resistance is the papaya, a tree which flowers after its first year (Tennant et al. 1994). Forest trees such as the European aspen need a decade or two to reach sexual maturity. A breakthrough for tree engineering was achieved by Weigel and Nilsson (1995) by introducing the LEAFY gene, which controls flower development in *Arabidopsis*, into the European aspen: they managed to get trees to flower in the first year of growth. If this strategy can be applied successfully to other perennial plants, too, genetic engineering and breeding could be accelerated dramatically.

The experts in the included countries agree that the elucidation of virus resistance mechanisms in perennial plants would be most appreciable (see table 5.48) and a great contribution to scientific knowledge. It might be due to the nature of this development, which constitutes an advance in basic research and does not directly imply an application, even rather critical experts do not feel much need to reject this statement. Furthermore, most experts probably welcome the prospect of transgenic virus-resistant trees being available on European markets one day, as this would open new possibilities for a sustainable, non-chemical pest control without yield losses.

Table 5.48: Personal attitude concerning scientific developments in plant production with indirect effects on the environment

	Genetic engineering of polygenic traits in plants (Statement 54)		
	Positive	Indifferent	Negative
Germany	43 %	34 %	21 %
Greece	69 %	9 %	20 %
Italy	61 %	17 %	16 %
The Netherlands	63 %	17 %	18 %
Spain	91 %	4 %	3 %
	Development of plant varieties resistant to drought/salinity (Statement 55)		
	Positive	Indifferent	Negative
Germany	80 %	11 %	8 %
Greece	90 %	4 %	5 %
Italy	86 %	5 %	8 %
The Netherlands	82 %	8 %	10 %
Spain	95 %	2 %	2 %
	New plant defence mechanisms against pathogen resistance (Statement 56)		
	Positive	Indifferent	Negative
Germany	73 %	14 %	13 %
Greece	81 %	4 %	13 %
Italy	77 %	13 %	9 %
The Netherlands	79 %	7 %	14 %
Spain	95 %	2 %	3 %
	Elucidation of virus resistance mechanisms in perennial plants (Statement 58)		
	Positive	Indifferent	Negative
Germany	92 %	6 %	2 %
Greece	96 %	2 %	2 %
Italy	99 %	1 %	0 %
The Netherlands	90 %	8 %	1 %
Spain	96 %	3 %	0 %

The first virus resistance genes in trees will probably be cloned by DNA-marker-based approaches by the year 2000. Afterwards, genetic, biochemical and structural studies will be needed to discover the functions of these genes and their interactions with other genes involved. After this basic work in selected model organisms, it will be necessary to clone the homologous genes in other economically important species and to establish similarities and differences in their function. With the completion of this stage, statement 58 would count as fully realized. Most experts expect that this would take six to ten years (see table 5.49). This seems to be quite an optimistic estimation, even if the generation time for trees can be shortened by genetic engineering.

Table 5.49: Time of realization of scientific developments in plant production with indirect effects on the environment

	Genetic engineering of polygenic traits in plants (Statement 54)				
	0 - 5 years	6 - 10 years	11 - 15 years	later than 16 years	No realization
Germany	3 %	27 %	36 %	28 %	3 %
Greece	41 %	46 %	9 %	2 %	1 %
Italy	15 %	54 %	21 %	4 %	0 %
The Netherlands	12 %	54 %	25 %	6 %	2 %
Spain	48 %	34 %	12 %	2 %	1 %
	Development of plant varieties resistant to drought/salinity (Statement 55)				
	0 - 5 years	6 - 10 years	11 - 15 years	later than 16 years	No realization
Germany	5 %	22 %	44 %	22 %	4 %
Greece	41 %	47 %	8 %	2 %	1 %
Italy	17 %	67 %	13 %	2 %	1 %
The Netherlands	9 %	47 %	25 %	9 %	9 %
Spain	51 %	38 %	8 %	2 %	0 %
	New plant defence mechanisms against pathogen resistance (Statement 56)				
	0 - 5 years	6 - 10 years	11 - 15 years	later than 16 years	No realization
Germany	14 %	41 %	27 %	12 %	1 %
Greece	38 %	44 %	14 %	2 %	0 %
Italy	26 %	63 %	6 %	2 %	0 %
The Netherlands	24 %	56 %	12 %	4 %	2 %
Spain	56 %	38 %	4 %	0 %	0 %
	Elucidation of virus resistance mechanisms in perennial plants (Statement 58)				
	0 - 5 years	6 - 10 years	11 - 15 years	later than 16 years	No realization
Germany	6 %	47 %	32 %	10 %	1 %
Greece	35 %	49 %	10 %	5 %	0 %
Italy	12 %	62 %	20 %	2 %	0 %
The Netherlands	10 %	58 %	25 %	3 %	1 %
Spain	26 %	63 %	7 %	1 %	0 %

5.4.3   Potential negative environmental impacts

Three statements are included in the Delphi questionnaire, which refer to possible negative impacts of modern biotechnology on the environment. These statements relate to the deliberate release of genetically modified organisms which result in the transfer and recombination of the introduced genes (statement 26), the environmental risks of the deliberate release and marketing of genetically modified microorganisms (statement 27), as well as the loss of traditionally used organisms in food production due to the use of genetically engineered organisms (statement 34).

These developments are clearly rejected by a broad majority of the experts in all five countries (see table 5.50). The rather heterogeneous response behaviour to statement 27 in the countries surveyed is mainly due to the fact that in this statement two different developments are linked with each other. These developments are: firstly, the environmental risks of the deliberate release and marketing of genetically modified microorganisms (like uncontrollable spread, disturbance of the natural balance), and secondly, a possible reaction to these risks in the form of a broad rejection of these activities. While it can be assumed that the environmental risks of genetically engineered microorganisms most probably will be rejected by most of the experts, current debates in other application fields of genetic engineering indicate that, in general, there is much more controversy about how to tackle such problems and what consequences have to be drawn.

Despite their negative personal attitude, the majority of the experts in Germany, Greece, Italy and Spain expect that in around ten to 15 years at the latest, negative environmental impacts will occur due to the use of modern biotechnology in the Agro-Food sector. By contrast, the Dutch experts express some confidence that it might be possible to prevent the mentioned negative impacts on the environment, since in each statement more than 40 % of the experts think that these developments will never be realized (see table 5.51). The estimation of the Delphi experts in Germany, Greece, Italy and Spain related to a significant transfer and recombination of the introduced genes (statement 26) is supported by findings that the introduced genes in oil-seed rape can be spread to *Brassica campestris*, the wild and weedy form of *Brassica rapa*, (Mikkelsen et al. 1996) and to *Hirschfeldia incana* (Williamson 1996), with effective gene flow from crop to weed. In addition, specifically designed field trials show that significant quantities of pollen travel long distances (Timmons et al. 1996) and might be transferred to wild relatives of the genetically modified plant. Whether the trait becomes initially established in the population depends more on chance effects than fitness, since the great majority of mutants are lost from the population even if the trait confers specific advantages (Gale 1990). In cases where no compatible wild or weedy relatives exist or no other cultivars grow where the genetically modified plants are grown, gene flow from these plants will not be identified as an environmental problem. In other cases the specific characteristics and nature of the introduced trait need to be considered (OECD 1993).

In addition to a rather self-evident cross relation to the influential factor ecology, mainly social and ethical acceptance is regarded as key factor for the occurring of negative environmental impacts following the use of modern biotechnology. This can be so interpreted that the experts participating in the Delphi survey have more confidence in public control of modern biotechnology than in strong legal rules and regulations, since the influential factors regulation and politics are only mentioned in some statements (mostly in Central Europe) as crucial for the future development. All in all, the answers of the experts clearly indicate the need for additional activities by the responsible political institutions and administrative authorities in order to reduce the risk of occurence of the anticipated negative effects on the environment due to the use of modern biotechnology in the Agro-Food sector or to minimize such effects if required.

Table 5.50:   Personal attitude to statements with potential negative environmental impacts due to modern biotechnology

	Deliberate release of genetically engineered organisms results in significant transfer and recombination of introduced genes with unintended negative impacts (statement 26)		
	Positive	Indifferent	Negative
Germany	2 %	12 %	85 %
Greece	4 %	1 %	93 %
Italy	1 %	7 %	90 %
The Netherlands	2 %	5 %	92 %
Spain	8 %	5 %	81 %
	Public debate about environmental risks of release and marketing of genetically engineered microorganisms leads to broad rejection of such activities (statement 27)		
	Positive	Indifferent	Negative
Germany	28 %	14 %	58 %
Greece	69 %	7 %	19 %
Italy	26 %	28 %	42 %
The Netherlands	17 %	18 %	64 %
Spain	6 %	8 %	83 %
	Use of GMOs aggravates loss of traditionally used organisms in food production (statement 34)		
	Positive	Indifferent	Negative
Germany	1 %	25 %	72 %
Greece	2 %	15 %	81 %
Italy	7 %	8 %	84 %
The Netherlands	2 %	28 %	69 %
Spain	3 %	17 %	75 %

Table 5.51:   Time of realization of statements with potential negative
                environmental impacts due to modern biotechnology

	Deliberate release of genetically engineered organisms results in significant transfer and recombination of introduced genes with unintended negative impacts (statement 26)				
	Up to 5 years	6 to 10 years	11 to 15 years	More than 16 years	No realization
Germany	28 %	38 %	10 %	4 %	14 %
Greece	9 %	57 %	20 %	6 %	1 %
Italy	26 %	32 %	11 %	6 %	8 %
The Netherlands	19 %	24 %	6 %	1 %	43 %
Spain	25 %	41 %	13 %	5 %	2 %
	Public debate about environmental risks of release and marketing of genetically engineered microorganisms leads to broad rejection of such activities (statement 27)				
	Up to 5 years	6 to 10 years	11 to 15 years	More than 16 years	No realization
Germany	81 %	10 %	1 %	0 %	5 %
Greece	24 %	55 %	15 %	2 %	2 %
Italy	77 %	14 %	1 %	0 %	1 %
The Netherlands	28 %	5 %	1 %	0 %	64 %
Spain	62 %	18 %	5 %	4 %	2 %
	Use of GMOs aggravates loss of traditionally used organisms in food production (statement 34)				
	Up to 5 years	6 to 10 years	11 to 15 years	More than 16 years	No realization
Germany	3 %	33 %	37 %	12 %	9 %
Greece	13 %	46 %	25 %	13 %	1 %
Italy	9 %	38 %	25 %	13 %	6 %
The Netherlands	7 %	29 %	8 %	3 %	49 %
Spain	13 %	49 %	15 %	6 %	0 %

5.4.4   Controversially discussed specific topics

Reduction of environmental pollution through cultivation of genetically engineered herbicide-resistant plants

The introduction of the genes of herbicide resistance in farm crops represents one of the earliest commercial applications of Agro-Food biotechnology. Since the first genetically engineered herbicide-resistant crops like soybeans or canola are already available e. g. in the USA and Canada and have got market approval in the EU as well, there has been an intensive public debate concerning the benefits and risks of herbicide-resistant plants in most member countries of the EU in recent years. In this context, scientists and agro-industrial companies often argue that with the help of genetically engineered herbicide-resistant plants currently used selective herbicides can be replaced by non-selective herbicides which are supposed to be more friendly to the environment than the traditionally used herbicides.

Most of the experts in all included countries appreciate a potential reduction of environmental pollution due to the cultivation of herbicide-resistant plants (see table 5.52). Concerning the realism of such a development, clear destinctions have to be made between Central Europe and the Mediterranean countries since in Germany and the Netherlands a considerable proportion of the experts regards this option as unrealistic, while in the other three countries most experts expect a short- to medium-term realization (see table 5.53). Extreme differences occur in the assessment of this statement between the different expert groups in Germany. More than 50 % of the biotechnology critics regard a 50 % reduction of environmental pollution in plant production after the cultivation of herbicide-resistant field crops as unrealistic, while more than 85 % of the experts of the industry/research group expect that such a development will be realized in the future. According to their comments, most of the critics argue that environmental damage in plant production will be increased by the cultivation of herbicide-resistant plants. Some of the respondents of industry and research doubt whether there will be a 50 % reduction of environmental pollution due to herbicide-resistant plants, but in general, they expect an improvement of the environmental situation in plant production.

The critical view of the experts in Central Europe is supported by the findings of scientific assessment studies which indicate that no clear picture can be painted whether and which agent used together with herbicide-resistant plants (like e. g. glyphosate, glufosinate, bromoxynile) may have environmental advantages compared to traditionally used herbicides (Sandermann and Ohnesorge 1994). Calculations of different application schemes for non-selective herbicides suggest that some reductions in herbicide quantities might be achieved for various herbicide-resistant crops. For sugar beets, up to 30 % reductions are expected (van den Daele et al. 1997).

Table 5.52:   Personal attitude towards reduction of environmental pollution due to the cultivation of herbicide-resistant plants (statement 29)

	Positive	Indifferent	Negative
Germany	76 %	14 %	8 %
Greece	85 %	5 %	10 %
Italy	76 %	8 %	15 %
The Netherlands	77 %	9 %	13 %
Spain	92 %	1 %	6 %

Table 5.53:   Time of realization of reduction of environmental pollution due to
the cultivation of herbicide-resistant plants (statement 29)

	Up to 5 years	6 to 10 years	11 to 15 years	More than 16 years	No realization
Germany	8 %	49 %	17 %	6 %	20 %
Greece	29 %	52 %	14 %	3 %	1 %
Italy	44 %	41 %	5 %	2 %	6 %
The Netherlands	11 %	51 %	10 %	4 %	22 %
Spain	45 %	50 %	4 %	0 %	0 %

Modern biotechnology and biodiversity

In 1992 the Convention on Biological Diversity was passed at the UNO Conference
on environment and global development in Rio de Janeiro. Major targets of this
convention are the preservation of the worldwide biodiversity as well as a
sustainable use of the biological and especially the genetic resources. In this context
an intensive discussion arose whether this convention mainly prefers industrial
countries, in which most biotechnology approaches are developed, while the genetic
resources of the developing countries might not be adequately protected by this
convention (Wolfrum and Stoll 1996). In this context a broad variety of concerns
have been mentioned regarding potential negative impacts of patents in the area of
biotechnology (Moufang 1995).

The influence of Agro-Food biotechnology on the biodiversity of agriculturally
influenced ecosystems as well as the use of traditionally used varieties is differently
assessed by the experts in Central Europe and the Mediterranean countries. While a
majority of the experts in all countries rejects possible negative impacts of modern
biotechnology on biodiversity (see table 5.54), the estimations of the experts differ
significantly in relation to the realism of these developments. A considerable
proportion of the German and Dutch experts expects negative impacts of modern
biotechnology on the biodiversity in agriculturally influenced ecosystems while this
opinion is almost negligible in the Mediterranean countries (see table 5.55). In
contrast, the Mediterranean respondents expect a loss of traditionally used varieties
in food production in the medium to long term, while almost half of the Dutch
experts regard such a development as unrealistic mainly influenced by acceptance
issues. In Germany the impact of modern biotechnology on biodiversity is assessed
extremely differential by the different expert groups. While more than 50 % of the
biotechnology critics regard no negative effects of the maintenance of biodiversity
in agriculturally influenced ecosystems despite the use of modern biotechnology as
unrealistic, more than 85 % of the experts on the industry and research group expect
that such a development will be realized in future. According to their comments,

most of the critics argue that modern biotechnology will have a negative impact on biodiversity.

Table 5.54:   Personal attitude towards interactions between modern biotechnology and biodiversity

	No negative effect on biodiversity in agricultural ecosystems can be discovered, despite the widespread use of modern biotechnology (statement 30)		
	Positive	Indifferent	Negative
Germany	76 %	13 %	8 %
Greece	63 %	5 %	22 %
Italy	65 %	16 %	14 %
The Netherlands	86 %	7 %	7 %
Spain	80 %	4 %	13 %
	Use of GMOs aggravates loss of traditionally used organisms in food production (statement 34)		
	Positive	Indifferent	Negative
Germany	1 %	25 %	72 %
Greece	2 %	15 %	81 %
Italy	7 %	8 %	84 %
The Netherlands	2 %	28 %	69 %
Spain	3 %	17 %	75 %

Table 5.55:   Time of realization of developments related to modern biotechnology and biodiversity

	No negative effect on biodiversity in agricultural ecosystems can be discovered, despite the widespread use of modern biotechnology (statement 30)				
	Up to 5 years	6 to 10 years	11 to 15 years	More than 16 years	No realization
Germany	8 %	37 %	18 %	5 %	25 %
Greece	14 %	60 %	13 %	4 %	4 %
Italy	24 %	26 %	19 %	10 %	5 %
The Netherlands	30 %	25 %	4 %	4 %	21 %
Spain	26 %	57 %	10 %	2 %	2 %
	Use of GMOs aggravates loss of traditionally used organisms in food production (statement 34)				
	Up to 5 years	6 to 10 years	11 to 15 years	More than 16 years	No realization
Germany	3 %	33 %	37 %	12 %	9 %
Greece	13 %	46 %	25 %	13 %	1 %
Italy	9 %	38 %	25 %	13 %	6 %
The Netherlands	7 %	29 %	8 %	3 %	49 %
Spain	13 %	49 %	15 %	6 %	0 %

Genetic engineering and organic farming

In recent years the number of farms which produce according to the rules of organic farming is growing in all member countries of the EU. This development goes parallel with an increased consumers demand for food which is produced in such a way. In the EU production of organic food is regulated in the regulation 2092/91/EEC, in which the use of genetic engineering approaches is not foreseen. However, in December 1997 the United States Department of Agriculture suggested standards for organic agricultural products, in which the application of genetically modified organisms would be allowed in organic agricultural practice (Kirschenmann and Kirschenmann 1998). More than 200,000 farmers, food-processing companies, retailers and consumers have reacted to these suggestions, mainly rejecting the use of GMOs and irradiation in US organic farming (Löhr 1998).

In the Delphi survey the experts are divided into those from the Mediterranean countries who express a more positive attitude and those from Central Europe who show a more negative attitude towards the integration of genetic engineering approaches in organic farming (see table 5.56).

Table 5.56:   Personal attitude towards integration of genetic engineering approaches in organic farming (statement 33)

	Positive	Indifferent	Negative
Germany	42 %	18 %	39 %
Greece	69 %	7 %	20 %
Italy	63 %	14 %	20 %
The Netherlands	56 %	24 %	18 %
Spain	77 %	11 %	11 %

The same answering pattern emerges in the time estimation of this statement as well, indicating a medium-term time horizon in the Mediterranean countries, whereas in Central Europe the experts react in a rather reserved way since around one quarter of the German and Dutch experts assume that this development will never be realized (see table 5.57). In these countries discrepancies in the assessment of statement 33 can be registered between critics/consumers and industry/research. Biotechnology critics and consumers mostly reject this development for ethical reasons and the low acceptance of genetic engineering among organic farmers in general. A relatively high proportion of the experts of these groups do not see a chance of realization for the integration of specific genetic engineering approaches in organic farming in Germany and the Netherlands, while respondents of industry and research see a much higher chance of realization, e. g. in the field of vaccines. The results of the Delphi survey are in line with findings of other studies, especially in Central and Northern Europe since e. g. a Danish study has shown that only 10 % of the Danish food industry finds GMOs and genetic engineering compatible with organic production, suggesting that the food industry is sensitive to consumer opinion (Kristensen and Nielsen 1997).

Table 5.57:   Time of realization of the integration of genetic engineering approaches in organic farming (statement 33)

	Up to 5 years	6 to 10 years	11 to 15 years	More than 16 years	No realization
Germany	7 %	42 %	17 %	7 %	24 %
Greece	17 %	48 %	23 %	8 %	0 %
Italy	20 %	49 %	11 %	1 %	7 %
The Netherlands	20 %	39 %	6 %	1 %	28 %
Spain	32 %	47 %	10 %	1 %	5 %

5.4.5   Framework conditions

Framework conditions in different countries

Concerning the framework conditions relevant for potential environmental impacts of Agro-Food biotechnology, a clear distinction has to be made between Central Europe and the Mediterranean countries. In Germany and the Netherlands mainly social and ethical aspects, ecology, policy and regulation are regarded as crucial, while in the involved southern countries R&D infrastructure, technology transfer, ecology, and information activities get a rather high weight (see table 5.58). Surprisingly, international collaboration is weighted rather low in both clusters.

The relatively high relevance of social acceptance and ethical aspects might be explained by the post-industrial cultural patterns of the northern countries and the potential of modern biotechnology to offer solutions for environmental problems relevant in the Agro-Food sector, which can only be put into effect if a favourable public opinion and environment exist for these approaches.

The Mediterranean countries perceive R&D infrastructure and mechanisms of technology transfer as rather important factors for the environmental impacts of Agro-Food biotechnology, i. e. this aspect is mostly related to the respective scientific and technological developments for these experts. The explanatory hypothesis is an antagonism in relation to the high relevance of social acceptance in the Northern countries. Since the Mediterranean countries are less industrialized, and therefore have less problems with pollution by industrial type waste, the environmental problems seem less anthropogenically provoked, technical solutions are preferred instead of more behavioural and socio-cultural ones. In other words, environmental issues are perceived more as objectifible problems and not as consequences of a complex interaction of a society with nature, and as such, they can be faced by technical means. It seems that the relations between modern biotechnology and environmental problems are regarded mainly from the potential "pollution perspective" in the Mediterranean countries compared to a "risk perspective" in Central Europe.

Policy, regulation and market variables are more pronounced in the northern countries. This result can be justified by using the rationale from the above mentioned argumentation lines. Policy sensitivity and market-oriented activities go together with societies, for which the question of social orientation is equally important with the forms of mobilization of resources and the capacities of production and commercialization of technical solutions. And, of course, market aspects are more pronounced due to concrete (present or future) activities of Agro-Food biotechnology as an established technico-economic sector.

Table 5.58:   Influential factors relevant for "environment"

Influential factors	Germany	Greece	Italy	The Netherlands	Spain
R&D infrastructure	5 %	23 %	14 %	15 %	16 %
Personnel	0 %	10 %	5 %	0 %	3 %
Technology transfer	6 %	10 %	15 %	4 %	16 %
Industrial innovativeness	8 %	2 %	5 %	10 %	9 %
Markets	10 %	3 %	3 %	7 %	4 %
Funding	4 %	13 %	4 %	5 %	1 %
Regulation/standards	9 %	4 %	16 %	8 %	10 %
Policy	11 %	4 %	4 %	13 %	5 %
Environment	14 %	10 %	15 %	13 %	13 %
Social/ethical acceptance	25 %	9 %	9 %	19 %	8 %
Information	7 %	10 %	10 %	5 %	14 %
International collaboration	1 %	2 %	1 %	1 %	1 %

Information as a framework condition is more pronounced in the Mediterranean countries, but is also of certain relevance in Central Europe. The following aspects may justify the importance attributed to information activities by the experts asked (Centerick 1997):

• Consumers do not easily associate food consumption with environmental issues.

- Consumers often correlate the environmental friendliness of a food product with its packaging. It is rather difficult for them to understand objectives such as efficient use of raw materials, minimization of emission of intermediary materials or products.

- Consumers make a distinction between information, attitude and practical use of environmentally benign food products, since often products with less environmental benefits are chosen, if they are of the same or better quality and have the same price or a lower price than environmentally friendly products.

- Producers and retailers do not practice adequate communication strategies for the environmentally friendly food products (e. g. insufficient labelling)

- Authorities have not succeeded in educating and guiding producers about the possibilities of cleaner technologies.

The above mentioned elements indicate that the need of information is important and the role of the information is critical in any integrated chain management of food products. The experts of the Delphi study have, precisely, recognized the significance of this influential factor.

Since genetic engineering is a modern, even global technology (Millstone 1997), it is rather surprising that international collaboration only obtains relatively low ratings as influential factor in the view of the experts asked in the Delphi survey. Agro-Food biotechnology has become global and respective environmental issues concern our planet. Not only the origin of the problems become more and more international but mainly, the solutions depend on international collaboration in order to obtain economies of scales, codification, cross-fertilization, co-ordination of skills and capacities, and articulation of dispersed expertise. It seems that environmental issues are perceived mainly in their local expression, and not in their international dimension.

Low relevance of "ecology" as influential factor

In all countries included in the Delphi survey, "ecology" is regarded of minor relevance for the future application of Agro-Food biotechnology, since less than 5 % of the mentioned influential factors relate to ecology (see figure 4.6 and chapter 4). Even if environmental aspects might be incorporated in other influential factors as well (like regulation or policy, because these activities often deal with safety issues of modern biotechnology which are often linked to environment), this relatively low relevance of ecology as driving force for future activities in the biotech field is rather surprising, since public debates in recent years were often focused on this aspect. A high impact of "ecology" is only seen in relation to the statements of the environment domain in all countries (see chapter 4). The results of the Delphi survey go hand in hand with the findings of the Eurobarometer survey in

1996 indicating that benefit/risk considerations gain in importance in assessing different biotechnology applications instead of pure risk-oriented approaches (European Commission 1997).

In analogy to international cooperation, the influential factor "ecology" can also be regarded as underestimated by the Delphi experts, taking into account that the results of other studies suggest that consumers do not automatically make the link between food consumption and environmental issues, they need more information about the real impact of production patterns on the environment (Centerick 1997). In addition, from the comments of the experts the general impression arises that many of them argue in favour of a pragmatic and practical approach to analyse and handle potential environmental impacts of modern biotechnology instead of a rather ideologically coloured discussion of these issues. Often the acceptance of Agro-Food biotechnology is linked with its capacity to solve environmental problems and to avoid major environmental complications or risks. This argumentation is substantiated by the response behaviour of the experts in relation to biotechnological approaches to reduce environmental problems in the Agro-Food sector, to relevant scientific developments in plant biotechnology as well as to potential negative environmental impacts of Agro-Food biotechnology (see chapters 5.4.1, 5.4.2, 5.4.3)

Win-win situation between ecology and economy

Due to small profit margin and a low research intensity in agriculture and the food industry, new application areas for biotechnical processes are innovative products and reduction of production costs. Biotechnological innovations which only reduce the environmental load do not offer sufficient incentives for an alteration of existing production procedures. In most cases the environmental effects of new biotechnological approaches, like specifically developed enzyme systems or organic waste management systems based on modern biotechnology, have not been the driving force for the adoption of the respective processes. The reductions of the environmental load are more or less so-called "desired external effects" of cost-saving strategies, more efficient production or processing processes, or novel products. Therefore, future potentials for biotechnical approaches lie in particular in innovations, where reduction of environmental problems can be coupled with economic advantages for the companies (Hüsing et al. 1998).

In table 5.59 the statements which achieve an index higher than 0.5 concerning the importance for the national economy as well as improvement of the environmental situation/sustainable development are summarized. At a first glance, there are significant differences between the countries involved concerning the assessment of the statements included in table 5.59 since e. g. all statements are marked in Germany and Spain and none in Greece. This is mainly due to the fact that in Greece especially the index values for "improvement of the environmental situation" of most statements are below 0.5, which was chosen as selection criteria for table 5.59. Taking into account the differing level of the indices in the involved countries, the same developments are characterized as being important both for the national economy and improvement of the environmental situation in almost all countries.

The statements mentioned in table 5.59 fulfil to a high extent the so-called "win-win criteria" between economy and ecology and are more likely to be realized in future than developments which only reduce environmental problems but do not offer any economic benefits. Biotechnological approaches which are regarded by the experts asked in the Delphi survey as important for the national economy and reduction of environmental problems refer to the following areas:

- Use of modern biotechnology to reduce environmental problems in animal production

- Environmentally friendly enzyme systems in food industry

- New biotechnology approaches in biological pest control

- Salt-/drought-resistant plant varieties

- Cultivation of renewable resources to a greater extent

Table 5.59:   Win-win situation between economy and ecology[1]

State-ment No.	Content	Germany	Greece	Italy	The Nether-lands	Spain
22	Due to the application of modern biotechnology, farmers produce renewable resources (e. g. bio-fuel, starch, fatty acids) which are used outside the food sector on 20 % or more of the arable land in your country.	X		X		X
25	Enzyme systems are specifically developed to improve the environ-mental performance of conventional food-processing procedures.	X		X	X	X
28	Genetically engineered microorga-nisms and plants producing biocides are widely used for biological pest control.	X		X	X	X
29	The widespread cultivation of genetically engineered field crops resistant to herbicides results in an approximate 50 % reduction of en-vironmental pollution in plant production.	X		X		X

State-ment No.	Content	Germany	Greece	Italy	The Nether-lands	Spain
31	Modern biotechnology is widely used to reduce emissions and waste from animal production (e. g. less manure due to enzymes as feed additives, improved anaerobic waste treatment, biofilter).	X			X	X
32	Modern biotechnology significantly contributes to the transformation of 40 % or more of the organic agricultural waste into marketable products (e. g. energy, secondary raw materials).	X		X	X	X
55	Genetically engineered plant varieties resistant to salinity and/or drought have significantly improved agricultural productivity in arid environments.	X		X		X
56	To prevent resistance against biocides in pathogens, new genetic engineering approaches for plant defence mechanisms have been developed (e. g. combination of different resistance genes, increase in pathogen tolerance).	X		X		X
58	The molecular basis of virus resistance mechanisms in economically important perennial plants like e. g. vine, olive and fruit trees is elucidated.	X		X		X
[1] All statements are included which achieve an index higher than 0.5 in both sub-categories.						

Interestingly, most of these developments are regarded as being very important for future knowledge creation in science and technology as well (see table 5.60). This means that on the one hand clear benefits can be expected if these developments are realized, on the other hand this implies that fundamental questions in science and technology are not yet solved in these areas. Due to the high relevance of these fields as well as the high expectations expressed by the experts if these developments might be realized in future, it should be checked whether additional activities from science, industry, politics or public institutions are required in order to raise the full potential of these approaches.

Table 5.60: Statements with high relevance for scientific knowledge, economy and environment

Stat.	Germany			Greece			Italy			The Netherlands			Spain		
	sc.&tech.	econ.	environ.	sc.&tech.	econ.	environ.	sc.&tech.	econ.	environ.	sc.&tech.	econ.	environ.	sc.&tech.	econ.	environ.
25	0.78	0.81	0.73	0.88	0.28	0.52	0.77	0.71	0.77	0.87	0.83	0.79	0.96	0.87	0.77
28	0.70	0.75	0.80	0.72	0.34	0.53	0.59	0.77	0.76	0.75	0.67	0.74	0.83	0.88	0.85
31	0.58	0.61	0.89	0.30	0.00	0.62	0.47	0.48	0.91	0.62	0.63	0.87	0.82	0.74	0.94
32	0.71	0.79	0.89	0.54	0.40	0.80	0.68	0.63	0.91	0.75	0.66	0.87	0.83	0.81	0.91
55	0.81	0.72	0.63	0.70	0.81	0.19	0.72	0.73	0.66	0.82	0.27	0.26	0.94	0.93	0.68
56	0.86	0.78	0.62	0.88	0.64	0.13	0.77	0.68	0.62	0.85	0.67	0.38	0.95	0.93	0.68
57	0.87	0.49	0.78	0.85	0.30	0.62	0.77	0.36	0.84	0.82	0.23	0.54	0.97	0.84	0.81
58	0.93	0.69	0.59	0.93	0.57	0.19	0.96	0.67	0.52	0.90	0.37	0.15	0.93	0.89	0.62

5.5 Health

5.5.1 Introduction

In recent years a scientific consensus has been reached that there are close relationships between food, nutrition and public health. First this relates to the quality of food and its impacts on individual health in terms of specific food compositions, e. g. the content of vitamins, minerals, essential amino-acids, etc. Eating habits and diets have an influence on health, but they are not the only factors. If we conceive health as a multivariable situation, then genetic-hereditary, environmental (e. g. occupation) and life-style factors contribute to the health status of persons and of populations. It is rather difficult to quantify the contribution of each factor as a causal agent for most of nutrition-related diseases and also to assess the synergetic causality.

Sociologists have well documented that health is a socio-cultural construction and its meaning and perception is related to many other aspects such as fitness, slimness, capacity to resist and conditioning by the social context (Agrafiotis 1988). That means that both "we live to eat" and "we eat to live" is valid in post-industrial societies, in which we have to face commercialization, co-modification and with a metaphor consumption of health and healthy lifestyles. So, there is a demand for diets which are healthy, safe and tasty.

On the other hand, there is scientific consensus, based on knowledge and evidence gathered over the past 40 years, that certain chronic non-communicable diseases are closely related to diet and aspects of lifestyle (e. g. dietary lipids play a causal role in the etiology and pathogenesis of obesity, cardiovascular diseases, hypertension). Heart attack and other diseases of the circulatory system account for nearly half for all causes of death in the EU and their highest risk factor is total fat consumption. The direct costs of non-healthy eating habits have been estimated to amount to over 5 % of the total costs of health care in industrialized countries. In Germany the costs of food and nutrition-related diseases was estimated at about 59 billion ECU per year (Schmitt 1997). This reason, amongst many others, explains why public health policy has become central at national and European level and, more than that, the section of "nutrition" has confirmed its place and importance in these policies.

All studies from different scientific fields (epidemiology, medicine, public health, nutrition) agree that on a European level the main problems related to nutrition are: energy intakes are too high, fat consumption is too high, intake of complex carbohydrate and dietary fibres is too low, ratio of polyunsaturated to saturated fatty acids in the diet is too low, protein intake is more than adequate, and some vitamins and minerals may be deficient. This general image has to be differentiated according to educational and socio-economic conditions. This diagnosis can explain the adverse impact of diets on health, but it is necessary to find out in each case how these factors influence obesity, tooth decay, coronary heart disease etc.

In addition, food and food habits represent an essential part of the culture and society in the European countries. In all EU member states food and food habits play an important role in the social interaction of the population and contribute in this sense to the "mental" health of the population.

In this landscape a new element has given a new "texture" in recent years. Modern biotechnology and genetic engineering can influence the interactions between nutrition, food and health in many ways. In this context the central questions are the following :

- Can these technologies contribute to reducing the adverse health effects provoked by food?

- Can modern biotechnology and genetic engineering contribute to a healthier, safer and more pleasurable nutrition?

- Which additional activities are required in order to use potential benefits of modern biotechnology and to minimize potential negative effects of these technologies?

5.5.2   Methodology, analysis design

In order to identify major health related factors existing in the view of the experts involved in the Delphi survey, a factor analysis was carried out based on the personal attitude of the experts. For this purpose the answers of 785 European experts have been analysed, thereby identifying four key factors as presented in table 5.61.

Table 5.61:   Health-related factors

Factor No.	Factor name	Included statement no.	Content	Factor Loading	Total Variance
1	Nutritional surveillance	42	The hygiene monitoring of food processing is significantly improved due to the widespread use of modern biotechnological analytical methods.	0.86409	29.8
		43	Techniques based on modern bio-technology are practically used for on-line control of quality parameters in food processing (e. g. content of micronutrients or harmful substances).	0.88095	
		62	Rapid test systems based on modern biotechnology are widely used for pathogen identification in plant production and animal husbandry.	0.59321	
2	Bio-enriched food	36	Allergy-sufferers are offered special food, of which the allergenic potential has been decreased by genetic engineering.	0.6462	18.0
		38	Food enriched with specific micro-organisms having positive health effects achieve approximately 25 % turnover share in their product group (e. g. yoghurt).	0.80785	
		39	A large variety of food and beverages of superior nutritional value (e. g. vitamin, protein and fibre content; fatty acid ratios) supporting dietary health requirements is produced with the help of genetic engineering.	0.83649	
3	Adverse health effects	37	The increased use of modern biotechnology in food production and processing results in additional allergies in people actively involved in these processes.	0.85279	11.3
		40	Unintended impacts on the health of consumers occur in some products or manufacturing processes due to the use of genetic engineering during food processing.	0.8463	
4	Information/ Knowledge creation	35	Food companies inform the consumers in detail about the specific health-relevant effects of food made with the help of genetic engineering.	0.73136	8.8
		41	The long-term health impacts of the use of Agro-Food products made with the help of genetic engineering on con-sumers, farmers and employees are investigated (e. g. by epidemiological surveys).	0.81928	

Factor 1 ("Nutritional surveillance") mainly comprises developments which focus on the improved monitoring of pathogens, food quality and the general hygienic status in food production and processing with the help of modern biotechnology. This factor has a rather high relevance, accounting for almost 30 % of the variance in the answers of the analysed experts (see table 5.61). Factor 2 ("Bio-enriched food") derives from the content of statements 36, 38 and 39 which deal with the application of modern biotechnology to improve the health-related quality of food in general or specific food products. For factor 3 ("Adverse health effects"), the identity derived from the content of statements 37 and 40. Their content converges towards the exposure to negative health impacts of people directly or indirectly involved in production processes, where modern biotechnology or genetic engineering is applied. Factor 4 ("Information/knowledge creation") consists of statements which mainly deal with accompanying measures in the health area, aiming at increased information of consumers on the health effects of gene food products as well as the investigation of the long-term health impacts of these products (statements 35, 41).

Based on the loadings of the ten statements to the four identified factors, as well as the individual response behaviour of the answering experts in the category "personal attitude", four factor scores were calculated for each of the 785 considered experts which facilitates classification of the experts in nine "attitude groups " including interviewees with a similar personal attitude related to a specific factor (table 5.62).

Table 5.62: Definition of "attitude groups" in health area

Attitude group			Number of included experts
Number	Name	Attitude	
1.1	Monitoring sceptics	Indifferent/negative concerning nutritional surveillance	54
1.2	-	Positive concerning nutritional surveillance	0
2.1	Health-food sceptics	Indifferent/negative concerning bio-enriched food	143
2.2	Health-food optimist	Positive concerning bio-enriched food	275
3.1	-	Negative towards adverse health effects of modern biotechnology	0
3.2	Do not care about adverse health effects	Indifferent towards adverse health effects of modern biotechnology	70
4.1	Do not care about information	Indifferent concerning enhanced information activities	81
4.2	-	Positive concerning enhanced information activities	2
5.1	Neutrals	No specific attitude towards one of the four factors	160
Total			785

Afterwards, the individual experts were sorted into the "attitude group", in which they had the highest value in the calculated four factors scores. This procedure resulted in eight attitude groups ranging from (1.1, 1.2) to (4.1, 4.2) and a ninth group (the so-called "neutrals") who had individual factors scores between -0.5 and +0.5 for each factor, which does not allow the definite classification of this expert group to one of the other eight attitude groups. The importance of the selected attitude groups is shown in Figure 5.22.

At first glance, a rather heterogeneous distribution of experts among the different attitude groups can be observed. More than one third of the experts express strong optimism towards bio-enriched food ("health-food optimist") and represent by far the largest attitude group in the health area. Around 18 % of the analysed experts have an opposite opinion towards the contribution of modern biotechnology to healthier food ("health-food sceptics"). The other three attitude groups which are related to a specific factor are rather small and always represent those experts who have a rather indifferent personal attitude towards the respective factor. "Monitoring sceptics" who are not totally in favour of developing new monitoring techniques and test systems based on modern biotechnology represent a rather small group (6.9 %) in the sample (see Figure 5.22). The same applies to attitude group 3.2, in which experts are included who are more or less indifferent about possible negative health effects of modern biotechnology in food products ("Do not care about adverse health effects"). Attitude group 4.1 which comprises experts who do not fully agree with accompanying measures in the health area ("Do not care information") represents a rather small group in the sample as well. By contrast, the attitude group of the so-called "neutrals" who show no strong personal attitude to any of the four identified factors includes around 20 % of the analysed experts (see Figure 5.22).

Attitude group 4.2 with experts who have a strong positive attitude towards enhanced information activities and the investigation of long-term health effects has only two members. Due to the small number this group will not be taken into account in subsequent analyses. Two possible attitude groups representing persons who are totally in favour of new monitoring systems based on modern biotechnology (attitude group 1.2) or experts who reject adverse health effects of modern biotechnology application in the Agro-Food sector (attitude group 3.1) could not be identified among the analysed experts. This does not mean that there are no experts who express the respective personal attitude, but that the experts show a specific answering pattern related to the other factors and, therefore, are classified in the respective attitude groups.

Figure 5.22: Relevance of different "attitude groups" in the health area

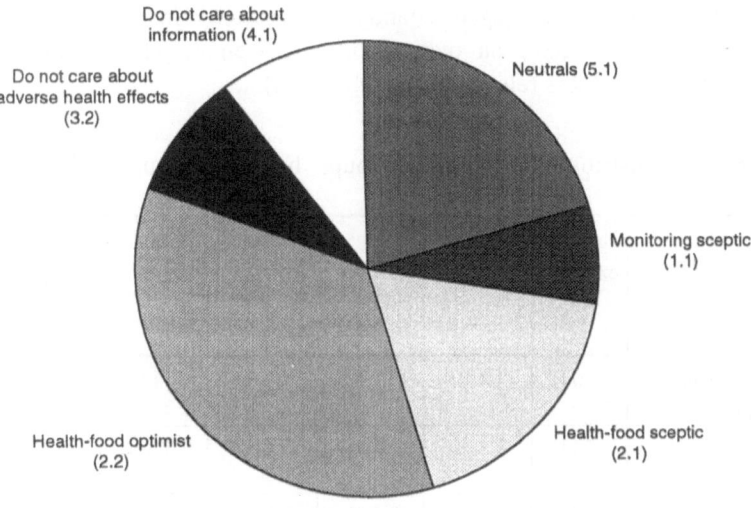

In table 5.63 and 5.64 the distribution of the attitude groups by country or expert group according to the original coding of the experts by the project teams are presented. The most important attitude group of health-food optimists is overrepresented in the Mediterranean countries (especially in Spain) and strongly underrepresented in Germany (see table 5.63).

Table 5.63: Distribution of "attitude groups" by country in the health area

Attitude group		No. of experts	Germany	Greece	Italy	The Nether-lands	Spain
No.	Name						
1.1	Monitoring sceptics	54	70.3 %	14.8 %	1.9 %	11.1 %	1.9 %
2.1	Health-food sceptics	143	70.6 %	11.9 %	2.1 %	11.9 %	3.5 %
2.2	Health-food optimist	275	33.1 %	21.8 %	14.9 %	12.0 %	18.2 %
3.2	Do not care about adverse health effects	70	54.3 %	17.1 %	11.4 %	7.1 %	10.0 %
4.1	Do not care about information	81	58.0 %	2.5 %	6.2 %	21.0 %	12.3 %
5.5	Neutrals	160	67.5 %	4.4 %	2.5 %	23.1 %	2.5 %
Total		785 [10]	53.9 %	13.5 %	7.9 %	14.9 %	9.8 %

[10] Two experts have been excluded from the analysis.

In relation to the expert groups only industry is overrepresented in this attitude group, at the expense of other experts (see table 5.64). The group of health-food sceptics shows a mirror-imaged country distribution. In this group mainly consumers and critics have an overproportional weight, whereas industry and research experts are of less relevance (see table 5.64).

Table 5.64:   Distribution of "attitude groups" by expert group in the health area

Attitude group		No. of experts	Industry	Re-searcher	Farmer	Con-sumer	Biotech-nology critics	Other experts
No.	Name							
1.1	Monitoring sceptics	54	11.1 %	16.7 %	14.8 %	7.4 %	35.2 %	14.8 %
2.1	Health-food sceptics	143	9.8 %	16.8 %	13.3 %	23.8 %	23.1 %	13.3 %
2.2	Health-food optimist	275	25.5 %	29.1 %	10.9 %	10.6 %	15.3 %	8.7 %
3.2	Do not care about adverse health effects	70	24.3 %	44.3 %	7.1 %	8.6 %	8.6 %	7.1 %
4.1	Do not care about information	81	22.2 %	30.9 %	11.1 %	8.6 %	11.1 %	16.1 %
5.5	Neutrals	160	19.4 %	31.2 %	8.8 %	5.6 %	16.9 %	18.1 %
Total		785[11]	19.9 %	28.2 %	10.8 %	11.3 %	17.3 %	12.5 %

In accordance with the results obtained in the different countries (see chapter 3), monitoring sceptics are overrepresented in Germany and underrepresented mainly in Italy and Spain (see table 5.63). In this attitude group critics have a highly overproportional weight, mostly at the expense of industry and research experts (see table 5.64). Experts who are mostly indifferent towards adverse health effects of gene food products are underrepresented in the Netherlands (see table 5.63) and among the farmers, consumers and especially critics. In contrast, experts particularly from research institutions have a rather high weight in this attitude group (see table 5.64). Taking into account the relatively high weight of information as an influential factor for the future development of modern biotechnology in Greece (see chapter 3.2), it is not surprising that "do not care about information" experts are only of small relevance in Greece. The attitude group of the "neutrals" is

---

[11] Two experts have been excluded from the analysis.

of underproportional weight in the Mediterranean countries and overrepresented in Central Europe (see table 5.63).

In the following sub-chapters the response behaviour of the different attitude groups will be analysed and discussed, taking into account the findings of other research activities. The analysis will be structured according to important issues related to the health impacts of modern biotechnology.

5.5.3   Promoting health by nutrition

Although in industrial or post-industrial societies life expectancy is the highest in human history, at the same time health-related issues are among the most important concerns of the population (Agrafiotis 1996). Health is seen more and more as an achievement, and not a destiny. In this perspective each determinant of health (e. g. housing, transport, socio-economic activities, ecology) has to be exploited or managed in such a way that their maximum contribution might be obtained. Food production, eating habits and diets should contribute to this general objective; therefore, the food supply to the population has to be of high quality and safe.

On the other hand, "healthy behaviour" depends not only on information or technical characteristics of diets but also on social norms, established practices, cultural legitimacies and socio-economic circumstances (e. g. financial possibilities, availability of time, traditions). Furthermore, studies on the social organizations of food have revealed that consumption of food is a consequence of socio-economic circumstances and the internal structures of the households and the division of labour in the family (Charles and Kerr 1988).

In spite of a lot of scientific studies, it is still not totally clarified how people adopt and follow a "healthy" behaviour in their eating habits. Different theoretical models have been tested (like "Health as a Locus of Control", "Health Belief Model") but all these psycho-social models cannot predict satisfactory behavioural change and, therefore, they have to be supplemented by other approaches (Agrafiotis 1996). All in all, it can be concluded that "healthy behaviour" depends on a lot of determinants of different nature and its adoption is not a linearly related phenomenon to information, which is of course an important factor, but not always the most crucial.

In the factor analysis, new types of food based on modern biotechnology have been identified as one of the key factors for the assessment of the future health impacts of this technology. This factor that we call "bio-enriched/targeted food" originates from statements 36, 38, 39 and refers to novel food, functional food or food with specific positive health effects due to the use of modern biotechnology.

Bio-enriched food is assessed differently by the different attitude groups, as shown in table 5.65. Per definition health-food optimists appreciate this type of food to a high extent. The other attitude groups with the exception of health-food sceptics and monitoring sceptics are in favour of food with positive health effects due to the use of modern biotechnology as well, and expect market introduction of this type of food in six to ten years (see table 5.65). In this context the situation on the food markets, the acceptance of bio-enriched food as well as information activities are regarded as key factors for the success of this type of food. Some experts mention technology-push factors like R&D infrastructure, technology transfer and industrial innovativeness as important factors, while indicating that there are still open scientific and technical questions in this area.

In contrast to the other attitude groups, health-food sceptics and monitoring sceptics regard bio-enriched food rather critically (see table 5.65). Therefore, they assess the development and market penetration of this type of food as medium- to long-term developments which will be realized in more than ten years. In addition, a considerable proportion of these attitude groups regard bio-enriched food as unrealistic (see table 5.65). In analogy to the other attitude groups, health-food sceptics and monitoring sceptics identify the situation on the food markets, the social and ethical acceptance of this type of food and information activities as decisive for the future development.

Table 5.65:   Assessment of factor 2: Bio-enriched food

Attitude group	Personal attitude			Time of realization			
	positive	indifferent	negative	0-5	6-10	11-15	no real.
Monitoring sceptics	39.6 %	25.1 %	32.1 %	22.2 %	35.3 %	14.4 %	17.0 %
Health-food sceptics	24.2 %	37.1 %	37.8 %	18.3 %	42.9 %	17.6 %	8.3 %
Health-food optimist	98.6 %	0.0 %	0.0 %	39.5 %	48.9 %	8.4 %	0.9 %
Do not care about adverse health effects	73.9 %	20.8 %	5.3 %	37.2 %	43.5 %	11.1 %	0.5 %
Do not care about information	66.9 %	25.6 %	5.0 %	29.5 %	44.4 %	18.0 %	1.7 %
Neutrals	64.9 %	34.2 %	0.0 %	14.1 %	61.4 %	16.6 %	1.9 %

An important element of this factor is functional food; that means food, which claims to have positive health effects due to specific functional components which go beyond their basic nutritional value (statement 38). Currently, for this type of food a number of questions related to classification (food or drug), proof of efficacy and labelling practice are still open. In addition, in several cases the scientific characteristics and mechanisms of this type of food are not totally clarified. Often the borders between food and pharmaceuticals are not totally clear in the area of functional food. This choice is not an easy one: if the product is a "pharmaceutical", then medical doctors could subscribe it; if it is a food product, it goes to market without the long assessment of drug evaluation institutions.

In addition to health-related aspects, marketing influenced fashion trends play an important role in the food markets. In USA "healthy", "natural" and minimally processed foods continue to grow in popularity, suggesting a reaction against the mass-produced and highly processed foods that, until recently, have comprised the typical US diet (Proops 1997, Martens 1997). In Japan, a growing need for introduction of functional food is proclaimed in order to compensate the poor nutritional content of convenience foods which replaced the traditional healthy diet, based on rice and fish (Martens 1997).

Socially and politically speaking, bio-enriched food represents a controversial subject and the experts in the Delphi survey clearly confirm this fact through their answers. By far the biggest attitude groups arise in relation to this factor with controversial assessment of this type of food. This could be interpreted in the sense that these kinds of foods are perceived as a social concern and problem and in this context, the question of effectiveness and information arises immediately.

5.5.4   Adverse health effects

Quality of food, safety of new food technologies and related health effects have become issues which provoke social debate and controversies, participate in political agendas and are reasons for NGO's advocacy. These issues call for new forms of social actions and put cultural norms to the test for their validity and relevance.

Some of these aspects have been filtered out in the factor analysis of the results of the Delphi survey as well. The assessment of factor 3 "adverse health effects" by the different attitude groups is documented in table 5.66. With the exception of attitude group 3.2 ("do not care about adverse health effects"), who are per definition mostly indifferent towards negative health impacts of modern biotechnology, all the other attitude groups strongly reject such developments (see table 5.66). Despite this negative personal attitude, a big majority of the experts expect that adverse health effects due to the use of modern biotechnology will occur in future. Concerning the time of realization, there are some differences in the assessment of the attitude groups ranging from a short- to medium-term realization by monitoring sceptics and health-food sceptics to a rather long-term realization by attitude group 3.2 ("do not care about adverse health effects") (see table 5.66). Health-food optimists, "do not care about information" and neutrals are the most optimistic attitude groups in relation to the prevention of adverse health effects due to modern biotechnology, because in these groups a considerable proportion of the experts (up to 26 %) regard such a development as unrealistic (see table 5.66). From all attitude groups mainly information of consumers and employees, regulation activities and to a lower extent R&D infrastructure are regarded as key factors related to adverse health effects of modern biotechnology.

Table 5.66: Assessment of factor 3: Adverse health impacts

Attitude group	Personal attitude			Time of realization			
	positive	indifferent	negative	0-5	6-10	11-15	no real.
Monitoring sceptics	1.9 %	4.7 %	88.8 %	45.0 %	26.0 %	9.0 %	4.0 %
Health-food sceptics	0.4 %	0.7 %	96.8 %	33.5 %	40.3 %	8.3 %	7.2 %
Health-food optimist	0.0 %	0.0 %	96.6 %	23.4 %	34.2 %	8.0 %	17.2 %
Do not care about adverse health effects	27.0 %	50.4 %	27.0 %	28.5 %	38.0 %	13.9 %	10.2 %
Do not care about information	0.6 %	12.0 %	85.4 %	21.9 %	27.8 %	8.6 %	25.8 %
Neutrals	0.0 %	0.0 %	98.7 %	19.6 %	39.9 %	5.9 %	22.2 %

The aspect that modern biotechnology is perceived by the experts as a strong force having something of inevitable, makes the adverse health effects due to applications of this technology a crucial criterion for its future development. The results of this Delphi survey correspond with those of Eurobarometer 46.1, in which European citizens consider the use of modern biotechnology in food production as one of the application areas of this technology with the greatest risk to society (European Commission 1997).

The health risks related to consumption of novel food are toxicological, additional allergies, and the resistance to antibiotics (transfer of resistance to antibiotic markers). The first and the last aspect have not been considered separately in the questionnaire, mostly due to the limited number of statements. The question of additional allergies due to the use of modern biotechnology is a central issue in the preoccupations of the experts. This confirms (on the level of their perception) the well-known fact that in industrialized countries the increase of allergic disorders is striking. In Europe almost one third of the population suffers from allergenic problems (Zwick 1997).

An important question in this context is to which degree food allergies and intolerances contribute to the overall increase of allergies. In this area, existing knowledge is very limited since basic scientific questions are not solved up to now (Knop and Saloga 1996). These questions relate to the pathological mechanisms of food allergies and intolerances, to necessary tools for the diagnosis of food allergies and intolerances, the relation of the results of these diagnoses to diets, the epidemiological evidence of the problem, as well as to an overall assessment of the phenomenon (Schäfer and Ring 1996, Wüthrich 1996).

In spite of the fact that in recent years food allergies and food intolerances are an area of public concern, very few studies have been conducted analysing the relevance of this problem in the general population in the industrial countries. Two prevalence studies based on the performance of proper double-blind, placebo-controlled food challenges (DBPCFC) have been conducted in the United Kingdom and in the Netherlands. In answering to a questionnaire, 20.4 % of the nation-wide sample complained of food allergy or intolerance in the United Kingdom. After a DBPCFC procedure the estimated prevalence of the population varied from 1.4 % to 1.8 % according to the definition used (Young et al. 1994). In the Netherlands, 12.4 % of population have reported allergies or intolerances to specific foods while the estimated prevalence based on DBPCFC varied between 0.8 % to 2.4 % (Niestijl Jansen et al. 1994). Other studies have been carried out for specific populations (children, blood donors), but without the DBPCFC procedure. In these cases there is a big difference between reported food allergies or intolerances and estimated prevalence as well.

Up to now no specific studies have been conducted in order to analyse the specific allergenic potential of novel food compared to traditional foodstuffs. Due to the limited knowledge in this field, general research activities are needed to understand better why some proteins are allergenic and others are not. This question includes the allergenic potential of novel food. In this context methodologies and investigation techniques have to be developed and tested, not only in a large population, but also in vivo and in vitro. For the latter aspect life sciences and modern biotechnology could be very useful (Madsen 1996). Ortolani et al. (1997) concluded as well that the identification of the major and intermediate allergens for the main foods should be followed by purification of the major allergenic proteins and their immunological study, aimed at identifying the epitopes involved in the immune response. Immunological research will then provide proteins produced by recombinant DNA technology and these can be employed for research into characteristics of their immunogenicity.

All in all, it can be concluded that modern biotechnology is seen by the respondents as a factor of aggravation, but at the same time, as a possibility to reduce allergies. There are strong convictions and expectations, especially for the contribution of genetic engineering to prevent allergies. But at the same time, there are clear concerns about harmful health effects of the same technology.

One possibility to avoid adverse health effects of modern biotechnology is the development of new monitoring and test systems which can be based on these techniques as well. This aspect has been filtered out in the factor analysis related to the health effects of modern biotechnology as well. The factor "nutritional surveillance" covers the biotechnological control and monitoring mechanisms necessary to assure the food safety and to prevent food-borne diseases, which are now a crucial subject for the citizens of industrialized societies. As documented in table 5.67, all attitude groups, with the exception of monitoring sceptics (who per definition express an indifferent attitude concerning this issue), highly appreciate this development and expect a rather short-term realization within the following five years (see table 5.67) mainly depending on funding possibilities and technology-push factors (like R&D infrastructure, technology transfer, industrial innovativeness).

Table 5.67:   Assessment of factor 1: Nutritional surveillance

Attitude group	Personal attitude			Time of realization			
	positive	indifferent	negative	0-5	6-10	11-15	no real.
Monitoring sceptics	31.2 %	50.0 %	17.7 %	27.9 %	46.8 %	9.7 %	4.6 %
Health-food sceptics	92.5 %	4.3 %	0.3 %	40.5 %	44.5 %	9.1 %	0.3 %
Health-food optimist	99.7 %	0.0 %	0.0 %	60.1 %	34.3 %	3.2 %	0.0 %
Do not care about adverse health effects	94.7 %	3.4 %	1.5 %	53.4 %	35.9 %	6.3 %	0.0 %
Do not care about information	90.0 %	8.3 %	0.4 %	51.1 %	39.8 %	6.3 %	0.5 %
Neutrals	99.8 %	0.0 %	0.0 %	44.8 %	46.7 %	6.1 %	0.0 %

Today the problems of food surveillance and food-borne diseases are not restricted locally, regionally or even on a national level, but extend to international, even global dimensions. The high acceptance of the experts (only low opposition against factor 1) projects the urgent and important need to establish a system to cope with this emerging situation. There is a strong demand for a future concept and a system of surveillance related to the dimensions of the problem. The national strategic plan of the USA to improve food safety is a good example in this direction. In this plan notable emphasis has been placed on prevention through enhanced Hazard Analysis and Critical Control Point (HACCP) measures, increased research and development of detection and diagnostic kits, a national early warning system that could isolate incidents and sources of infection and public education (Yeaton Woo et al. 1997). HACCP is a possible answer, a possible methodology, to the problem of assuring an overall approach to food safety from production to consumption. In this system, modern biotechnology is supposed to provide new instrumental tools for identification or detection of pathogens and support innovative strategies of control.

5.5.5   Information on health effects of food

Since adverse health effects due to modern biotechnology have been identified as a possible future problem by the experts, publicly available information on health relevant aspects of food are one possible activity in order to minimize these negative effects. This aspect has been filtered out in the factor analysis as well, in the sense of information activities of food companies on the health effects of gene food products as well as the investigation of the long-term health impacts of these products on consumers, farmers and employees (statements 35, 41). As shown in table 5.68, all attitude groups with the exception of "do not care about information" (who per definition express an indifferent attitude concerning this issue) highly appreciate information activities and expect a medium-term realization (see table 5.68). However, a considerable proportion of the experts (mainly of the monitoring sceptics, health-food sceptics and do not care about information-group) are hesitant, whether the mentioned developments will be realized at all. All attitude groups regard a broad set of influential factors as crucial, including market conditions, social and ethical acceptance, political and regulation activities and funding possibilities.

Table 5.68:   Assessment of factor 4: Information/Knowledge creation

Attitude group	Personal attitude			Time of realization			
	positive	indifferent	negative	0-5	6-10	11-15	no real.
Monitoring sceptics	86.5 %	9.9 %	2.7 %	31.7 %	26.9 %	11.5 %	21.2 %
Health-food sceptics	97.5 %	2.5 %	0.0 %	25.8 %	35.5 %	14.7 %	14.7 %
Health-food optimist	99.1 %	0.0 %	0.0 %	48.8 %	36.9 %	7.3 %	3.0 %
Do not care about adverse health effects	85.3 %	10.5 %	4.2 %	36.4 %	42.7 %	11.2 %	2.8 %
Do not care about information	37.6 %	42.3 %	14.6 %	33.8 %	37.5 %	11.3 %	11.3 %
Neutrals	99.4 %	0.0 %	0.0 %	39.1 %	42.5 %	9.5 %	4.9 %

The experts of the five countries express strong demand for epidemiological surveys in order to investigate the long-term health effects of gene food products. This demand, however, has its own specificities due to the following challenges which have to be faced by the epidemiologists and other public health specialists:

• Since new processes, new products, new types of food and new life styles have to be taken into account, some of the "diseases" or health problems are only partly known or even yet unknown.

• Definitions of the variables or factors contributing to the above mentioned "diseases" are not well established or codified.

• The elaboration of definitions and codifications (concerning the collection and treatment of the data) have to face one more obstacle relating to the fact that conditions, mechanisms and determinants work both in linear and non-linear ways.

• It is necessary to develop new corresponding time scales in order to determine the dynamics of the diseases, health harms or disorders.

• Specific attention has to be paid to studies in working places of the Agro-Food sector.

The comments of the experts of the Delphi survey as well as the different estimations concerning the "time of realization" show a certain scepticsism concerning the effective realization of the mentioned developments. In this context, the missing interest of the food industry in extended information activities, especially in such a sensitive field like health, are mentioned. In addition, some experts express some doubts concerning the objectivity of the transferred message.

All in all, the answers given by the experts impose a serious demand on the information released by firms and food companies; along with their products they should also provide arguments and proofs about the quality, safety and non-harmfulness (to the health of consumers) of their products. Since labels on food packages are only one possible source of information of consumers in relation to food and nutrition (besides others like TV/radio, magazines, newspapers, health professionals, advertisement) which have only a limited impact in this field (Schmitt 1997), a communication strategy of companies and public institutions on health-relevant aspects of gene food products should aim at involving different target groups (like health professionals, mass media, consumer organizations) and try to adapt the information as well as possible to the needs of the consumers, i. e. it should not be the target to distribute all types of information as widely as possible among the consumers but to focus information relevant for allergies on the persons who suffer from this type of disease.

Our study has already proposed a unified element for the "acceptance" issue. This is, the potential of genetic engineering to offer solutions to concrete, precise problems of healthy food or environment. It will be interesting to observe the capacity of this aspect to shape the acceptance in different socio-cultural settings by implementing long-term socio-cultural research.

5.6 Scientific and technological development

5.6.1 Introduction

In the following chapter the results of the Delphi survey with respect to the scientific and technological development in four subfields will be discussed. These subfields are:

- Enzymes (chapter 5.6.2)
- Genetic engineering of polygenic traits in plants (chapter 5.6.3)

- Reproduction techniques and genetic engineering in farm animal breeding (chapter 5.6.4)

- Diagnostics and analytical methods based on modern biotechnology (chapter 5.6.5)

5.6.2   Enzymes

Introduction

Enzymes are nature's catalysts. These biocatalysts are proteins which are synthetized by living cells. Their natural function is to speed up biochemical reactions considerably. The enzymes are neither converted nor consumed in this reaction. The catalytic properties of enzymes can be exploited in technical systems and are used in a wide range of industrial applications. The origin of "classical" biotechnology was to use enzymatic reactions for food processing, and even today, innovations in enzyme technology have a major impact on the food industry.

Out of the 7,000 enzymes known in nature, approximately 3,000 enzymes are known which catalyze a large variety of different reactions. Only a small fraction is commercially available (140 enzymes in 1972, 250 enzymes in 1985 (Crueger and Crueger 1989)), and an even smaller fraction of only 75 enzymes is of industrial relevance (Dixon 1994). The latter are mostly hydrolases which catalyze the hydrolytic degradation of proteins, fats and carbohydrates. Therefore, there is a huge gap between the nearly unlimited potential of enzymes on the one hand and their limited application in industry on the other hand.

The global market for industrial enzymes is estimated at approximately 1.5 bill. US$, with an average annual growth rate of 10 % over the last decade (Cowan 1996). This growth was mainly due to the market introduction of new enzyme products and applications and to the acquisition of new customers in other industrialized countries. Europe holds a strong position in industrial enzymes: approximately 60 % to 70 % of the world supply of industrial enzymes are prepared in Europe, 15 % in North America and 12 % to 15 % in Japan (Godfrey and West 1996). Approximately 5,000 people are employed in the European enzyme industry (Amfep 1995). Moreover, European enzyme producers establish production facilities in China, South-East Asia, Latin America and Eastern Europe. These countries are considered to be interesting emerging markets with annual growth rates far above global average due to their growing economies.

Within the global market for industrial enzymes, enzymes for the food and beverages industry have the largest share of approximately 50 % (750 mio. US$).

Taking into account that enzyme costs are usually only 0.1 % to 3 % of the final food product value (Uhlig 1991) this corresponds to a food product value of up to 750 bill. US$. Figure 5.23 gives an overview of the most important applications of enzymes within the Agro-Food sector: these are starch processing, followed by dairy applications (e. g. cheese making), alcohol production and fruit and vegetable processing (e. g. juice production). The addition of enzymes to animal feed is still a small, albeit fast-growing market.

Figure 5.23:   Market share of the most important applications of enzymes within the Agro-Food sector

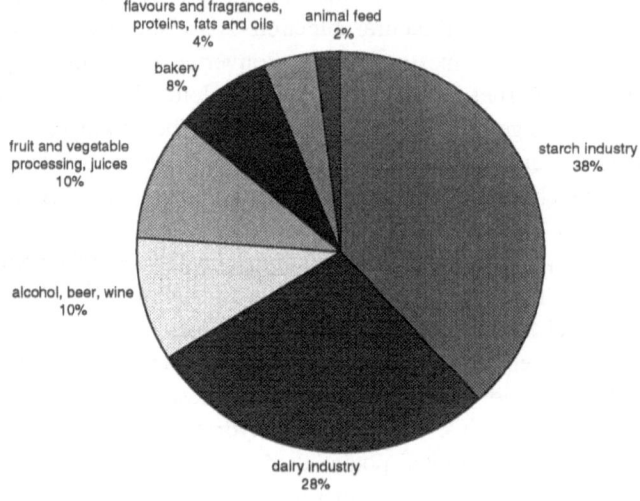

Source: Hüsing et al. 1997

Present state of the art and future development

Traditionally, the following steps led to the discovery, development and commercial production of industrial enzymes:

- Screening for interesting enzymes in source organisms such as bacteria, yeast, fungi, plants, and animals[12]
- Enzyme isolation and characterization of the enzyme properties
- Development of the source organism to a production strain, e. g. by mutation

---

[12] However, the range of plant and animal enzymes is small, due to their limited availability and relatively high costs.

- Development of a production process and a downstream processing procedure in order to obtain maximum enzyme yields
- Formulation of the enzyme preparation

This technology has been developed to sophistication in the past (Falch 1991). A significant bottleneck in this classical route was the development of a production process. Therefore, screening for interesting enzymes was mainly confined to groups of organisms which promised a realistic chance of developing a cost-efficient production process on a large scale. The introduction of genetic engineering into enzyme technology has significantly changed this classical route. From the point of view of enzyme-producing companies, genetic engineering offers the following possibilities:

- Significant cost reductions in the development and production process of enzymes
- The exploitation of new types of enzymes and new source organisms (e. g. even enzymes from non-culturable organisms)
- Drastic shortening of development times from screening to market
- Improved product safety and fewer production risks, due to the production of enzymes from a wide variety of different source organisms in a small number of well-characterized enzyme production organisms of GRAS[13] status
- Genetic engineering is a prerequisite for the optimization of enzyme properties by protein engineering

In the last years, strategic decisions have been made by enzyme-producing companies to use genetic engineering as a core technology, together with protein engineering and rational design, irrational design and high-throughput screening activities, and to exploit synergies between the different methodological approaches, in order to stay internationally competitive. At present, more than 60 % of the industrial enzymes are produced by genetically engineered organisms (Cowan 1996). Enzymes from these sources are - however - primarily applied in non-food applications, e. g. detergents, textile industry, pulp and paper industry. The introduction and use of enzymes produced by genetically modified organisms has been much later and slower in the food industry than in other industrial sectors: at present, only few food enzymes produced by genetically engineered organisms are commercially available. Among them are chymosin for cheese production, amylases for starch liquefaction and the conversion of starch into sugars, for brewing, for anti-staling of bread and several animal feed additives (e. g. phytase) (Hüsing et al. 1997).

---

[13] GRAS: abbreviation for "generally recognised as safe".

Which role will food enzymes produced by genetically engineered organisms play in the future? According to the view of the experts asked in the Delphi survey, there is hardly any doubt that approximately 90 % of the enzymes used in the food industry - that is, virtually all food enzymes - will be produced by genetically engineered organisms (statement 50) in future. Only a significant share (13 %) of the Dutch experts are sceptical that such a high percentage will ever be realized - they think the percentage is too high. Approximately two thirds of the Delphi experts are convinced that enzymes from genetically engineered organisms will have replaced traditionally produced enzymes in the food sector within the coming decade - realization within an even shorter term is mainly perceived by the Spanish and Greek experts.

It has already been shown on laboratory/greenhouse scale that the production of industrial enzymes is not only possible in bacteria, yeast and fungi (as is common practice today), but also in genetically engineered crop plants (Goddijn and Pen 1995). The development of large-scale production of industrial enzymes by genetically engineered field crops (statement 52) seems feasible to most Delphi experts, but will still take ten to 15 years. This seems plausible, given the fact that no field trials with such plants have been performed so far. However, several experts expressed doubts in their comments whether industrial enzyme production by crop plants will ever be cost-competitive with enzyme production by fermentation. In fact, downstream processing from plant biomass is generally assumed to be difficult and expensive. However, the economics of purifying proteins from plant biomass has not been investigated in detail yet (Goddijn and Pen 1995). This problem might be circumvented by concentrating on applications where the need for purification and formulation of the enzymes is completely obviated (e. g. the expression of an alpha-amylase in the potatoes or corn used for starch liquefaction or the expression of phytase in animal fodder plants (Pen et al. 1992, 1993)).

Enzymes are traditionally used by the food-processing industry. So why has the adoption of enzymes produced by genetically engineered organisms been so late and will proceed more slowly than in other industry branches? The food-processing industry is in a quandary: on the one hand, they heavily depend on enzyme suppliers who have decided to use genetic engineering as a core technology. On the other hand, they also depend on large retail companies and finally, on the consumers, who - in high percentages - reject food produced with the help of genetic engineering. As a consequence, most food-processing companies adopt a wait-and-see strategy towards the use of enzymes produced by genetically engineered organisms: on the one hand, food enzyme suppliers offer enzymes produced either conventionally or by genetically modified organisms in parallel, as an interim solution. For example, in the mid-90s Dutch milk processing companies voluntarily agreed on a moratorium for the use of officially approved chymosin from genetically engineered organisms for fear of adverse consumer reactions. Such moratoria, however, begin to erode – especially, because enzymes produced by genetically engineered organisms are not covered by the "Novel Food Regulation" and therefore do not have to be labelled as such.

Despite this quandary, the Delphi experts see a danger in the reluctance to apply recombinant enzymes in food processing: they see the broad use of enzymes produced by genetically engineered organisms or crop plants as crucial for the competitiveness of the economy - these statements are among the top ten which could contribute most to economic competitiveness. These enzymes could enhance competitiveness significantly, because they offer the potential to process food much more cost-efficiently and to develop new products with specific characteristics, promising high profit margins. As a consequence, a lack of industrial innovativeness, followed by deficits in technology transfer, is seen as the main hindrance for this development by the Delphi experts.

From the scientific point of view, the next logical step following the use of enzymes produced by genetically engineered organisms is the use of enzymes which have additionally been tailored for specific applications by protein engineering (statement 49). This technology makes it possible to deliberately change small parts of the amino-acid chain of a cloned enzyme, thus also altering the enzyme properties. It has been shown that substrate specificity and affinity, pH and temperature dependence, long-term stability and activity in organic solvents can successfully be engineered by this approach (Hüsing et al. 1997). The prerequisite for this "rational design" of enzyme properties is its production by genetic engineering and detailed information on structure-function relationships. As can be seen from the comments, obviously a large share of the Delphi experts, however, confused "protein engineering" with "genetic engineering" or with the optimization of proteins by classical mutation and screening for optimized variants.

Up to now, tailor-made enzymes optimized by protein engineering have only been commercially available for the highly competitive market of detergent enzymes. However, the market leader Novo Nordisk has announced the market introduction of the first protein-engineered food enzyme, an amylase for the conversion of starch to sugars, for 1998 (Heldt-Hansen 1998). Amylases for starch hydrolysis and chymosin for cheese production are enzyme best-sellers, since the starch and dairy industry have a share of almost 70 % of the entire food enzyme world market (Figure 5.23). From this aspect, it can be concluded that enzyme suppliers use their core technologies (genetic and protein engineering) preferably for enzymes with a high potential turnover and not in niche markets.

Enzymes are applied in food processing for the following reasons:

- Increasing the efficiency of food-processing steps (e. g. higher yields, flavour development, fewer by-products)

- More economic production (e. g. shortening of processing or ripening times, use of cheaper raw materials, improvement of food-processing properties, use of by-products)

- Increasing the process safety and controllability (e. g. reduction of losses or charges which have to be discarded)

- Increasing the reproducibility of the food-processing process, achieving more uniform and homogeneous products

- Development of new products

- Improvement of certain quality aspects of food (e. g. nutritional value, texture, shelf life, flavour, "naturalness" of food)

- Replacement of chemical additives (e. g. colorants, preservatives)

- Replacement of harsh processing conditions by milder ones

Most of these applications offer an advantage for the food-processing company (e. g. improvement of food-processing properties, more cost-efficient production), and some applications also offer a direct benefit for the consumer (e. g. new products, improvement of certain quality aspects of food). It was beyond the scope of this Delphi survey to cover all these developments by specific statements. With the two statements "the wide use of genetically engineered organisms and enzyme systems to improve the processing quality of food (e. g. prolonged shelf life of biological products, synchronization of the ripening process)" (statement 46), and "the specific development of enzyme systems for the improvement of the environmental performance of conventional food-processing procedures" (statement 25), two developments are included in the survey which should primarily be beneficial for the food industry.

Approximately 80 % to 90 % of the Delphi experts estimate that these two developments will be realized within the next decade[14] and both are seen as very important for the competitiveness of economy. With respect to the hindrances for these developments there is a clear difference between the countries: for the Mediterranean countries Greece, Italy and Spain deficits in infrastructure, technology transfer and personnel are the main hindrances for both developments. Experts from Germany and the Netherlands, however, do not assess both developments similarly but differentiated between them: for the improvement of the

---

[14] 90 % of the Dutch experts (in comparison to around 50 % in all other countries) think that the improvement of the environmental performance of food processing processes will be realized within the coming five years: in the Netherlands, environmental problems arise from the disposal of large amounts of liquid manure in agriculture. Therefore, the maximum load of phosphorpentoxide per hectare has been restricted by law since 1990. Farmers with higher manure emissions are charged with a tax. As a consequence, the Netherlands have become the most important market for phytase. If this enzyme is given as a feed additive, the manure has a lower phosphorus content - in addition to significant cost-savings in feed costs.

processing quality of food by genetic engineering (statement 46), severe hindrances are seen in the markets due to the lack of acceptance. This is in line with the finding that applications of genetic engineering in food production are rejected to a higher extent by consumers if benefits primarily arise for the producer, but not for the consumer (Hamstra 1991), as is the case in statement 46.

For the improvement of the environmental performance of conventional food-processing procedures (statement 25), however, an even more important hindrance is seen in a lack of industrial innovativeness by the German and Dutch experts. At the same time, the experts see the improvement of the environmental performance as one of the crucial factors for the competitiveness of the economy: due to their highly specific working under mild reaction conditions - at ambient temperatures and pressures and in aqueous solutions - biotechnical processes have the potential to increase the conversion efficiency and yield of industrial production processes, reduce the consumption of raw materials and energy and minimize the production of undesired side products. The adoption of these techniques by food processors could therefore lead to significant cost savings, more competitive production processes, and new products (Hüsing et al. 1998), thus combining ecological and economic advantages.

However, comments on this statement indicate that this concept of production-integrated environmental protection and its likely competitive advantage is not yet widely seen in the Mediterranean countries, especially in Spain. Moreover, comments from critical groups indicate that they see the environmental protection aspect as a "fig leaf" which hardly hides the economic interests, that environmental protection can also be reached without genetic engineering, and that any environmental benefit is due to an increase in efficiency, but that they favour the much broader, encompassing concept of sustainability.

Summary and Conclusions

Within the world market for industrial enzymes, food enzymes are the largest market segment with a share of approximately 50 % (750 mio. US$). Two thirds of all food enzymes are used by the starch and dairy industry. Growth rates above average are, however, expected in presently rather small markets such as enzymes as feed additives in animal husbandry, baking enzymes, enzymes for protein hydrolysis or enzymes for taste and flavour development.

The use of enzymes in food processing is considered crucial for the future competitiveness of this industry by the Delphi experts: enzymatic processes have the potential to contribute to a more cost-efficient production, to the reduction of environmental loads (which are also costly to dispose of) and to the development of new products with high profit margins. Innovations in food enzyme technology, however, require extensive knowledge in core technologies, such as genetic engineering, rational protein design and protein engineering, efficient and innovative screening procedures for new enzymes, knowledge in directed evolution and irrational enzyme design (Kuchner and Arnold 1997), fermentation, downstream processing and a close interaction between the different methodological approaches in order to exploit synergies between them.

Food enzymes are predominantly supplied by a few large enzyme companies which operate globally in a highly competitive market. In the last years, the market leaders of the enzyme supply companies have invested heavily in the above mentioned core technologies and have taken strategic decisions in order to realize a cost-efficient mass production of industrial enzymes and to develop highly innovative pioneering enzymes in order to stay competitive.

Food-processing companies are the main customers for food enzymes. In Europe, small and medium-sized companies prevail within this industry branch. The economic situation of the food industry is characterized by relatively low profit margins because the food market in the EU is hardly growing and growth in one segment is compensated for by loss in another. At the same time, the dependency on retail companies is also increasing. There is intensive competition among food-processing companies and small and medium-sized companies can maintain their competitiveness only if they specialize, use highly innovative production processes or develop unique products. However, a traditional lack of innovativeness is a severe hindrance in the food-processing industry which is characterized by a very low research intensity. Innovations and the adoption of new technologies are predominantly realized by the purchase of machinery or other products of the supplying industry and not by own R&D. Know-how in such complex areas as biotechnology, genetic engineering and enzyme technology is not generated in the food-processing companies themselves, so that there is a technological dependency on supplying companies.

The Delphi survey clearly shows that enzymes can considerably contribute to the maintenance and increase of the competitiveness of the food industry. However, the main hindrance to the exploitation of this potential is the lack of industrial innovativeness. Especially in the Mediterranean countries Spain, Italy and Greece, but to a lesser extent also in the Netherlands and Germany, improvements in R&D infrastructure, technology transfer and qualified personnel could help to overcome these hurdles.

At present, only few food enzymes produced by genetically engineered organisms are commercially available. However, their share will increase considerably. Delphi experts think that nearly all food enzymes will be produced in this way within the coming decade. This seems plausible, since enzyme suppliers have already strategically decided to rely fully on genetic engineering and food-processing companies are technologically dependent on their enzyme suppliers.

However, consumers' and retailers' view of enzymes from genetically engineered organisms is very critical. Therefore, food processors are in a quandary because they depend economically on retailers and consumers. At present, they adopt a wait-and-see strategy and there are no uniform positions with respect to the use of enzymes from genetically engineered organisms within this sector. This ambivalence is also reflected in the fact that the respective Delphi statements show the highest share of experts, who are indifferent with respect to this development. Due to these difficulties, it is necessary to proceed in a very sensible way and to make transparent how individual enzymes have been produced.

5.6.3 Genetic engineering of polygenic traits in plants

Introduction

Since October 1991, when the Council directive 90/220/EEC on the deliberate release of GMOs into the environment and placing on the market came into force, approximately 1,000 field releases of transgenic plants have taken place in EU member states (OECD database of field trials). Figure 5.24 gives an overview of the genetically engineered traits of the crop plants released in 1996. As indicated in figure 5.24, in the overwhelming majority of transgenic crop plants currently investigated in field trials traits have been introduced which are based on the function of a single gene which gives a dominant phenotype (Briggs and Koziel 1998): herbicide resistance is conferred by single resistance genes against the herbicides glyphosate or glufosinate, insect resistance is based on the introduction of a *Bacillus thuringienis* endotoxin gene, and altered compositions, e. g. tomatoes with altered ripening characteristics, potatoes with modified starch composition or oil seed rape with altered fatty acid composition, are also mostly due to the action of a single gene.

However, many traits targeted by plant breeding programmes are naturally determined by the interaction of several genes. Consequently, the transfer of individual genes implicated in a certain trait has often failed to result in the desired phenotype. Using present technology, genetic engineering of polygenic traits is principally possible: several transgenes can be accumulated through breeding of different transgenic organisms and/or through the repeated or simultaneous introduction of several genes. However, such approaches become increasingly difficult when the number of genes involved exceeds a certain limit, which often renders them too labour- and time-intensive to be economically viable (Gelvin 1998).

Figure 5.24:   Overview of genetically engineered traits in EU field release applications of transgenic crop plants filed in 1996 (N=234)

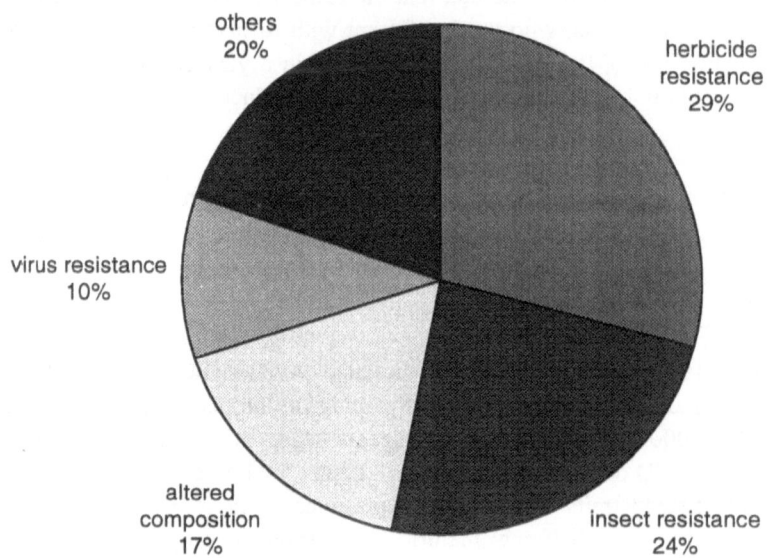

In order to assess the prospects of engineering polygenic traits, we analysed six statements of the Delphi survey dealing either directly or indirectly with future technical progress in this field, thereby covering developments in pest control, stress tolerance and the alteration of plant product composition (table 5.69).

Statement 54 addresses the engineering of polygenic traits in crop plants directly. Approximately two thirds of the Delphi experts personally appreciate this development, one fifth do not like it. The Spanish experts express a more positive attitude than the average and Germany is characterized by fewer supporters, but a large share of experts who are indifferent about this development. There is a large variation between the countries with respect to the time of realization: while two thirds of the German experts think that the realization is at least ten to 20 years ahead, 80 % of the Spanish experts are convinced that it can be realized within the coming decade. The other countries are placed in between. This large variation may be due to the fact that technological advances are in a steady flux and it is difficult to define a stage by which this development would count as realized. The molecular basis of many economically important polygenic traits is being elucidated, but the stage of scientific progress varies, depending on species and traits. The combinatorial expression of genes in transgenic plants is practised already, but limited to a few genes with current techniques. It is difficult to estimate how long it will take for revolutionary breakthroughs, such as the use of laser microsurgery for genetic engineering, to facilitate this difficult task.

Table 5.69: Statements related to genetic engineering of polygenic traits in crop plants included in the Delphi questionnaire

Statement No.	Content
General approaches	
54	Genetic engineering approaches are developed which allow the alteration of polygenic traits in most economically important plant species.
Developments in pest control	
28	Genetically engineered microorganisms and plants producing biocides are widely used for biological pest control.
56	To prevent resistance against biocides in pathogens, new genetic engineering approaches for plant defence mechanisms have been developed (e. g. combination of different resistance genes, increase in pathogen tolerance).
58	The molecular basis of virus resistance mechanisms in economically important perennial plants like e. g. vine, olive and fruit trees is elucidated.
Stress tolerance	
55	Genetically engineered plant varieties resistant to salinity and/or drought have significantly improved agricultural productivity in arid environments.
Alteration of plant product composition	
60	The content of erucic acid in rapeseed reaches 65 % or more, due to breeding systems based on genetic engineering.

According to the Delphi experts, developing genetic engineering approaches which allow the alteration of polygenic traits in most economically important plant species will be of major importance to knowledge creation in science and technology, followed by contributing to the competitiveness of the economy. This reflects that on the one hand this goal can only be reached through basic research and highly original approaches, and that on the other hand most economically important traits, such as yield, stress tolerance, or quality parameters, are polygenic ones. However, all countries perceive significant deficits for such research in the R&D infrastructure. There are hardly any other developments in the Delphi survey which are hampered by infrastructure to such an extent. Moreover, all countries point out a lack of funding for this kind of research. Deficits in technology transfer and in the availability of skilled personnel pose additional problems in Greece, Spain and to a lesser extent in Italy, while Germany and the Netherlands perceive lack of acceptance.

In the following paragraphs we will investigate how this general assessment of the alteration of polygenic traits through genetic engineering is modified by the Delphi experts for specific applications.

Developments in pest control

Much effort using transgenic approaches in agriculture is put into the improvement of crop protection. Aiming for cost-effective, flexible and efficient weed control, various crops resistant to non-selective herbicides have been engineered. Now a new "wave" of transgenic plants is heading for the market, which have been made resistant against insects, bacterial, fungal or viral diseases (Beachy 1997, Shah 1997). This development is seen as an additional tool for the control of crop pests and could offer certain advantages over conventional pesticides, such as

- Financial savings
- More efficient targeting of pests protected within plants
- Greater resilience to weather conditions
- Fast biodegradability
- Reduced operator exposure to toxic chemicals
- A reduction in the use of broad-spectrum pesticides, thereby extending the useful life of these compounds and reducing the ecological damage they cause

However, there is concern that the constitutive expression of anti-pest agents will encourage the selection of resistance to such products in pest populations (Schuler et al. 1998). Moreover, this raises the question of potential toxicity or allergenic risks to humans and animals when transgenic plants are consumed. The main question concerning environmental impacts is the potential effect on the natural microflora (e. g. in and on the plants or in the rhizosphere) by non-target effects, the risk of gene flow and the emergence of new races of the pathogen and the spread of the disease (Mourgues et al. 1998).

All in all, the three statements covering genetic engineering for pest control purposes (table 5.69) receive more support than the genetic engineering of polygenic traits without specification of the purpose, and thus are seen in general as positive developments. Whether the individual expert perceives these developments as positively or negatively depends on his terms of reference: compared to "conventional" agriculture, these developments are seen as clear improvements over the present situation, and the experts trust that such plants will find their position within integrated farming practices, provided that sound risk assessments were carried out and that special attention is given to possible long-term adverse effects. Experts who favour "organic farming" judge these developments as the "wrong approach", which will not contribute to sustainable agriculture.

Several crop plants producing biocides (statement 28) have already reached the marketplace. However, the approval of insect-resistant maize expressing an insecticidal toxin from *Bacillus thuringiensis* has been highly controversial in Europe. This issue was put on the agenda mainly by organic farmers, who have been applying solutions of B.t. toxins successfully for years as a means of biological pest control, and who fear the widespread emergence of resistant insect strains due to the careless cultivation of such transgenic plants. Experts from all countries see "ecology" as a very important influential factor for this development. Therefore, special attention has to be paid to sound pest management procedures and careful monitoring of the possible emergence of resistant insect strains.

Avoiding pathogen resistances, touched in the discussion of statement 28, is a clear case for polygenic and combinatorial approaches which are addressed in statement 56. In general, advances in this field are supported by the broad majority of experts. However, a minority of experts, mainly from critical groups, point out that such an approach based on genetic engineering would neglect basic biological principles, thus only creating new resistances in pests with every attempt to circumvent this problem. Additionally, most experts from all countries agree that the development of resistant-proof plant defence mechanisms would be very beneficial for the economy. However, concerning the time of realization the estimation of the experts is more controverse. The successful prevention of pathogen resistances by new genetic engineering approaches is expected in the coming decade and thus in the same time range as biocide-producing genetically engineered microorganisms and plants (statement 28). However, as already pointed out, there is considerable variance between the countries: it will take about twice as long according to German experts as compared to the Spanish experts' opinion.

Genetic engineering of perennial plants, in particular trees, has obtained relatively little attention until recently. Along with huge genomes and large sizes of trees, the main obstacles are slow growth and long generation times (Moffat 1996). According to the general appreciation of new plant defence mechanisms against pathogen resistances, the majority of the experts agree that the elucidation of virus resistance mechanisms in perennial plants (statement 58) would be most appreciated as well and a great contribution to scientific knowledge. Most experts think that this would take six to 15 years - a quite optimistic estimation, even if the generation time for trees can be shortened by genetic engineering (Fladung 1998). The estimations for the economic impact of this development correlate with the importance of perennial plants for the national economies: the Netherlands as a country with few forests and horticulture mainly based on vegetables rate the importance of this development far lower than the other countries which grow and export fruit, olives (olive oil) and wine to a larger extent.

Stress tolerance

Environmental stress causes significant crop losses. The stresses are numerous and often crop- and location-specific. They include drought, high salinity, temperature extremes, restricted oxygen supply in waterlogged and compacted soil, mineral nutrient deficiency, metal toxicity, pollutants and increased UV-B radiation (Smirnoff 1998). The most common environmental stress factor limiting plant growth and crop yield is water deficiency – not only during drought, but also as a consequence of elevated temperature and high salinity. Because of the strong link between transpiration and photosynthesis, this trait is extremely difficult to tackle. Due to the polygenic nature of water tolerance, the transfer of individual genes in these cases resulted in marginal increases in water stress tolerance – enough to reveal a function but too little to produce marketable strains. However, molecular markers are already used in breeding for drought resistance (Quarrie 1996). It seems clear that engineering of efficient tolerance will require the transfer of multiple genes or even several pathways.

The successful use of transgenic water-stress resistant plant varieties (statement 55) is generally welcomed and the expected impact on knowledge creation is rated high. The expected benefits for the national economy depend on the climate and the crops preferably grown: in the Netherlands, a country with high rainfall, excessive water resources and extensive greenhouse cultivation using irrigation, the use of such plants would be relatively limited and as a consequence, the Dutch experts only expect minor contributions to their economic competitiveness. On the contrary, Spanish experts who have experienced water scarcity in their country attribute great economic importance to this development.

Despite the overall positive appraisal of this development, a considerable proportion of experts also express their ambivalent or sceptical feelings about it: although they feel morally obliged to help poor countries which suffer most from drought and salinity, they doubt that these countries would be in the position to make use of such a technical development, be it for financial, political or educational reasons. Some experts point out in their comments that such a technical solution is only a "repair of symptoms" which may lead to devastating long-term effects (e. g. the excessive exploitation of water resources or the spread of transgenic plants into natural habitats), but does not tackle the cause of the problem. However, the ecological situation is not pointed out as a major influential factor by any country.

The realization of this development seems to be well underway. Several approaches are pursued and the first promising results have been obtained under laboratory conditions (Bohnert and Jensen 1996, Smirnoff 1998). The biggest open question is how long it will take to achieve a compromise between water-stress tolerance and productivity which is economically viable in arid areas - this would be a requirement for their marketability and the improvement of productivity in arid environments. A realization within the next decade, as predicted by the Greek and Spanish experts, is not impossible, but the estimation of the other countries of more than ten years is probably more likely.

Alteration of plant product composition

In 1995, the first transgenic crop with a modified seed composition – a rapeseed with a lauric acid content increased from 0.1 % to 40 % - was cultivated for commercial use. Rapeseed is among the genetically modified crops with the highest number of field trials worldwide. Reasons for the rapid progress in rapeseed engineering include the early development of efficient gene-transfer techniques and the possibility of influencing seed-oil production by relatively simple genetic modifications (Murphy 1996).

The production of erucic acid in rapeseed (statement 60) receives by far the least appreciation of all statements dealing with plant genetic engineering and the lowest value for scientific knowledge creation, although the economic gains expected are only slightly below average. From the comments of the experts it is very obvious that many participants did not understand why one would want to have a content of 65 % or more erucic acid since breeding programmes so far tried to eliminate this substance from rapeseed used as food.

Erucic acid is important for industrial uses only and is not intended to enter the food chain. It has been shown that the insertion of a single exogenous gene can alter the seed-oil composition dramatically and contents of 30 % to 40 % of the total seed-oil have been obtained for several single fatty acids. Addition of a second gene may lead to the accumulation of 60 %. However, even 60 % of a desired single fatty acid in the total seed-oil may not be a sufficiently high concentration for its commercial development. Especially for erucic acid in rapeseed-oil, economically feasible production depends on the ability to raise the present levels of about 50 % to approximately 90 % in the total seed-oil. This clearly calls for the introduction of several genes into rapeseed. Transgenic rapeseed varieties with erucic acid levels of up to 90 % have been mentioned in the scientific literature and field trials might be started as early as 1998 (Murphy 1996). Even if one took this stage as part of the realization of statement 60, it seems liable to happen in the very near future. This is recognized by a large number of experts from all countries, who suggest that this development will be realized within the next five years. German and Dutch experts see the main obstacle in the marketability of erucic acid. This reflects the fact that most of the transgenic rapeseed varieties currently under development are substitution products aimed at existing markets and the economics of such substitutions are far from clear (Murphy 1996).

Summary and conclusions

Approximately 1,000 field trials with transgenic crop plants have been carried out in the European Union since 1991, when the Council directive 90/220/EEC on the deliberate release of genetically modified organisms into the environment was put into force. In the overwhelming majority, the altered traits of these plants are due to the action of single genes which have been introduced into the crop plants through genetic engineering.

However, most economically important plant traits are encoded by several genes which closely interact. Moreover, traits conferred by a single gene are much more vulnerable to adverse effects, such as e. g. loss of the trait, emergence of resistance, transfer of the gene to other organisms etc. Therefore, the genetic engineering of polygenic traits is - on the one hand - a scientific and technical challenge which requires a broadening of our knowledge of the genetic basis of agronomically important traits. On the other hand, genetic engineering of crop plants will remain restricted to certain applications and will not play a significant role in the plant breeder's toolbox unless the genetic engineering of polygenic traits will become possible.

The Delphi experts' answers reflect the notion that this goal can only be reached through basic research, highly original approaches and technological breakthroughs. There is, however, no consensus among the Delphi experts on how long it will take to reach this goal – the time scale of realization ranges from a few years to two decades. However, all countries uniformly point out significant deficits in the R&D infrastructure. There are hardly any other developments in the Delphi survey which are hampered by infrastructure to such an extent. Additional funding is also seen as necessary by all countries for this type of research which would significantly contribute to knowledge creation in science and technology.

While the contribution to knowledge creation in science and technology is widely acknowledged, the question of how to apply this knowledge in crop plant breeding causes controversy among the Delphi experts. Support seems to depend on three factors: the purpose of the genetic engineering or the traits altered, respectively, the expert's personal terms of reference, and the frame conditions under which the transgenic crops will be grown commercially. This can be illustrated by the genetic engineering of polygenic plant traits in order to make plants more resistant or less vulnerable to plant pathogens or pests. These pest control approaches are rated more positively than e. g. the alteration of crop plant compositions. In addition, it depends on the individual expert's terms of reference whether these approaches are perceived positively or negatively: when compared to conventional agriculture, these approaches are seen as clear improvements over the present situation of extensive pesticide use. When compared to organic farming, genetic engineering of pest-resistant crop plants is highly disapproved of, because it is seen as incompatible with sustainable agriculture. This conflict is not likely to be solved in the near future.

However, it is remarkable that the experts' assessments become more differentiated and critical for close-to-market applications than for basic research applications, and more emphasis is given to the frame conditions, under which the transgenic crops will be grown. This reflects the notion that transgenic plants are not the "magic bullet" which will solve all problems of present agriculture, but that their problem-solving potential can only be exploited when their use is embedded into broader concepts, e. g. sound pest management concepts or strategies for sustainable use of scarce water resources. Further elements of such concepts are careful risk assessments of transgenic plants prior to their commercialization, monitoring programmes for possible adverse, indirect or long-term effects (e. g. emergence of resistant pest strains) which cannot be detected during risk assessments, and educational programmes (e. g. for retailers and plant growers).

5.6.4   Reproduction techniques and genetic engineering in farm animals

Introduction

Techniques based on modern biotechnology and genetic engineering are gaining increasing influence in farm animal breeding and animal husbandry. However, most of the techniques currently applied are biotechnological reproduction techniques which must be distinguished from genetic engineering techniques. As can be seen from table 5.70, the Delphi survey does not cover these biotechnological reproduction techniques. The statements focus on cloning techniques which have recently gained much publicity due to the birth of the cloned lamb "Dolly" as well as on transgenic animals.

Table 5.70: Overview of selected developments based on modern biotechnology in livestock reproduction and breeding, and their coverage in the Delphi survey

Development	Covered by statement in the Delphi survey
Reproduction techniques	
Artificial insemination	-
Embryo transfer	-
In vitro fertilization	-
Cloning techniques	
Cloning by embryo splitting	The use of cloned embryos of cattle, sheep and goats is a widespread technique for the reproduction of farm animals (statement 65).
Cloning by nuclear transfer	
Transgenic animals	
Alteration of monogenic traits by genetic engineering	Monogenic traits are specifically and stably altered in fish important for aquaculture by genetic engineering as a matter of routine (e. g. anti-freeze gene, growth hormone gene) (statement 69).
	The elucidation of principles underlying the developmental processes of oocytes leads to the genetic engineering of early stages of embryo cells from farm animals (statement 66).
	Breeding procedures based on genetic engineering are practically used for those farm animals (cattle, pigs, sheep, goats) which are of major economic importance for food production in the EU (statement 63).
Alteration of polygenic traits by genetic engineering	The elucidation of the principles regulating the expression of polygenic traits in farm animals allows the alteration of such traits by genetic engineering with regard to time, extent, quality, and localization (e. g. formation of milk similar to human milk in cows) (statement 64).

Reproduction techniques

Although not covered by statements of the Delphi survey, this chapter will start with an overview of biotechnological reproduction techniques. They represent a lasting element in animal breeding and have had an ever increasing influence on the breeding process since the implementation of artificial insemination in the 1940s. In livestock breeding biotechnological reproduction techniques aim at

- Optimizing the efficiency of selection in breeding

- Extensive use of individuals with a preferable genotype

- Control of the genetic structure of populations and families

- The extension of gene migration, e. g. across species barriers (Geldermann 1997)

Several reproduction techniques are available which are currently being applied to different extents. These techniques are:

- *Artificial insemination.* Artificial insemination represents a traditional and widely used basic technology in reproduction techniques. Currently, the proportion of artificial insemination is between 60 % and 95 % in cattle in Europe; in pigs this technique shows an increasing tendency and in sheep artificial insemination is most widespread in the sheep-breeding countries of the Southern hemisphere. Despite its technical maturity, artificial insemination is practically used in the breeding of horses or goats only under specific circumstances (Merkt 1980). Artificial insemination is often used in combination with other biotechnological reproduction techniques like e. g. embryo transfer (see below). A new variant of artificial insemination that deposits sperm deeply into the cow's uterus opens up the possibility to use sexed sperm to produce sex predetermined calves. Before the development of this technique, sperm sexing technology could only be used in combination with in vitro fertilization, embryo development and embryo transfer, which are impractical procedures in large dairy and beef operations (Potera 1996).

- *Embryo transfer.* One major target of embryo transfer techniques is the extended use of the stock of germ cells in the ovary of a female farm animal by inducing super ovulation. Up to ten oocytes mature during this process which are inseminated artificially. So far, embryo transfer is applied almost exclusively in cattle for the breeding of elite cows (e. g. high number of descendants from elite cows, nucleus breeding progammes, embryo bank). So far, the costs of this technique are too high to be used for breeding of farm animals on the production level. Worldwide about 250,000 embryo transfers are carried out per year (Geldermann 1997). At present, the main difficulties of embryo transfer techniques are high costs, variable reactions of female animals to superovulation, low amount of transferable embryos, as well as reduced pregnancy rates if conserved embryos are used. In addition, kryo-conservation of embryos in pig and horse is not successful and satisfactory respectively (Geldermann 1997).

- *In vitro fertilization.* A technique associated with embryo transfer is in vitro fertilization, in which fertilization of oocytes takes place outside the maternal organism. At present, the use of this technique is limited by low efficiency and high costs. However, an increase of in vitro fertilization is expected in future because in vitro fertilization has advantages at the breeding and production level. These advantages are increased selection intensities, shortening of the generation interval and use of sexed sperm. Moreover, in vitro fertilization can successfully be applied in connection with cloning and genetic transfer techniques (see below, Geldermann 1997).

All these reproduction methods have in common that only few sires or dams are used in the breeding practice. This is connected with the risk of increased inbreeding in the population along with possible negative consequences on genetic variability, fitness and performance of the animals.

Cloning techniques

Cloning is defined as non-sexual reproduction of descendants from a part of an organism. Cloned individuals possess the same or nearly the same genetic information as their ancestors. Subsequently, cloning techniques just aim at multiplicating desired genes, but they do not contribute to a modification of the gene setting of an individual animal. Two different techniques are applied for the cloning of farm animals: embryo splitting and cloning by nuclear transfer. Both techniques require the in vitro intervention into very early stages of embryonic development and the subsequent transfer of the cloned embryos into surrogate mothers until birth (embryo transfer, see above).

- *Embryo splitting.* Embryo splitting makes it possible to produce identical twins (or even more genetically identical embryos) by dividing blastomers or blastocytes into two or more embryos with the help of microsurgery. One major problem of the embryo splitting technique is its low efficiency if blastomers or blastocytes are divided into three or four parts (Niemann and Meinecke 1993). Moreover, quality requirements for the splitting process are only met by 30 % of the embryos planned to be used for this purpose. Although this technique was developed to maturity 15 years ago and is a routine procedure in cattle, sheep and goats, it is not generally applied in practical animal husbandry so far (Epping et al. 1997). At present, embryo splitting is restricted to cattle and aims at the production of additional calves on the production level. In research activities, the production of identical twins by embryo splitting is of importance for the analysis of inheritance of multifactorial dependant traits. In embryos cloned in this way only the properties of the maternal and paternal animals are known, but not the precise gene setting of the embryos themselves - which phenotype results from the mixture of the parental traits can only be tested after the birth of the clones.

- *Embryo cloning by nuclear transfer*. In contrast to embryo splitting, the precise gene setting of embryos obtained by nuclear transfer is known - these embryos have only one parent and are therefore genetically identical to this parent animal. Embryo cloning by nuclear transfer is carried out by transferring a diploid nucleus into eggs, from which the DNA was removed. Usually, nuclei derived from blastomeres (undifferentiated embryonic cells) are used for this purpose. However, recently, nuclei derived from adult tissue have also been successfully used as was reported for the first time in February 1997 by Wilmut et al. (1997), who announced the birth of the lamb "Dolly". Initial doubts whether really adult tissue was used (e. g. by Sgaramella and Zinder 1998) have been put to rest by careful re-examination of the procedure (Solter 1998, Ashworth et al. 1998, Signer et al. 1998) and papers currently in press show that Dolly-style nuclear transfer does not only work in sheep but also in mice (Wakayama et al. 1998), cows (Fox 1998) and primates (Stokes 1998).

The Delphi statement concerning the widespread use of cloning techniques for the reproduction of cattle, sheep and goats as farm animals (statement 65) does not differentiate between these two different cloning techniques. However, the second round of the Delphi survey was conducted simultaneously with the intensive public debate on cloning which followed the announcement of the birth of "Dolly" in February 1997 and the answers were certainly influenced by this debate. At that time a certain enthusiasm for the possibilities and prospects of cloning was noticeable in scientific publications: the use of cloning techniques as a substitute for conventional breeding with non-transgenic animals will pay off when having "nearly" identical animals is necessary (MacQuitty 1997). This would be the case for producing strains with certain desired gene combinations in specific breeding programmes. Furthermore, cloned animals could be of use for basic research activities on the regulation of genes, complex traits and the reasons for variability of important traits (Geldermann 1997). In addition to commercial purposes, cloning might be valuable in the breeding of endangered species. The most important advantage of this technique is the potential to derive animals from cells that have been genetically modified. Cells bearing the desired genetic modification will be selected and characterized in culture before nuclear transfer into recipient eggs. Using this technique would reduce time and production costs and increase the efficiency of producing transgenic animals (Rudolph 1997a).

However, this enthusiasm is not reflected in the results of the Delphi survey. The widespread use of cloned embryos in reproduction of farm animals is regarded rather critically across all countries. Cloning is welcomed by a majority only in Spain. On the other hand, the share of cloning opponents reaches 70 % in Greece. They derive from the expert subgroups of farmers, critics and consumers, whereas the attitude towards cloning of experts from research institutions is much more positive. In Germany and the Netherlands a negative to indifferent attitude towards all biotechniques described in the animal statements is predominant, with cloning being the least desired of all. It is outstanding that 20 % of the Dutch experts believe that the widespread cloning in farm animals will not be realized at all. This might reflect that the Netherlands is currently the only country in Europe where the law demands that decisions about experimental cloning procedures must involve ethical advice. As a consequence, the Dutch government introduced a temporary ban on cattle cloning experiments planned by a Dutch company in February 1998. The reason given by the Dutch Minister of Agriculture for this decision was that "the use of nuclear transfer is undesirable and can only be permitted if it will lead to a substantial advantage for society and if nuclear transfer has no impact on animal health and welfare" (Peerenboom 1998).

Moreover, the cloning statement provoked a large number of comments. Experts expressed their ethical concerns with regard to the cloning of farm animals, most of them pointing out the slippery slope towards the cloning of human beings. Other concerns often expressed were the inherent risk of loss of genetic diversity among farm animals and adverse impacts of inbreeding, e. g. increased susceptibility to diseases. Moreover, many experts are of the opinion that cloning will be confined to specific applications, especially in basic research and in cloning transgenic animals for pharmaceutical purposes, but that it offers too few advantages and is too costly for livestock reproduction on a large scale.

The widespread use of cloning is expected, or (on the basis of ethical misgivings) rather feared to be realized within the coming ten years or even earlier. This assessment is in accordance with former Delphi surveys in Germany (Fraunhofer-Institut für Systemtechnik und Innovationsforschung 1997a, b) and the United Kingdom (Loveridge et al. 1995) and obviously preoccupied with successful cloning experiments lately discussed in public.

Currently, efficiency of embryo cloning is rather limited. The biggest clones in cattle consist of seven to eleven animals (Petzoldt 1998). In the case of nuclear transfer of differentiated adult cells, "Dolly" was the sole survivor of the production of 277 sheep embryos cloned from a single adult mammary cell (Brower 1997). Only low pregnancy rates of 25 % to 35 % are achieved, even if morphologically intact embryos are used (Geldermann 1997). The loss of embryos happens at every stage of the process and most of the reasons for these problems are not elucidated so far. One more problem reported with cloned animals is high birthweight which endangers the life of the surrogate mothers during birth (Brower 1997). Regarding the technical problems still existing in the cloning process, the costs and not least the intensive adaptation processes necessary in breeding organizations, the expectations for widespread use in livestock reproduction seem to be rather unrealistic. Most of the biotechnological methods will not be applied on farms but on specific artificial insemination centres, laboratories, scientific institutes or commercial breeding companies (Geldermann 1997).

Genetic engineering of farm animals

All techniques described so far can be considered as "enabling techniques" for the production of transgenic animals whose gene setting has been deliberately altered by genetic engineering. The initial goals of commercial transgenesis were to increase the meat and production characteristics of food animals. The first working model was the pig, whose relatively short generation time and large litters made it ideal for this purpose (Rudolph 1995). Different growth hormone gene constructs were introduced into pigs, but also into fish, cattle, sheep and poultry. However, the resulting transgenic animals often suffered from negative side effects, such as lower birth weight, reduced appetite, lethargy, reduced fertility and other severe disorders (Niemann 1997). In this context it is important to know that almost any foreign protein can be produced in mammary glands of cows. Strategies have been suggested for transgenic cattle with altered milk composition to enhance cheese yield, to reduce the energy costs of milk production, to reduce the microbial load in milk and to eliminate some bovine milk protein genes and to replace them by human ones, so that the resulting milk could be used as a supplement or replacement for infant formula (Wall et al. 1997). To improve the properties of cow milk for human neonatal nutrition, it has been suggested to introduce the genes encoding human lactoferrin and lysozyme, which are thought to provide antibacterial activity in human milk. A co-expression of both genes might also result in a reduced risk of intramammary infection or mastitis. Additionally, the gene for human lipase could be transferred in order to increase the digestibility of lipids (Karatzas and Turner 1997). But fact is that the necessary experiments are still at an infant stage. All in all, attempts to apply genetic engineering techniques to produce livestock with improved production characteristics have been discouraging to date, largely because of the unintended introduction of undesirable characteristics on other tissues or organs (Langer 1998), but also due to technical problems.

Currently, genetically modified farm animals are produced by pronuclear injection, involving the physical injection of 200 to 300 copies of a gene into recently fertilized eggs. One major difficulty of this technique is its low success rate and efficiency. Only about 2 % to 3 % of the born animals are transgenic and - in addition - only a small proportion of them express the gene products deriving from the added genes at high level (Stokes 1998). Due to these technical problems with the generation of transgenic animals, high R&D costs and low profit margins in the food industry, priorities were shifted to the production of protein pharmaceuticals in the mammary gland of transgenic sheep, goats and cows (Rudolph 1995).

At present, three pharmaceutical proteins from transgenic animals are in human clinical trials: human antithrombin III from transgenic goats for the treatment of acquired antithrombin III deficiency (phase III), alpha-I-anti-trypsin from transgenic sheep for the treatment of cystic fibrosis (phase II), and alpha-glucosidase from transgenic rabbits for the treatment of Pompe's disease (phase I expected in 1998) (Rudolph 1997b). Other therapeutic proteins in the "transgenic animal R&D pipeline" are human serum albumin, alpha-1-proteinase inhibitor, prolactin, malaria proteins, calcitonin, factor VII, factor VIII, factor IX, fibrinogen, collagen, protein C (Hicks 1998), and transgenic pigs as source animals for xenotransplantation (Cozzi and White 1995).

Altering traits in farm animals by genetic engineering, as addressed in the statements 63, 64, 66 and 69 in the Delphi survey (see table 5.70), is uniformly rated more negatively by the Delphi experts than using these techniques in plant or microorganism breeding in all countries. However, there are large differences between the countries: approximately 80 % of the Spanish experts have a positive attitude towards transgenic animals, whereas up to 50 % of the Dutch and German experts disapprove of these developments. Although the four statements which deal with transgenic animals are assessed differently, no uniform pattern for all countries can be observed - the acceptance patterns seem to be country-specific. Reasons for the personal rejection of transgenic animals given in the comments concern ethical aspects and the risk of loss of genetic variability.

The majority of experts think that transgenic fish, genetic engineering in the breeding of farm animals and the genetic engineering of early stages of embryo cells from farm animals will become a reality within the next decade - only the targeted alteration of polygenic traits by genetic engineering will take some more years to realize. Given the minimum of seven years it takes today to generate a herd of transgenic cows and the considerable amount of breeding required to produce a viable stock which stably carries only three transgenes, this could easily take one human generation. Thus, the experts' relatively cautious estimations seem to be very realistic.

For all developments, German experts are the most cautious, favouring a rather late realization, while Spanish and Greek experts think that an earlier realization will be possible. It is striking that up to 40% of experts in the Netherlands consider the realization of genetic engineering in farm animal breeding as "not possible". This is especially remarkable as the Netherlands are the homebase of one of the few biotech companies in Europe which focus on transgenic animals.

Most important influential factors for the application of genetic engineering in farm animal breeding are social and ethical acceptance on the one hand, and R&D infrastructure on the other hand. The need for additional research activities seems to be highest for the elucidation of the principles underlying the developmental processes of oocytes and the possibilities of genetic engineering of early stages of embryo cells from farm animals (statement 66) compared to the other developments. As a consequence, the infrastructure in R&D is considered to be the most limiting factor, even in countries like Germany and the Netherlands with comparatively favourable infrastructure. Broadening the knowledge in the area of embryonic cells is crucial, because full benefits of cloning will only be achieved in combination with this technique.

In Germany and the Netherlands the public debate on the use of genetic engineering in agriculture has a rather long history compared to the Mediterranean countries. Apart from acceptance which is mentioned as most influential factor, emphasis is given to the marketing of products deriving from biotechnological production methods as well as to political activities in the Central European countries. Obviously, the implementation of these production methods is not primarily limited by scientific and technical constraints, but by a missing public consensus often giving more weight to risks and ethical concerns than to the respective benefits.

In the Mediterranean countries acceptance as limiting factor is inferior. Biotechnical reproduction techniques and their application are considered very positively, especially in Spain. But it has to be recalled that basic conditions for these techniques like infrastructure in R&D as well as technology transfer are considered to be the most limiting factors. For this reason, the question arises whether the realization of these breeding techniques is possible in the relatively short period of time assessed by the experts in the survey, if the respective knowledge and techniques are developed domestically.

Summary and conclusions

The present situation is characterized by the more or less isolated application of different reproduction, cloning and genetic engineering techniques in livestock breeding and production. Each technique has its unique technical drawbacks which prevent a broad application from the purely technical point of view alone. In addition to ethical concerns, the application as major routine techniques in the commercial area can lead to reduction of genetic variability in farm animal populations if few clones only would be spread on the production level in future.

However, with the cloning of the sheep "Dolly", it has become obvious even to the non-specialist that combinations of the different techniques can give a considerable impetus to the whole field. Indicators for the expected synergies from a combination of different techniques are recent commercial alliances between companies with cloning know-how and companies with transgenic animal know-how. The first transgenic lamb and calves obtained by nuclear transfer cloning technique have already been born (Schnieke et al. 1997, Anonymus 1998), expressing the genes for the blood clotting protein factor IX and human serum albumin, respectively.

Pronounced differences in attitude towards biotechnological applications in animal reproduction and breeding are observed between the participating countries in the survey. Whereas in Germany and the Netherlands technical constraints are comparatively low, the discussion on the possibilities, the risks and the benefits of genetic engineering in general and the use of biotechnological reproduction techniques has left the academic circles and has been carried out in public. The public discussions led to differentiated attitudes towards the use of biotechniques and reservations, especially in animal reproduction. Researchers and politicians will have to concern themselves with this attitude as acceptance is regarded as the most limiting factor for the use of the new techniques.

In Greece the attitude towards biotechnological reproduction techniques is determined by traditional values. Especially cloning is rejected, which might partly be due to detailed discussions in the public after the successful cloning of sheep "Dolly" in February 1997. In contrast to Greece, the attitude towards biotechniques in reproduction of farm animals in Spain is rather enthusiastic and does not differ from attitudes towards the scientific and technical development in plants or food. Therefore, insufficient infrastructure and technology transfer procedures are regarded to have highest influence on the future development of biotechniques in animal reproduction. The results of the Delphi survey give no clear indication whether with increasing use of these techniques the public debate will change dramatically towards a more critical assessment of these techniques in Spain or not. In analogy to the former public discussions in the Central European countries, the conclusion can be drawn as well that public attitudes might switch to a more critical view, as has been the case in Germany or the Netherlands.

Acceptance turned out to be a key issue in the debate on scientific and technical development in animal reproduction. Beside other ethical aspects concerning the protection of animals, it is mainly feared that results obtained in animals will be transferred to humans. Therefore, the neutral, objective and profound information of the public and additionally, a further intensive discussion on the ethical aspects and implications of the biotechnological methods are indispensable requirements. If certain subfields of animal biotechnology turn out not to be tolerated in society, it should be checked whether legislation should be amended accordingly. This is of particular importance for commercial applications as well: biotechnology firms need a clear legal framework to invest capital in this field and to establish their business activities.

5.6.5   Diagnostics and analytical methods

Introduction

The demand for analytical and diagnostic methods in the Agro-Food sector is considerable, in a large variety of different applications. Table 5.71 gives an overview of these applications and their coverage in the Delphi survey. Standard tools for analytics and diagnosis are physical and chemical methods. However, new biotechnological methods for analytics and diagnosis have been developed in recent years which are now being added to the toolbox, which sometimes even replace conventional methods, or make completely new applications possible. These new biotechnological methods are:

- *Enzymatic assays.* Enzymatic assays have been used for decades in the food industry for the quantification of major and minor food constituents. Usually, co-enzyme-dependent enzymatic reactions are employed for the spectrophotometric quantification of the analyte (Bergmeyer 1977). More recent applications are assays based on bioluminescence which are used for the detection of invisible dirt and stains on surfaces, e. g. of food-processing equipment in order to control the efficiency of cleaning and to keep them in a hygienic state.

- *Biosensors.* Biosensors are sensors which incorporate a biocomponent, e. g. an enzyme, an antibody, microbial cells, receptors or a nucleic acid probe. Most biosensors have been developed for clinical applications, but many of them could also be applied in the food sector. Only few biosensors have been commercialized and none is exclusively dedicated to the food industry. Most biosensors for food applications that have been reported in the literature have been developed for the determination of carbohydrates, mostly glucose. However, there is a large gap between academic biosensor developments and practical application: many promising biosensors fail miserably during their first contacts with food samples, largely due to the high complexity of the samples (Luong et al. 1997).

- *Immunoassays.* Among the outstanding achievements of modern biotechnology have been monoclonal antibodies (Milstein 1980). Today, monoclonal antibodies are the functional part of a large variety of immunoassays, the majority being so-called ELISAs[15]. Immunoassays are based on the concept of linking antibodies to enzymes or other compounds that give a detectable signal. The antibodies then bind very specifically to the analyte, and this binding is detected by the coupled tag (Self and Cook 1996, Wrotnowski 1997). In the Agro-Food sector, immunoassays are commercially available which detect e. g. pesticides or dangerous food pathogens (Larkin 1996).

---

[15] ELISA: "enzyme-linked immunosorbent assay".

- *Nucleic acid based assays.* Nucleic acid based assays are used to detect certain DNA sequence information in the sample. These types of assays are important to all industries and applications which rely upon, utilize, or modify biological organisms (Abramowitz 1996). In the Agro-Food sector, important applications are e. g. in plant and animal breeding, in monitoring of infectious agents in crop production, animal husbandry and food processing, in identity control of production strains and starter cultures, in controlling the pedigree of farm animals, in controlling the origin of raw materials and components of processed food (e. g. whether a pork meal really contains meat from pork, whether raw materials from genetically engineered organisms have been processed), in the assessment of biodiversity, and in monitoring the fate of genetically engineered organisms deliberately released into the environment. The importance of nucleic acid based assays is thought to increase significantly in the coming years, because enormous amounts of DNA sequence information are expected to become available from genome sequencing efforts currently underway. Currently, technologies are being developed which will make automated, easy-to-use, inexpensive DNA assays widely available (Abramowitz 1996).

The Delphi survey does not differentiate between the different types of assays and sensors; in all relevant statements they are addressed as "test systems based on modern biotechnology". This is reasonable because it depends on the specific analytic or diagnostic problem that has to be solved whether one of these methods has comparative advantages over another. The Delphi experts' answers in the field of analytics and diagnostics will be assessed with respect to two questions:

- Which role will biotechnical analytical and diagnostic methods play in the Agro-Food sector in future?

- Does the use of genetic engineering in the Agro-Food sector create a demand for additional analytical and diagnostic examinations?

Table 5.71:   Overview of applications of analytical and diagnostic methods based on modern biotechnology in the Agro-Food sector, and their coverage in the Delphi survey

Applications of analytical and diagnostic methods based on modern biotechnology		Covered by statement in the Delphi survey
Plant and animal breeding	Total sequencing of the entire genome	The genomes of most economically important bacteria used in food production are completely sequenced (statement 44)
	Identification of quantitative trait loci and their mapping	
	Identification of carriers of desired traits for further breeding on the basis of molecular markers	Genetic maps and molecular test systems are practically used for the identification of economically important traits and marker-assisted breeding of fish (statement 68)
	Identification of undesired (recessive, heterozygous) gene variants	
	Identity and pedigree controls in livestock breeding	
	Control of the stable inheritance of genes newly introduced by genetic engineering	
Breeding/ ecosystem	Monitoring and assessment of biodiversity with the aim of its maintenance	
Environment	Monitoring of the fate of genetically engineered organisms and their genes after deliberate release into the environment	New monitoring and control techniques for genetically engineered plants in open fields have significantly improved the reliability of risk assessment (statement 57).
Plant and animal production	Diagnosis of diseases, identification of pathogens	
	Monitoring of pathogens, e. g. monitoring the effectiveness of disease treatment, monitoring the effectiveness of hygienic conditions, safe guarding the absence of pathogens (e. g. in milk, meat).	Rapid test systems based on modern biotechnology are widely used for pathogen identification in plant production and animal husbandry (statement 62).
	Determination of chemical contaminants (e. g. pesticides, heavy metals) in soil, groundwater	

Applications of analytical and diagnostic methods based on modern biotechnology		Covered by statement in the Delphi survey
	Determination of biological contaminants in feed and products (e. g. mycotoxins in feed, antibiotics in milk and meat)	
Food processing	Quality control: determination of the chemical composition of raw materials and products	
	Quality control: determination of the hygienic status of raw materials and products	The hygiene monitoring of food processing is significantly improved due to the widespread use of modern biotechnological analytical methods (statement 42).
	Quality control: assessment of the nutritional value and quality parameters of raw materials and products	Techniques based on modern biotechnology are practically used for on-line control of quality parameters in food processing (e. g. content of micronutrients or harmful substances) (statement 43).
	Quality control: control of the origin and identity of raw materials	
	Process control and regulation	
	Hygiene monitoring	
Food surveillance authorities	Compliance with regulations	The responsible authorities in [your country] practically use specific technical equipment and standardized methods to identify food and beverages which were produced with the help of genetic engineering (statement 11).
	Hygiene monitoring	

Present state of the art and future development

In the coming years the need for diagnosis and analytics will increase in the Agro-Food sector for several reasons:

- Genome-sequencing projects currently underway will yield enormous amounts of information which form the knowledge base for the development of new diagnostic and analytical possibilities.

- There is increased concern about food composition and safety (Luong et al. 1997). A major problem are food-borne diseases. Moreover, there is concern with respect to contaminants in food, e. g. hormones, antibiotics, pesticides or toxins and with respect to additives in food. In addition, there are new emerging diseases, e. g. the recent bovine spongioform encephalopathy (BSE) outbreak in the UK, the fear of adverse health impacts of new technologies such as genetic engineering, and the increase of food allergies.

- These concerns result in increased regulatory action and enhanced consumer awareness. Many food products require detailed labelling of their constituents. Consumers wish to be informed about the origin of their food and about the production and processing methods applied (e. g. genetic engineering). The compliance with regulations and standards has to be monitored by independent authorities from time to time.

Therefore, food industry and surveillance authorities need suitable methods for process, product and quality control which give reliable, quick, specific and cost-effective information about the physical, chemical and biological characteristics of food. It is estimated that the food industry spends, on average, 1.5 % to 2 % of the value of its total sales on quality control and appraisal (Luong et al. 1997). The global food safety testing business is estimated at about 1 bill. US$ per year (Larkin 1996).

How are these trends reflected in the Delphi experts' answers? Until 1998, 20 prokaryotic and one eukaryotic[16] genomes have been completely sequenced and the sequencing of additional 45 prokaryotic and 17 eukaryotic genomes is underway (Magpie genome sequencing project list 1998). However, most of the organisms are model organisms or pathogens. At present, there are only very few genome sequencing projects on microorganisms of economic importance for the food industry: among them are baker's yeast (*Saccharomyces cerevisiae*), *Schizosaccharomyces pombe*, *Clostridium acetobutylicum*, *Bacillus subtilis*, *Escherichia coli*, *Aspergillus nidulans*, and several viruses or phages which are pathogenic for plants, starter cultures or animals. The German-based company Degussa has announced the commencement of a *Corynebacterium glutamicum*

---

16 Genome of baker's yeast, *Saccharomyces* cerevisiae.

genome sequencing project (Hodgson 1998). The majority of Delphi experts estimates that it will take up to ten years to totally sequence the genomes of most economically important bacteria used in food production (statement 44), a substantial share of approximately 15 % thinks that it will take even longer.

The largest benefits from total genome sequencing are expected for knowledge creation in science and technology – in all five countries this is one of four statements which rank highest with respect to knowledge creation. However, neither economic nor environmental importance is attributed to it. The response behaviour with respect to influential factors shows that the Delphi experts see the genome sequencing efforts as mainly basic research which require an appropriate R&D infrastructure, adequate funding, trained personnel and should be performed in international collaboration. Nevertheless, the comments given by the Delphi experts also echo a certain scepticsism which has often been expressed with respect to all genome sequencing efforts: whether the foreseeable results are worth the effort and that the sequence information alone is not useful unless it is correlated with gene function and phenotypic traits (functional genomics).

On the other hand, the practical use of genetic maps and molecular test systems for the identification of economically important traits and marker-assisted breeding of fish (statement 68) is – according to the Delphi experts – a little less important for knowledge creation in science and technology, but a bit more important for the competitiveness of the economy than whole genome sequencing and will be realized a few years later. The most important influential factors are the same as for the whole genome sequencing. Moreover, lack of acceptance is also mentioned, especially by German and Dutch experts, who refer to the social rejection of transgenic animals. However, the use of genetic maps and marker-assisted breeding are used as additional tools in conventional breeding so that animals bred in this way are not transgenic and cannot be distinguished from animals bred without these tools. For other animals than fish, several diagnostic tests based on genetic maps and molecular markers are already used in practical breeding: among them is a test to identify and exclude pigs from breeding which are susceptible to malignant hyperthermia. This syndrome results in stress sensitivity of the animals and a poor pork quality (PSE[17]) (Fuji et al. 1991, Otsu et al. 1991). Moreover, gene defects resulting in the bovine leucocyte adhesion deficiency (BLAD), a lethal hereditary disease in Friesian cattle or the Weaver syndrome, another bovine hereditary disease, can be detected by molecular test systems (Shuster et al. 1992, Georges et al. 1993). In order to breed cows with favourable milk protein composition, molecular methods are applied in the screening for bulls with desired milk protein variant genes (Lindersson et al. 1995).

---

[17] PSE: pale, soft, exudative meat

Despite the availability of veterinary pharmaceuticals and pesticides, animal and plant diseases still cause considerable yield losses in agriculture or reduce the product quality. Food-borne infectious diseases are a major health concern in industrialized countries: according to the World Health Organization, each year 130 mio. Europeans (15 % of the total population of the WHO European Region) are affected by episodes of food-borne diseases, ranging from mild gastrointestinal infections to severe gastroenteritis or even death (WHO, European Centre for Environment and Health 1995). The implications of this high incidence in terms of population health as well as direct and indirect costs for the social system are obvious (Schmidt 1995). Therefore, it is important to introduce farming and food-processing practices which help to reduce the direct and indirect detrimental effect of plant, animal and human pathogens.

The statements 62 and 42 address the question whether diagnostic methods based on modern biotechnology can contribute to solving these problems in plant production and animal husbandry (statement 62) and in food processing (statement 42). Statement 44 addresses the use of modern biotechnology for the on-line determination of quality parameters of food. All three statements are assessed in a very similar way by the Delphi experts – there are hardly any significant differences in the response behaviour between the three statements. These developments are among the top ten statements which receive the highest degree of approval by the Delphi experts, who express a very positive attitude. According to the majority of experts, all three developments will be realized within the coming five years. This estimation is certainly due to the fact that a few applications are already commercialized: for example, immunoassays are commercially available which detect e. g. pesticides and organic chemicals in soils, water and food products. In addition, immunoassays are on the market which detect dangerous food pathogens such as *Salmonella*, *Listeria*, *Escherichia coli* O157:H7 and *Campylobacter* in raw materials and processed food (Larkin 1996).

However, there are a number of hindrances for the broader use of analytical and diagnostic methods based on modern biotechnology: the Delphi experts point out deficits in infrastructure, technology transfer, industrial innovativeness, and a lack of skilled personnel, especially in Spain, Italy and Greece, inadequate funding especially in Germany and the Netherlands and assess the regulatory situation in the food industry as unfavourable. Because the principle of analytic methods based on modern biotechnology is different from conventional microbiological, physical or chemical methods, personnel will have to be trained to use the new tests – obviously a deficit, especially in the Mediterranean countries. In general, new tests are primarily developed and applied in specialized laboratories and there is a lack of knowledge transfer and diffusion to possible users. This is especially true in agriculture, but also in the food sector. Moreover, because no widely applicable standards have been defined for routine analyses (e. g. for mould detection in a wide variety of foods), the validation of a new test system is needed in each specific situation. This requires R&D investments which are frequently not affordable for smaller companies. However, it has to be mentioned that this also provides a unique market niche for small diagnostics companies which offer custom-made test systems for special analytical problems.

Within a quality control concept in the food industry it is necessary to validate the hygienic and aseptic aspects of the process equipment, the process components and the production systems. Requirements and test procedures are standardized by the European Hygienic Equipment Design Group (EHEDG) and the European Committee for Standardization (CEN). However, many of the standardized methods are conventional ones, thus providing unfavourable conditions for newly developed ones based on modern biotechnology.

These deficits and hindrances also point to R&D needs which are being addressed in order to make the tests more user-friendly and to allow their broad application:

- Develop test systems simple to use for customers on-site as well as in the laboratory

- Develop low-cost, labour-saving tests which can also be used by less specialized personnel

- To provide not only rapid methods for qualitative screening and monitoring tests, but also develop standardized, validated quantitative, confirmatory methods

There is considerable concern about risks associated with the deliberate release of transgenic organisms into the environment and about the question whether compliance with labelling regulations for Novel Food can effectively be controlled. In both cases analytical methods, ironically based on modern biotechnology, can be applied for control and monitoring purposes. The significant improvement of the reliability of risk assessment by new monitoring and control techniques for genetically engineered plants in open fields (statement 57) is seen as one of the most important developments in the Delphi survey for knowledge creation in science and technology, as well as for the protection of the environment/sustainable development. According to the Delphi experts, this is not important for the competitiveness of the economy, which may reflect the notion that the commercialization of transgenic crops has not been significantly hindered by the risk assessment and that the risk assessment and its improvement is not a private, but a public task. The latter is also supported by the fact that the Delphi experts see major barriers to the improvement of the reliability of risk assessment in an insufficient infrastructure, a lack of qualified personnel for this highly interdisciplinary task and a lack of funding. In Spain, Italy and Greece experts perceive also deficits in technology transfer. That this issue is a very controversial one where dogmatic positions clash can be seen from the comments: reactions to this statement range from "the statement implies that risk assessment is unsafe and this is definitely not true" to "the possible damages associated with the release of transgenic plants are so large that the risk must not be taken; any improvement in the methodology cannot prevent the imminent catastrophe". However, a moderate position prevails among the Delphi experts who see an urgent need for improving the knowledge base for risk assessment. Moreover, this is perceived as an important prerequisite for social-ethical acceptance.

After years of intensive and controversial discussion the "Novel Food Regulation" came into force in May 1997 – shortly after the completion of the second round of this Delphi survey. According to the "Novel Food Regulation", labelling is compulsory for food which is no longer equivalent to its "conventional" counterpart (see chapter 5.2). With respect to food produced with the help of genetic engineering, however, the "Novel Food Regulation" does not specify by which methods and tests the genetic modification has to be analysed and whether the analysis has to be performed on the level of the individual food ingredient or the final product. Guidelines which specify which data and information are required and by which analytical tests these data should be generated still have to be developed.

Will the responsible authorities in the respective countries practically use specific technical equipment and standardized methods to identify food and beverages which were produced with the help of genetic engineering (statement 11)? The large majority of the Delphi experts welcomes this development and sees its realization within the next five to ten years. This is rather late, given the fact that several methods[18] have already been developed, standardized and validated and are available to surveillance authorities, at least on a national level (see e. g. Engel et al. 1995, Zagon et al. 1998). In addition, several small service companies already offer such tests on a commercial basis. However, up to one fifth of the Dutch and German experts think that statement 11 will not be realized at all. They argue that the detection of all possible genetic modifications will neither be scientifically nor practically feasible: only known genetic modifications can be detected and genetic modifications may be detectable in some foods (e. g. soya protein), but not in others (e. g. soya oil). Others are convinced that the detection of genetic modifications will be scientifically feasible, but they see practical hindrances: qualified personnel is lacking in all countries except the Netherlands, where deficits in infrastructure are perceived, the costs are thought to be prohibitive and a lack of funding is pointed out and, above all, a lack of political will to press for new methods and for the development and the implementation of uniform standards on regional, national and international level.

---

18 Among these methods are the detection of defined genetic modifications in potatoes, in *Lactobacillus curvatus* and *Streptococcus thermophilus* from sausage and yoghurt, and in glyphosate-resistant soy beans.

Summary and conclusions

There is a considerable need for diagnostics and testing in the Agro-Food sector. It is estimated that the food industry alone spends, on average, 1.5 % to 2 % of the value of its total sales on quality control and appraisal (Luong et al. 1997). The global food safety testing business is estimated at about 1 bill. US$ per year (Larkin 1996). Not yet included in this estimation are assays which are performed in the Agro sector, e. g. in plant and animal breeding, in health monitoring of crops and livestock and in environmental monitoring.

Modern biotechnology adds new tools to the analysts' toolbox: enzymatic assays, biosensors, immunoassays and nucleic acid based assays which have comparative advantages and disadvantages over existing microbiological, chemical and physical methods. Among the scientific-technological developments addressed in this Delphi survey, analytical and diagnostic methods based on modern biotechnology are among those which are perceived as generally favourable and raise the fewest concerns.

Among the most powerful methods are nucleic acid based assays which are at present at the threshold of revolutionizing analytics and diagnostics in human health care, because inexpensive, automated, easy-to-use DNA chips and related technologies are very close to the market (Hoheisel 1997). This principle could also be transferred to the Agro-Food sector, and recent strategic decisions from multinational life science companies such as Monsanto and Novartis which invest heavily in DNA-based technologies in the Agro-Food sector show that these multinationals have already set the points in this direction.

However, the Delphi survey indicates that the European agriculture and food sector may be ill-prepared for these future developments. Most statements relating to the development and use of analytical and diagnostic tools based on modern biotechnology show that personnel is not sufficiently trained to use and adapt the tests, that there is a lack of transfer and diffusion from specialized laboratories to the "ordinary user" and that the regulatory environment is unfavourable for new bioassays in the food sector. Moreover, research and development is required to develop test systems which are simple to use on-site as well as in a laboratory, which are inexpensive, labour-saving and which can be used by less specialized personnel. Moreover, efforts should be devoted to the goal not only to provide rapid methods for qualitative screening but also to develop quantitative, confirmatory methods.

5.6.6 Summary and conclusions

At present, there are relatively few products produced with the help of modern biotechnology in the European Agro-Food marketplace or in broader practical use in the EU Agro-Food sector. Examples are several food enzymes produced by genetically engineered microorganisms, such as amylases or chymosin for cheese production, transgenic maize, soybean or rapeseed with altered monogenic traits, (e. g. tolerance to non-selective herbicides, resistance to certain pests or altered fruit or seed composition, changed ripening characteristics), as well as several analytical and diagnostic tests (e. g. for pathogen detection in food, for detection of recessive hereditary diseases in livestock breeding). There are no transgenic animals bred for food purposes yet. However, modern biotechnology plays a role in animal husbandry and breeding, e. g. in the production of feed additives or veterinary pharmaceuticals, in reproduction and in diagnostics.

The results of the Delphi survey confirm that the number of modern biotechnology-derived products and methods is likely to increase significantly within the next ten years, due to scientific-technological advances and due to the synergetic combination of different approaches, methods and technologies:

- *Enzymes.* Market-leading companies producing food and feed enzymes have made strategic decisions in recent years to use genetic engineering as a core technology, together with efficient and innovative screening procedures for new enzymes, rational and irrational protein design, fermentation and downstream processing, thus reengineering their R&D activities substantially. Investing in these core technologies and exploiting synergies between them makes it possible for the enzyme producers to drastically shorten the time to market for new enzymes and to make a much larger variety of different enzymes commercially available than before.

- *Plants.* Large multinational "life science companies" such as Monsanto and Novartis invest heavily in "green biotechnology". Their strategic goal is to transfer the genomics approach from the pharmaceutical sector to the Agro-Food sector. Relevant scientific information stems e. g. from genome sequencing projects of plant genomes (e. g. the model plant *Arabidopsis*) and from research elucidating plant-pathogen interaction and plant development and metabolism on the molecular level. Methods still have to be refined in order to allow the engineering of polygenic traits in crop plants.

- *Animals.* At present, artificial insemination, embryo transfer, in vitro fertilization and cloning by embryo splitting are established reproduction techniques which are practically used in animal breeding and animal husbandry. Genetic engineering of farm animals and cloning by nuclear transfer is mainly in the research stage and, if applied, is predominantly targeted at the pharmaceutical sector. In the coming decade it will most likely become possible to combine these different approaches, thus overcoming present technical difficulties and exploiting synergies between the individual technologies.

- *Analysis and diagnostics based on modern biotechnology.* Research efforts in elucidating the molecular and genetic basis of economically important characteristics of food, crop plants, livestock and food-relevant microorganisms will provide a multitude of information in the coming decade, which can be used for new analytic and diagnostic tests in the Agro-Food sector. Moreover, technologies such as DNA chips, presently at the threshold of revolutionizing analytics and diagnostics in human health care, can also be transferred to the Agro-Food sector. This may result in the broad application of inexpensive, easy-to-use and automated assays in the Agro-Food sector.

Although scientific-technological advances are an absolute requirement, other factors also play a significant role in the future introduction of products and methods based on modern biotechnology in the Agro-Food sector.

Both for enzymes and for analytic and diagnostic methods based on modern biotechnology the most important influential factor is a lack of industrial innovativeness on the side of the users of these enzymes and bio-assays. In the European food industry small and medium-sized companies predominate. Their economic situation is not very satisfactory, because the food market is more or less stagnant and growth in one segment is compensated by decreases in other segments. Competition among the food companies is considerable, not least due to increased pressure from large retail companies. In this environment, small and medium-sized companies can maintain their competitiveness only if they specialize or if they use up-to-date production processes and market innovative products. It is a drawback for the latter strategy that the research intensity of the food industry is below industrial average. Innovations and the introduction of new technologies are often realized by purchasing new equipment, rather than by own R&D activities. In the case of enzymes and bioassays, technological competence in core technologies remains mainly confined to the supply industry or research institutions and is not broadly available in the food companies themselves. So there is a certain technological dependency of small food companies on the supply industry or R&D institutions. The required expertise can only be acquired via cooperation and participation in appropriate networks.

Therefore, the further diffusion of assays based on modern biotechnology in the Agro-Food sector will depend significantly on whether personnel can be qualified, especially in the Mediterranean countries, whether technology transfer between "assay developers" and "assay users" can be improved, whether additional funding can be made available, especially in Germany and the Netherlands and whether the regulatory environment can be made more favourable for assays based on modern biotechnology in the food area. For enzymes produced with the help of genetically engineered organisms, the future prospects are different: their use will most likely increase if the enzymes enable the food company to reduce costs significantly or to develop innovative products with high profit margins. This might require intensive customer counselling by the enzyme supply industry. The results of the Delphi survey indicate as well that the cost-saving potentials of enzymes if employed in production-integrated environmental protection are not yet widely known in the food industry. Therefore, there seems to be a need for education and for the dissemination of information on companies which have successfully integrated enzymes in their production process.

Moreover, the dependency of food companies on retailers is increasing, due to a concentration process among the retailers. The use of enzymes produced with the help of genetically modified organisms is disapproved of by consumers and retailers, although to a lower extent than genetically engineered crops or animals. Therefore, food companies which technologically depend on "pro-genetic engineering" enzyme suppliers and economically depend on "contra-genetic engineering" retailers and consumers find themselves in a difficult situation. At present, many food processors have not yet defined their own position in this conflict and adopt a wait-and-see strategy. Therefore, it is necessary to act very sensibly and to readily provide information on how the applied enzymes have been produced.

There is considerable uncertainty about how long it will take to reach the goal of genetically engineered polygenic traits in plants – the time scale of realization ranges from a few years to two decades. This uncertainty might stem from significant deficits in the R&D infrastructure, which are uniformly pointed out by experts from all countries. This is certainly true if one compares the Agro-Food sector and the pharmaceutical sector. In the last years, a whole new industry with a multitude of strategic alliances between pharmaceutical companies, biotech companies and research institutions has evolved in the field of genomics. Similar strategies are pursued by large multinationals such as Monsanto and Novartis in the Agro-Food sector, but relevant R&D projects and cooperations are still at a relatively infant stage, especially in the publicly funded area. Against this background, improving the infrastructure and providing additional funding for broadening the knowledge of the molecular basis of economically important crop plant traits are required.

On the other hand, public resistance to genetically engineered crop plants and food derived from them is considerable, at least in the Central European countries, as can be seen e. g. from the opposition to field trials with transgenic plants. Moreover, the delay in putting into force clear and practicable labelling regulations for food produced from genetically engineered crops does not conform with consumers' needs and does not at all contribute to transparency. It seems unlikely that scientific-technological development can be pushed much further and at a quick pace if the public's concerns are not taken seriously. This requires among other things the improvement of our knowledge base for risk assessments, careful risk assessments of transgenic plants prior to their commercialization, sound monitoring programmes for rare adverse or long-term effects after their commercialization and the integration of transgenic plants into broader concepts, such as e. g. sound pest management strategies.

A significant bottleneck in the genetic engineering of livestock seems to be a lack of understanding of the early stages of embryonic development. Full benefits of cloning will only be achieved in combination with genetic engineering. If the commercial application of these techniques in farm animal breeding and reproduction is desired, support should be given to an advancement of R&D infrastructure in this field.

However, ethics is a key issue in the debate, on which scientific and technical developments in animal biotechnology are socially required and acceptable. Ethical aspects concern animal welfare, but it is even more feared that results obtained in animals will one day be transferred to humans. Therefore, the full and objective information of the public and an intensive discourse on the ethical aspects and the implications of the biotechnological methods for animals as well as for man is an indispensable requirement. If certain subfields of animal biotechnology turn out not to be tolerated in society, it should be checked whether legislation should be amended respectively. This is also of importance for commercial reasons: biotechnology firms need a clear legal framework to invest capital in this field and to establishing their business activities.

6.   Conclusions and recommendations

In the following chapter the most important findings of the project will be combined according to the thematic fields covered in the Delphi survey. This relates to future scientific and technical developments, their impacts on the health of consumers and the environment, the future acceptance of Agro-Food biotechnology, and regulatory aspects as well as economic implications. Finally, the suitability of the applied methodology is discussed. Based on these results, recommendations on how to use the obtained results for politics, industry and other social actors are worked out.

6.1   Which scientific and technical developments can be expected in future?

According to the answers of the experts the following fields of Agro-Food biotechnology can be regarded as being of specific interest from a scientific and technical point of view in future:

- Analytical approaches based on modern biotechnology

- Use of optimized enzymes in the food industry

- Genetic engineering of polygenic traits in plants

- Biotechnological reproduction techniques and genetic engineering of animals

Analytical approaches based on modern biotechnology

Modern biotechnology offers enzymatic assays, biosensors, immunoassays and nucleic acid based assays for diagnostic and analytic purposes in the Agro-Food sector. These methods can be employed in quality control and process control and regulation in the food industry, in plant and animal breeding and production, in monitoring of biodiversity in plant and animal breeding, in health monitoring of crops and livestock and in environmental monitoring, to name but a few possible application fields.

Analytical and diagnostic methods based on modern biotechnology are among those scientific-technological developments addressed in this Delphi survey which are perceived as generally favourable and raise the fewest concerns. However, there is considerable need for R&D in order to make the available test systems simple to use on-site as well as in a laboratory, inexpensive, labour-saving and usable by less specialized personnel. Moreover, efforts should be devoted to the goal not only to provide rapid methods for qualitative screening but also to develop quantitative, confirmatory methods.

Hindrances for the development and use of analytical and diagnostic tools based on modern biotechnology in the Agro-Food sector lie in personnel not sufficiently trained to use and adapt the tests, in a lack of transfer and diffusion from specialized laboratories to the "ordinary user", and in the regulatory environment, which is unfavourable for new bioassays in the food sector.

Use of optimized enzymes in the food industry

Traditionally, the application of enzymes plays an important role in this industry branch. However, the use of isolated enzyme preparations is a relatively new issue. In recent years these enzyme preparations have been produced by classical biotechnology (fermentation technology). Currently, new technological trends revolutionize this process with respect to speeding up the discovery/R&D process, broadening the scope of available enzymes as well as engineering enzyme properties. These trends refer to techniques and approaches such as genetic engineering, screening of new enzyme sources (e. g. extremophiles, non-culturable organisms), high throughput screening, combinatorial approaches (irrational design), protein engineering (rational design), cost-efficient bulk enzyme fermentation and cost-efficient downstream processing.

The use of optimized enzymes in the food industry has been identified as a key development for the future competitiveness of the European food industry by the experts. However, only few food enzymes produced by genetically engineered microorganisms are commercialized at present. This process has progressed much slower than would have been technically feasible. The main reasons for this development are reluctance of food-processing companies as well as lack of consumer acceptance. As the results of the Delphi survey indicate, this lack of acceptance differs significantly between the countries involved. While in the Mediterranean countries a positive attitude and rather short-term time expectations arise for the use of recombinant enzymes in the food industry in general, as well as the application of protein engineering for its optimization in particular, the German and the Dutch experts express some hesitations, especially related to the latter aspect. Nevertheless, in all countries a short-term use of protein-engineered enzymes in specific sectors of the food industry is expected. This estimation of the experts seems to becoming true, since the first food enzymes optimized by protein engineering are announced to be introduced into the market by Novo Nordisk in 1998.

The future innovativeness related to food enzymes greatly depends on the know-how in the above mentioned platform-technologies and in the close interaction/exploitation of synergies among the different technologies. In most cases the know-how in the platform-technologies has been built up in the supply industry, not in the food-processing industry itself. Therefore, the food-processing industry depends on innovations developed in other industry branches. In this context, it is crucial to keep in mind that often SMEs in the food industry are confronted with large multinational suppliers. In addition, SMEs in the food-processing industry are often not yet aware of the benefits which could arise from the use of enzymes produced with the help of genetically modified organisms (e. g. more cost-efficient production, production-integrated environmental protection).

Although the food-processing industry is reluctant to apply food enzymes produced by genetically engineered organisms, most food enzymes will be produced this way in the near to mid-term future, as the results of the Delphi survey indicate. Taking into account the differing interests of the involved actors, the food-processing industry is in a dilemma since mostly economic reasons support the use of this type of enzymes, while on the other hand often the request for "no genetic engineering" is expressed on the demand side (e. g. retailers, consumers).

Modification of polygenic traits in plants

The overwhelming majority of transgenic crop plants which are grown worldwide have been genetically engineered in traits which are conferred by the action of single genes. However, most economically important traits in plants, such as yield, stress tolerance or quality parameters, are encoded by several genes which closely interact. Whether genetic engineering will play a significant role in future crop plant breeding therefore heavily depends on whether the engineering of polygenic traits will become possible.

According to the Delphi experts this is an ambitious, both scientifically and technically challenging goal, which requires basic research, highly original approaches and technological breakthroughs. There is considerable uncertainty how long it will take to reach this goal – the time scale of realization ranges from a few years to two decades. This uncertainty might stem from significant deficits in the R&D infrastructure which are uniformly pointed out by experts from all countries. This is certainly true if one compares the Agro-Food sector and the pharmaceutical sector. In the last years, the new field of genomics has emerged in the pharmaceutical sector, aiming at exploiting the information from genomics programmes for the development of innovative or improved diagnostics and pharmaceuticals. As a consequence, a whole new industry with a multitude of strategic alliances between pharmaceutical companies, biotech companies and research institutions has evolved. The approach of genomics is also applicable to the Agro-Food sector, aiming at the development of diagnostics, agro-chemicals and plant and animal breeding. However, relevant R&D projects and cooperations are still at an infant stage. Against this background, improving the infrastructure and providing additional funding for broadening our knowledge of the molecular basis of economically important crop plant traits are required.

Much more controversial than basic research concerning the molecular basis of economically important traits is the question, how to apply this knowledge in crop plant breeding. Support seems to depend on at least three factors: the purpose of the genetic engineering or the traits altered, respectively, the expert's personal terms of reference, and the frame conditions, under which the transgenic crops will be grown commercially. Transgenic crop plants will only display a potential to solving present problems in the Agro-Food sector if their use is embedded into broader concepts, such as sound pest management concepts or sustainable use of scarce water resources etc. Further elements of such concepts are careful risk assessments of transgenic plants prior to their commercialization, monitoring programmes for possible adverse, indirect or long-term effects (e. g. emergence of resistant pest strains) which cannot be detected during risk assessments and educational programmes (e. g. for the plant growers).

Biotechnological reproduction techniques and genetic engineering of animals

Another area of specific interest represents the use of modern biotechnological reproduction techniques as well as genetic engineering approaches in animal husbandry. Currently, artificial insemination, embryo transfer, in vitro fertilization and cloning by embryo splitting are established reproduction techniques which are used - with differing relevance depending on the species, country and technique - in animal breeding and animal husbandry. By contrast, genetic engineering of farm animals and cloning by nuclear transfer is still at the research stage. The full potential of these techniques can only be exploited if different approaches are combined (e. g. in vitro fertilization, genetic engineering, cloning and embryo transfer). However, this integrated approach will take at least ten years from now, and requires additional research activities as can be concluded from the results of the Delphi survey.

The widespread use of cloning techniques for the reproduction of farm animals is expected in the medium term by the majority of experts. On the other hand, clear concerns related to the ethical justification of this technique arise in the expert panels and underline the need for an overall consideration of the social and ethical justification of animal biotechnology. The majority of the experts in Germany, Italy and the Netherlands expect the practical use of genetic engineering approaches in breeding of farm animals within six to ten years or even later, while their colleagues in Spain and Greece expect this development earlier. Basic research activities targeted to the development of oocytes for genetic engineering of farm animals are regarded by the majority in all expert panels as a medium- to long-term development, mainly depending on enhanced research activities. The above mentioned developments are mainly influenced by the consideration of social and ethical acceptance issues in Germany and the Netherlands, while in the Mediterranean countries especially infrastructure in R&D and technology transfer are regarded as crucial. The main concerns expressed by the experts refer to animal welfare and the possibility that the methods and techniques used in animal biotechnology may be transferred to humans, resulting in the "production of human beings".

6.2 Impacts on health of consumers and the environment

In the public debate on Agro-Food biotechnology in the EU, potential risks of this technology related to the health of consumers as well as to the environment play a prominent role. In addition, the measures foreseen in official risk assessment activities as well as regulation procedures before marketing of gene food products mainly focus on these aspects.

Health effects of Agro-Food biotechnology

In the Delphi survey four main factors have been identified which are decisive for the assessment of the future health impacts of modern biotechnology. One factor relates to the development and marketing of food with specific positive health effects due to the use of modern biotechnology ("bio-enriched food"). The majority of the experts are in favour of this type of food and expect market introduction in six to ten years, mainly depending on demand-pull factors like market conditions, and acceptance of these products, as well as accompanying information activities. Some experts regard technology-push factors as important as well, because there are still several open scientific and technical questions in this area. A minority of the experts express some doubts concerning the proclaimed positive health impacts of these new products and expect a later realization or even regard "bio-enriched" food as an unrealistic scenario.

As could be expected, almost all experts strongly reject adverse health effects (e. g. additional allergies) due to the use of modern biotechnology on consumers, users and employees. Despite this negative personal attitude, a big majority of the experts expect that these effects will occur in future, even if there are differing estimations concerning the time of realization. A considerable proportion of the experts (in particular from the Netherlands) regard such a development as unrealistic, i. e. they seem to be optimistic that these adverse health affects would be avoided. Both groups mainly regard information of consumers and employees as well as regulation activities as key factors.

One possibility to avoid adverse health effects of modern biotechnology is the development of new monitoring and test systems which might also be based on biotechnology. The experts express strong expectations in these systems and expect a short- to medium-term realization, mainly depending on funding possibilities and technology-push factors (R&D infrastructure, technology transfer, industrial innovativeness).

Another key factor for preventing adverse health impacts of modern biotechnology are enhanced information activities as well as knowledge creation in this field. Information of consumers concerning possible health impacts of modern biotechnology, as well as the investigation of the long-term health effects, are highly appreciated by the majority of experts. Especially for the latter aspect a mid-term realization is expected, while information activities are expected to be realized in the coming five years. Despite their strong demand for epidemiological studies as one possibility to investigate the long-term health effects of modern biotechnology, the experts see some difficulties in realizing these surveys mainly due to the (at least partly) missing methodology, a broad range of variables to be included, as well the handling of the time aspect. In addition, several experts doubt that industry and partly public institutions are really interested in carrying out this type of studies.

All in all, it can be concluded that modern biotechnology is seen by the respondents as a factor of aggravation but at the same time as a possibility to reduce allergies. There are strong convictions and expectations, especially for the contribution of genetic engineering to prevent allergies. But at the same time, there are clear concerns about harmful health effects of the same technology. In total, the asked experts express higher expectations in technical solutions (e. g. the development of new monitoring and test systems) than in organizational approaches (like enhanced information activities) in order to reduce or prevent such adverse health effects.

Environmental impacts of Agro-Food biotechnology

The experts in the Delphi survey express an ambivalent opinion concerning the future environmental impacts of Agro-Food biotechnology. On the one hand, they regard modern biotechnology approaches as one opportunity to reduce environmental burdens, while on the other hand they see the possibility of additional, possibly long-lasting environmental damage from the use of these techniques.

The use of modern biotechnology to reduce organic waste in animal husbandry as well as its conversion into marketable products is strongly appreciated by the expert panels and regarded as highly beneficial both for the environment and the national economy. This also relates to the development of specific enzyme systems to improve the environmental performance of conventional food-processing procedures. However, this rather optimistic experts' view is counterbalanced by the estimation that a complex set of technically and demand-/market oriented activities are required to realize these developments in the medium to long term.

In addition, the experts express rather high expectations in different scientific developments in plant production, of which they expect direct or indirect positive impacts on the environmental situation. This relates in particular to the development of salt- or drought-resistant, transgenic plant varieties, the elucidation of the principles of virus resistance mechanisms in perennial plants, the use of biotechnological approaches in biological pest control, as well as an increased cultivation of optimized renewable resources. Most of these developments are regarded to be realized in the medium to long term, indicating the complex character of these innovations.

In contrast, all expert panels highly reject negative environmental impacts of modern biotechnology like horizontal gene transfer, negative impacts on biodiversity or the loss of traditionally used varieties and organisms. Despite this negative personal attitude, a majority of the experts expects that these effects will occur in future, although there are differing estimations concerning the time of realization. However, a considerable proportion of the experts (in particular from the Netherlands) regards such developments as unrealistic. Mainly political and regulation activities, public acceptance and information measures are seen as key factors to influence potential negative environmental impacts due to modern biotechnology in future.

In addition to the above mentioned fields with rather clear estimations of the experts, there are highly disputed areas. These relate in particular to the environmental impacts of herbicide-resistant plants, the effects of modern biotechnology on biodiversity in agriculturally influenced ecosystems, as well as the use of specific genetic engineering approaches in organic farming. In the first two areas there are controversial opinions whether modern biotechnology will positively or negatively contribute to the environmental situation while in the latter field the debate focuses on the aspect whether the fundamental rules of organic farming are compatible with this kind of techniques. Especially in Germany and the Netherlands at least some of the experts do not see any possibility that the controversy in these fields will be settled in the coming years.

In recent years, environmental impacts have been considered one of the most important risks of modern biotechnology applications in the Agro-Food sector, resulting in several regulations on national and international level focused on this area. In contrast to this opinion, but in line with other actual surveys, the experts asked in the Delphi survey give a relatively low importance to "ecology" as an influential factor. This rather surprising result might be interpreted in a sense that most experts argue in favour of a pragmatic handling of this issue, trying to avoid negative impacts on the environment due to the use of modern biotechnology instead of an "ideologically" coloured discussion. With respect to the future design of biotechnology regulations, these results point towards a possible shift in the underlying policies: a need for a strong focus on ethical concerns as driving forces for the design of regulatory frameworks might arise in addition to the present risk-driven approach.

6.3 Future acceptance of Agro-Food biotechnology

The results of the Delphi survey show that the degree of acceptance of Agro-Food biotechnology varies considerably from one country to another. This is an indication that cultural factors play a major role in shaping the personal attitude towards modern biotechnology. In line with the results of the three Eurobarometers on Biotechnology 35.1, 39.1 and 46.1, in this expert panel the most "optimistic" countries are Spain and Italy, while Germany confirms its very low degree of acceptance and the Netherlands appears to have an intermediate position. Only the position of Greece does not fit entirely in the results of the most recent Eurobarometer, according to which Greece would have the lowest degree of acceptance of biotechnology within the EU, while in the Delphi survey it appears more optimistic than the Netherlands.

There are clear differences between Central Europe and the Mediterranean countries concerning the personal attitude of the involved expert groups. While in the latter only limited differences are observed between the expert groups, a rather polarized response behaviour emerges in Germany. The extreme poles represent the rather optimistic experts from industry and research institutions on the one hand, and the relatively sceptical consumers and critics on the other. The response behaviour of farmers is placed between these two poles, with clear tendencies towards the consumers/critics cluster. In the Netherlands moderate differences in personal attitude are registered between the analysed expert groups.

In line with previous opinion surveys, the genetic modification of microorganisms appears slightly more accepted than genetic modification of plants. Both application areas of modern biotechnology are much better accepted than genetic modification of animals which is generally considered very negatively (except when it is used for medical purposes). Negative attitudes toward genetic engineering of farm animals are much more frequently observed in Central European countries and in Greece than in Italy and Spain.

Differently from most opinion surveys on biotechnology which are conducted on the general public, the Delphi survey has been carried out with a particular methodology on a selected sample of well-informed respondents (experts), in most cases with a relatively high degree of knowledge (though not necessarily specialized) of the subject. This could be expected to result in a different pattern of perception and acceptance of Agro-Food biotechnology. As mentioned above, the differences in the basic trends are not dramatic. The results of the Delphi survey seem to confirm what has been found by previous population surveys in the EU. This finding is surprising to some extent, as it shows that the level of knowledge and the familiarity with the issues concerned are not so decisive in shaping the main attitudes. Instead, cultural factors seem to be prevailing.

The expert panels in each country express rather high expectations in future developments aiming at clear regulations of market approval and penetration of gene food products as well as enhanced information activities in this area. Such developments are expected to be realized in the short term. A large majority of respondents in each country seems to be convinced of the need of informing the public and expresses a positive attitude toward the creation of ad hoc institutions and bodies which provide information on modern biotechnology to all interested persons and institutions. On the whole, the respondents seem to have significant expectations concerning proactive initiatives by public institutions, as well as by private companies, in the field of public information and regulation. However, the existence of a direct link between information and public acceptance is doubted at least by a considerable proportion of the respondents. In addition, widespread concerns about the objectivity of information provided by governmental institutions are expressed in the comments.

The majority of the experts appears concerned about the "freedom of choice" of the consumer, i. e. the consumer should be made aware of the available options through proper information and be allowed to make a rational choice on the basis of cost-benefit considerations. All developments involving a broad diffusion of gene food products receive a moderate consensus. Positive attitudes seem to be linked to the indication of clear and specific benefits for human health and the environment. This confirms that the existence of ethical purposes is a key factor in order to obtain the social acceptance of genetic modification. This expectation, however, is accompanied by a feeling of inevitability of the final adoption of genetic engineering techniques in food production and processing.

The most supported applications of Agro-Food biotechnology are those concerning monitoring and control activities and those involving some benefit for the environment. The perception of possible risks is clearly indicated by the very high support given to epidemiological research on possible long-term impacts on human health due to the application of modern biotechnology. Together with the high share of positive attitude towards the statements in the domain "regulation", this suggests that the social acceptance of biotechnology is also linked to satisfying the public demand for control.

The attachment of consumers to their national food traditions is seen as an important factor in the process of acceptance of food biotechnology. A rejection of the products of modern biotechnology resulting from such attachment is seen as less likely in the Mediterranean countries compared to Central Europe. This relates in particular to specific food products laden with symbolic meanings (like beer) than to food in general. However, on the whole, the final broad introduction/adoption of the new food (bio)technologies is seen as inevitable.

6.4   Regulatory issues

After several years of intensive debate, the "Novel Food Regulation" of the EU came into force on May 15th, 1997. This regulation is valid for food which contains genetically modified organisms (GMOs) or consists of GMOs. In addition, food produced from GMOs but not containing them, is covered in this regulation as well. In the "Novel Food Regulation" it is foreseen that the characteristics of the "Novel Food" have to be stated on a label explaining the missing equivalence with conventional food. In addition, existing GMOs in the food product have to be mentioned. Up to date, the "Novel Food Regulation" has not been implemented in the EU member countries. Details concerning the text of the labels of gene food products as well as scientific procedures which are required to identify these products have not been clarified so far.

An important aspect filtered out from the results of the Delphi survey is the efficacy of the existing regulations related to Agro-Food biotechnology. Many experts state that single regulations already in force are not fully implemented, i. e. the existing legal framework does not work and is not efficient enough. This situation raises the scepticism and erodes the credibility in the governmental institutions' will and ability to install a system to control biotechnological products. Mistrust in public management of potential risks might be a result of this development.

According to the experts' opinion, the existing EU legal framework which provides regulations on the marketing of genetically engineered food and beverages - experts mention e. g. the "Novel Food Regulation" - is not very specific and open to various interpretations. This situation makes it rather difficult to standardize the extent of norms and control activities: Which products should be controlled and labelled? Only the ones which are genetically engineered, or all food products which contain some biotechnological component?

In case compulsory labelling is chosen for all food products which contain biotechnological components, experts argue that this option is not technically feasible, as some of these components are not detectable in the final food product. In addition, this way is regarded as rather costly. Following this argumentation, it is not very useful either that public institutions use specific equipment and methods to identify such products, as they will only be able to identify some of them, leaving the rest to reach the market "uncontrolled".

All in all, it can be concluded that regulations already in force are not precise enough. An effort could be made to specify more precisely their extent and to speed up their implementation time schedule. Important issues in this context are the labelling procedures as well as relevant criteria for the market approval of gene food products. If there are technical handicaps to implementing existing regulations, the norms' aims and content should be adapted in order to make them realistic.

The mentioned technical problems of identifing all gene food products pose the question: What are these regulations for? The first answer which comes to mind is that they may guarantee the quality and safety of biotechnological food products, whatever the standards are. In this context the experts see the problem that potential negative impacts or side effects have not been proved yet and, as they state, to predict them involves great difficulty, or is even impossible. The mentioned difficulties to assess potential impacts of genetically engineered food and beverages could be addressed by promoting epidemiological studies concerning possible health impacts of modern biotechnology, as some experts suggested.

The consideration of social impacts and ethical concerns of Agro-Food biotechnology are differently assessed by the involved expert groups. In the words of a German expert, their assessment is "ideological", since rather subjective terms are relevant compared to e. g. the "objective" or quantifiable risks of technical or scientific processes. In this regard, experts mainly from consumer and environmental organizations are concerned that the "public" involved in debate and decision-making concerning Agro-Food biotechnology are going to be representatives of strong pressure groups, especially big companies.

Although compulsory labelling is very welcomed by consumers and other non-technical experts, it is seen as a very problematic measure by researchers and entrepreneurs because they believe the public's knowledge does not allow them to make an informed decision about the implications of their consumption behaviour regarding food and beverages. In this sense, experts see enhanced need for information and education campaigns to increase the scientific literacy or public understanding of biotechnology in the population. Otherwise, the possibility is mentioned that the right to know becomes a way of discriminating biotechnological products.

The public's lack of information is also seen as a handicap to their participation in decision-making, a handicap which, on the other hand, is not only present on biotechnological grounds, but affects other areas of political activities as well. As a conclusion it could be said that education, information and transparency seem the best elements to avoid discrimination and decrease of competitiveness as well as to foster consumers' responsible behaviour.

Labelling and as a consequence "No ..."-labelling has been an issue since the EU

Novel Food regulations were introduced. The Delphi survey shows that in the short term we can expect food companies and retailers to have non-GMO labels for food and beverages. However, most experts assume that these labels will be restricted to specific market niches.

6.5   Economic implications of Agro-Food biotechnology

An important argument for the promotion of modern biotechnology in general is the expectation, mainly expressed by governmental institutions, that the application of this technology will contribute to additional employment possibilities and the competitiveness of European industry in a global market. In accordance with the results of (the few) other studies on this subject, the results of this survey indicate that in the next five to ten years only limited job growth is to be expected in the specialist biotech service companies and in supplying companies of the food industry. This means that additional employment possibilities due to modern biotechnology most probably will affect small and medium-sized companies (SMEs) to a higher extent than large multinational companies. However, it is also expected that a decrease in the traditional jobs in agriculture will take place, partly due to the widespread use of modern biotechnology, but also as a result of structural changes in this sector which are independent of the use of modern biotechnology.

Small and medium-sized companies (SMEs) can directly profit from the new biotechnological developments. The Delphi survey shows that in the medium and long term SMEs might be able to introduce a large variety of biotech products and processes and even small farmers are expected to use new transgenic plants and animals. However, especially in the Mediterranean countries, the realization of such scenarios highly depends upon successful technology transfer from research institutions to industry. In most EU member states governments have specific programmes to stimulate SMEs to get hold of new technologies. The importance of these programmes is underlined by the results of this Delphi survey.

The expert panels expect that modern biotechnology will have a considerable effect on the production costs of agricultural bulk products like cereals and milk. They estimate a significant decrease of these costs in future, although some experts think that a decrease of 30 % probably will be too high. However, the increase in productivity in agricultural production will only result in additional income for farmers if it is not overcompensated by decreasing price trends for agricultural products in the coming years.

In accordance with the results of other studies, the experts asked in the Delphi survey express some uncertainties and concerns about a strong market penetration of gene food products. This may result in delays of the market approval and diffusion of food products based on genetic engineering. Nevertheless, the experts expect considerable market shares and a kind of "habituation" of the consumers to this type of products in around ten years. In this context, food products which offer clear benefits to the consumers are best suited for taking the lead, because they are the best opportunity to overcome existing acceptance barriers. One example in this direction are probiotic foods or other food products supporting health requirements. In contrast, products which cause specific emotions (like e. g. beer in Germany) are less suitable as future pilot products.

The experts asked in the Delphi survey have very high expectations in the market opportunities of biotechnology-based products for non-food purposes. This relates to renewable resources used in the chemical industry and for energy production, to pharmaceutical substances produced with the help of transgenic animals or plants, as well as to the use of biotechnology approaches to reduce agricultural waste. In general, the social and ethical acceptance of these products or approaches is far higher than those of the use of modern biotechnology in the food chain, i. e. non-food products based on modern biotechnology could represent excellent pilot products also in countries with a considerable opposition to the use of these technologies, like Germany and the Netherlands.

On the other hand, the results of the Delphi survey indicate as well that most non-food applications of modern biotechnology are regarded as medium- to long-term developments. In addition, a complex set of influential factors is seen as decisive for their realization, including both the technically oriented as well as demand-/market-driven aspects, i. e. this type of innovation can only be speeded up in a "concerted action" of science, industry and politics, taking into account the interests of other groups (e. g. farmers) as well. In this context, the future division of labour between public research institutions, multinational companies and SMEs, which are often regarded as being more innovative and flexible, is an open question.

A clear regulation of the market approval procedure for gene food products is required for economic reasons as well. The majority of the experts appreciate such a development and expect a short-term realization of compulsory labelling activities of gene food products and an official case-by-case procedure for market approval. On the other hand, the results of the Delphi survey indicate as well that a considerable proportion of the experts sees little chances for clear regulation procedures, for practical reasons and other intentions of industry and politics. This expectation may lead to a kind of fatalism and result in additional frustration and resistance against genetic engineering approaches in the Agro-Food sector, especially by militant opposition groups. In order to prevent such a development, an implementation of the "Novel Food Regulation" in each member country is strongly recommended as fast as possible, thereby ensuring clear control and monitoring activities. Industry and political institutions play a key role in this context. In the past they have been aiming frequently at liberalizing existing standards and laws in order to facilitate the applications of biotechnology. There are some indications in the results of the Delphi survey (especially in the comments of the experts) that parts of industry may change their argumentation from "deregulation" in the direction of a compromise with opposing groups in a sense that a certain regulatory framework including the labelling of gene food products would be accepted.

The use of genetic engineering techniques in organic farming is a controversial issue. On the one hand, new biotechnological techniques for developing transgenic resistant plants in order to decrease the use of pesticides and herbicides as well as the use of "natural" biocides and animal vaccines can be assessed positively from the environmental point of view. On the other hand the fundamental rules on which organic farming is based do not accept this kind of techniques. The Delphi survey shows that this controversy most probably will not be solved in the coming years. Although there is a positive attitude towards these developments, especially in Germany and the Netherlands a considerable part of the panel foresees no integration of genetic engineering techniques in organic farming.

Biotechnology in fish farming and fishery is a very unfamiliar subject to the panels in the five countries, although fish is often eaten. Although the industrialized production of fish is growing and also biotechnological techniques are being introduced for breeding and processing purposes, this has not been a subject of great public attention so far.

6.6   Suitability of the Delphi methodology

All in all, it can be concluded that the applied Delphi approach proved to be suitable for the analysis and forecast of future developments in complex and controversially discussed thematic areas like the application of modern biotechnology in the Agro-Food sector. This relates to all new features of this project, namely the international comparability of the results of this survey, the inclusion of framework conditions in the statements and answering categories of the questionnaire as well as the involvement of several social groups (like farmers, consumers, biotechnology critics), which are not considered in conventional Delphi surveys. Even if these groups do not reach the required number of answers in each country to be statistically analysed separately and some members of these groups expressed some difficulties in filling in the "scientific and technical" part of the questionnaire, the analysis of the degree of knowledge of the involved groups as well as the response behaviour of these groups (which with the exception of "personal attitude" and to a certain extent "time of realization" does not differ significantly from industry and research experts) indicate that the involvement of these groups enlarges the existing basis of knowledge, experience and creativity which should result in a more sound estimation of the future developments. In addition, this approach allows analysis the differences in the response behaviour between the "developer" of a new technology (like e. g. public researchers, industry) and its final user (like e. g. farmers or consumers).

However, some minor modifications of the Delphi approach are suggested in order to adapt this methodology to the specific topic addressed in this project. The first modification refers to the selection of the experts. While the way used in this project proved to be successful in Germany and the Netherlands, some difficulties arose in the Mediterranean countries. This was mostly due to the fact that Delphi surveys are rather unfamiliar in these countries, as well as that the use of modern biotechnology in the Agro-Food sector was not an important issue in the public debate in these countries before 1997. Therefore, only a limited number of experts are available in these countries who intensively deal with the application of modern biotechnology in the Agro-Food sector and its impacts.

Against this background a kind of "co-nomination system" might be more suitable for the selection of the experts, i. e. in this system a small number of selected specialists working in the relevant field are asked to provide names of other experts. The persons who are mentioned additionally are asked to deliver names of relevant specialists as well. This system is carried out in several rounds. Besides the identification of additional experts, the main target of this approach is to analyse the overlap in the answers of the asked persons in order to identify a kind of "core group" of experts to be asked in the survey. Taking into account the activities required for intensive personal contacts, which seems to be an important instrument to motivate the experts to participate in the Delphi survey, the suggested way to select the experts seems to be appropriate in countries in which the Delphi methodology is rather unknown and the thematic area addressed in the survey is not intensively discussed by the public.

Additional suggestions for modification refer to the development of the questionnaire which represents a sensitive and difficult step in each Delphi survey. Taking into account the clear differences in the results between the involved countries, statements including national particularities should have more weight in the questionnaire in order to specify the answering pattern of each country more clearly. In contrast, statements containing rather "visionary" developments/ situations should be given only limited space in the questionnaire, because a relatively high proportion of the experts has difficulties to answer these statements.

The project team was assisted by the national expert committees in the identification of the most important future developments relevant for the application of modern biotechnology in the Agro-Food sector in the questionnaire. In this sense, the final questionnaire represents a compromise between a broad range of relevant aspects with a lot of interactions among each other and a limited number of statements and answering categories (which are mainly restricted by the time necessary to fill in the questionnaire). In order to include the most relevant aspects in the questionnaire, some statements were formulated in a way that several developments were linked with each other and were included in one statement. In some cases these statements created some difficulties, both when filling in the questionnaire and when interpreting the results of the survey. Therefore, it is recommended to include only one major topic per statement and special attention should be paid to the clear and simple formulation of the statements. In order to get a sufficient number of filled-in questionnaires, the number of statements included in the questionnaire should not exceed 100. In this context, additional emphasis has to be given to the number and complexity of the answering categories, since they influence the time required to fill in the questionnaire to a high extent.

In accordance with other surveys, the results of this survey indicate that the Delphi methodology is suitable for the analysis and forecast of future developments in highly socially debated research areas as well. In this context, specific emphasis should be given to the rather differentiated response behaviour of the experts of the Delphi survey which clearly contrasts with the "black and white" argumentation pattern often emerging in the public debate, at least in some of the included countries. In this sense, the results of the Delphi survey represent a sound basis for follow-up activities of all interested social actors who are an important indirect target of this approach.

6.7   Recommendations

Based on the results of the Delphi survey, the following recommendations are formulated in order to allow decision-makers in politics, industry and society to adapt to emerging problems in time or to be prepared for future possibilities:

(1)   The acceptance of modern biotechnology is largely influenced by ethical and cultural issues, also in the Agro-Food sector. Therefore, public institutions and private enterprises should play a proactive and responsible role, always giving an adequate consideration to these issues at different levels: research, public policy, private initiatives, training, education and information, public debates.

(2)   The current regulations in the field of Agro-Food biotechnology focus on the potential risks of this technology related to the health of consumers, users and employees as well as to the environment. However, the results of this Delphi survey and other studies indicate that ethical and cultural factors gain increasing importance in shaping the acceptance of modern biotechnology. With respect to the future design of biotechnology regulations, these results point towards a possible shift in the underlying policies: a need for a strong focus on ethical concerns as driving forces for the design of regulatory frameworks might arise in addition to the present risk-driven approach.

(3) In this context, further research activities are recommended, focusing on the development of practicable ways to include ethical aspects in regulation and decision-making concerning Agro-Food biotechnology.

(4) A clear regulation of the market approval procedure of gene food products is required for safety, social and economic reasons. Therefore, it is recommended to implement the passed "Novel Food Regulation" in each EU member country as fast as possible, thereby ensuring clear and realistic control and monitoring activities.

(5) Fostering public participation on decision-making seems to be a way of increasing positive attitudes towards - and confidence in - government action and companies' performance. In consequence, it is recommended to initiate the development of activities for involving the public in the decisions that affect them directly, like the ones regarding food.

(6) The establishment of an independent institution - not exclusively related to economic interests for approving the marketing of genetically engineered food - may strengthen public confidence in the control and the guarantee of the quality and safety of food products. In this context, it should be checked whether regulations on gene food products should incorporate a norm to make the release of information and the transparency principle for companies and governmental institutions compulsory.

(7) The institutional and regulatory framework plays a fundamental role in influencing the acceptance of biotechnology in the Agro-Food sector. Public institutions should not underestimate the importance of keeping a major role in biotechnology regulations; they should be able to respond to the expectations asking for a neutral and objective position. In addition, public institutions should guarantee the freedom of choice and prevent that biotechnology products becoming the only possible choice, by supporting also alternative production technologies.

(8) The creation of ad hoc institutions and bodies in order to inform the public about the application of modern biotechnology in the Agro-Food sector, including representatives of various interest groups and viewpoints, is warmly recommended. Besides information activities, such institutions could aim at increasing public participation in the decision-making process in this area.

(9) Education campaigns and making information available and understandable to the general public would help to make public participation a realistic option, and would increase consumers' ability to make well-balanced decisions.

(10) Research and development policies aiming at social, economic and environmental benefits of modern biotechnology can contribute to increasing the acceptance of this technology in the Agro-Food sector. Particular attention should be paid to environmental and human health monitoring and control activities in the respective research programmes.

(11) The estimations of the experts concerning possible adverse health effects of modern biotechnology are a clear message to politicians and the responsible authorities at EU and national level to take effective measures in order to minimize these effects in future. These activities are not necessarily limited to regulations and the legal framework conditions, but should incorporate the development of appropriate scientific methods and techniques for monitoring and analysing health issues, stimulation of relevant research programmes, as well as public information activities to consumers and employees, in order to raise the awareness of this problem within these groups and to promote appropriate preventive measures.

(12) Taking into account the clear demand of the experts for enhanced information activities as well as knowledge creation concerning the interactions between gene food and health of consumers and employees, public research policy should focus to a higher extent on the analysis, especially of the long-term impacts of modern biotechnology. This relates to the development and/or the adaptation of existing instruments (e. g. the methodology of epidemiological surveys to the specific environment in the Agro-Food sector), as well as enhanced integration of these issues in publicly funded research programmes. In addition, public support mainly from supranational organizations like the EU to the arrangement, management and carrying out of transnational epidemiological surveys analysing medium- and long-term health impacts of modern biotechnology is warmly recommended.

(13) In order to promote the use of biotechnological approaches with environmental benefits (e. g. in animal husbandry), integrated activities of science, industry and politics are required, which should be focused on specific target fields which are selected according to their environmental performance as well as possible economic implications.

(14) The concept of production-integrated environmental protection, i. e. production processes designed in a way that environmental pollution is prevented from the onset instead of cleaning them up by end-of-pipe processes, is not yet widely known. Modern biotechnology could - among other technologies and management concepts - play a substantial role in pollution prevention. It is recommended to broadly disseminate information on this new concept of environmental protection and the role of biotechnology within this concept.

(15) The estimations of the experts participating in the Delphi survey show a clear demand to politicians and the responsible authorities on EU and national level to undertake additional activities in order to minimize the expected negative environmental effects of modern biotechnology in future. These activities are not necessarily limited to regulations and the legal framework conditions, but should aim at initiating and stimulating relevant research activities as well.

(16) At present there are no sufficient instruments available to measure the isolated employment effects of modern biotechnology in the Agro-Food sector. This is especially true for the employment effects in the food industry. Therefore, new instruments should be developed in order to get a close and more realistic view of this aspect. In addition, sound scientific studies should be initiated by national governments and the EU Commission analysing the employment effects of modern biotechnology in the Agro-Food sector in particular. Applying such new concepts would help to substantiate political arguments to stimulate modern biotechnology for employment reasons.

(17) There is a general readiness to accept specific biotechnology products, but a strong need for justification is expressed. Any new product should be accompanied to the market paying special attention to its possible social, economic and environmental benefits.

(18) Since acceptance barriers against gene food and their reasons differ between the analysed countries, a differentiated marketing strategy and adopted accompanying measures (like communication and information activities) are strongly recommended, if gene food is to be introduced into the market. Although at a first glance, countries with a rather favourable environment for gene food products seem to be ideal candidates for the first market approval of these products in the European Union, it should be taken into account as well that only gene food products which are physically available on the markets allow consumers to have their concrete and individual experience with this type of products, which represents one major prerequisite to overcome the reservations against these products.

(19) Taking into account the hesitations at least of parts of the experts concerning "bio-enriched" food, food marketing strategies, which influence to a considerable degree consumers dietary habits, should not be limited to technical food characteristics, but they should also incorporate the needs proposed by nutrition professionals as well as clear information concerning the claimed positive health effects.

(20) Most non-food applications of modern biotechnology can only be speeded up in a "concerted action" of science, industry and politics, taking the interests of other groups (e. g. farmers) into account as well. In this context the future division of labour between public research institutions, multinational companies and SMEs is an open question. For these reasons further research is recommended, aiming at analysing and evaluating the market potential and market conditions of non-food biotech products in the Agro-Food sector. In this context, the question of how SMEs could participate and benefit from such innovations deserves particular attention.

(21) In order to stimulate consumers' and producers' information of fish and fish products, a specific project is recommended, in which the impacts of modern biotechnology on fish farming and fishery are analysed. The results of this

study should be made available to the public, consumers, producers and research organizations in each member state.

(22) The future innovativeness related to food enzymes highly depends on the know-how in platform-technologies like genetic engineering, high throughput screening, combinatorial approaches or protein engineering and in the exploitation of synergies between the different technologies. In most cases the know-how in the platform-technologies has been built up in the supply industry, not in the food-processing industry itself. Therefore, the food-processing industry depends on innovations developed in other industry branches. In this context it should be checked if the network between the food industry, its suppliers and relevant research institutions is sufficient or if building up or optimizing such a network should be actively supported. In addition, close interrelations between these technologies should be taken into account in future R&D programmes at national and international level.

(23) Although the food-processing industry is reluctant to apply food enzymes produced by genetically engineered organisms, most food enzymes will be produced this way in the near to mid-term future. Taking into account the differing interests of the involved actors in this controversial field, the food industry should aim at strategically positioning itself (instead of staying passive as in recent years), define its particular interests and opinions and act proactively.

(24) Especially SMEs in the food industry are not yet aware of the benefits which could arise from the use of optimized enzymes which are produced with the help of genetically modified organisms (e. g. more cost-efficient production, production-integrated environmental protection approaches). Therefore, specifically targeted research in this area should be initiated in order to check how the adoption of these techniques by industrial companies can be supported (e. g. workshops, brochures, personnel transfer, demonstration projects).

(25) Since social and ethical acceptance is identified as one key factor for the use of animal biotechnology, an indispensable requirement for the application of these techniques is neutral and objective information of the public and a broad discussion of the ethical aspects in this context. The main purpose of these activities should aim at differentiating the acceptance level between different applications of animal biotechnology as well as the identification of application areas or purposes which are not tolerated by society. Assuming that there are interactions between social acceptance or risk perception and the existing regulations, it should be checked whether the legal framework should be adjusted accordingly.

(26) Taking into account the existing technical constraints in animal biotechnology expressed in the Delphi survey, emphasis should be given to an advancement of R&D infrastructure in this field, if the use of modern biotechnology in

reproduction of farm animals and potential commercial applications is desired by society.

(27) Most crop plant traits of economic and agronomic importance are of polygenic nature. However, genetic engineering of polygenic traits is an ambitious, both scientifically and technically challenging goal. If this goal is felt worthy of being pursued, it is recommended to consider whether the R&D infrastructure should be improved and additional funding for broadening the knowledge of the molecular basis of crop plant traits should be provided.

(28) Analytical and diagnostic methods based on modern biotechnology can be used in quality, hygiene and process control in the food industry, in plant and animal breeding, in health monitoring of crops and livestock, and in environmental monitoring. It is recommended to support R&D with the aim to provide simple to use, inexpensive, labour-saving assays and equipment, to provide rapid methods for qualitative screening and quantitative, confirmatory methods. In parallel, attention should be paid to the problem that the existing regulatory environment may be unfavourable for innovative methods. Measures should be supported which aim at the qualification and training of personnel to use and adapt these test systems and at the transfer of these methods from specialized laboratories to the "ordinary user".

7.   Summary

Although the first Agro-Food products based on modern biotechnology have entered the EU markets, the application of this technology is still being intensively discussed in the European Union. Recent opinion polls indicate as well that consumers' acceptance of genetically engineered food and agro-products is still relatively low, at least in some member states of the EU. By contrast, representatives from politics and industry underline the necessity to apply modern biotechnology in the Agro-Food sector as well, mainly to ensure the competitiveness of the EU agriculture and food industry and for employment reasons.

Against this background there is a need for a scientific analysis of the future impacts of modern biotechnology in the Agro-Food sector of the EU. Recent studies trying to analyse this issue usually comprise extrapolations of status-quo analyses. What has not been exploited so far in this context are systematic technology forecasting approaches which include more than one singular country, thus obtaining information on an international level. Therefore, in this project which was financially supported by the Commission of the EU (DG XII), the impacts of modern biotechnology on the Agro-Food sector in five member countries of the EU (Germany, Greece, Italy, the Netherlands, Spain) have been analysed with the help of the Delphi methodology. The Delphi approach is based on a questionnaire which contains possible visions of future developments which are assessed by selected experts. The results of the first round are included in the questionnaire of the second round, thereby providing a feedback mechanism. The specific features of this project are, on the one hand, the consideration of the scientific and technical developments in Agro-Food biotechnology as well as the development of framework conditions for food production and consumption and, on the other hand, the involvement of different social groups (like farmers, consumers, and opponents of biotechnology) who are not asked in traditional Delphi surveys.

Realization of the Delphi survey

The questionnaire used in this Delphi survey has been developed in an interactive process between the different project teams and national expert committees specifically set up in the five countries during the first six months of 1996. 71 statements representing different possibilities of future developments or situations in the Agro-Food sector have been defined for this questionnaire, which should be assessed by five answering categories (degree of knowledge, personal attitude, time of realization, influential factors, importance to knowledge creation, competitiveness of the economy and protection of the environment). In addition, the experts had the opportunity to submit personal comments. 30 statements of the questionnaire are dedicated to scientific and technical developments in the biotech area. 41 statements refer to the development of the framework conditions in the Agro-Food sector in the areas acceptance, regulation, the economy, environment and health.

In contrast to traditional Delphi surveys, the definition of "expert" is extended to additional social groups in this project, since the impacts of biotechnology applications have a rather high relevance for such groups as well. Therefore, experts from industry, research institutions, farmers' organizations, consumer and user organizations, groups critical of modern biotechnology as well as other experts (e. g. from the policy, administration and educational sector, journalists, biotechnology advisers, patent lawyers) have been involved in the survey. The project teams in the different countries screened various information sources (like electronic databases, handbooks, lists of participants of relevant conferences) for experts in this field and contacted important organizations and institutions in the Agro-Food sector, as well as specific key persons in order to deliver member lists or addresses for the relevant groups. With the exception of Spain, no major difficulties in identifying a sufficient number of experts in the involved countries occurred.

The databases of the experts were completed in August/September 1996. In total, more than 7,800 experts were asked to participate in the first round of the Delphi survey. The size of the expert panels range from around 1,200 experts in the Netherlands and Spain to almost 2,400 experts in Germany. The questionnaires of the first round were mailed to the experts in the involved countries during autumn 1996. In each country specific activities (like reminding letters, answering questions of the experts, personal phone calls) were undertaken to raise the number of respondents. The mailing of the questionnaires of the second round was continued in March/April 1997 in each country. Especially in Spain and Italy a lot of personal communication efforts were undertaken by the project teams to motivate the members of the panel to fill in the questionnaire a second time.

The response rates differ significantly between the Mediterranean and the Central European countries. In the latter, a surprisingly high response rate can be registered (especially in Germany), whereas a slower and lower rate of response occurred in the Mediterranean countries. Nevertheless, a sufficient number of respondents could be achieved in total in all involved countries. While in Germany each expert group could be statistically analysed separately, in the other countries some expert groups (e. g. consumers and critics) had to be merged, for statistical reasons.

Differences between the involved countries

In a first step the results of the Delphi survey were analysed on a national level. In this analysis considerable differences occur between the involved countries. Regarding their personal attitude the experts of the Central European countries (Germany, the Netherlands) are more critical than their colleagues of the

Mediterranean countries, whereby among the latter the Spanish experts are by far the most positive ones. In total, more than half of the statements are appreciated, while less than one fifth of all statements are opposed by the experts of the five European countries.

Only in ten statements do no highly significant differences occur in the personal attitude of the experts in the five countries. These statements mainly refer to biotechnological approaches which contribute to reduce health problems like allergies or to develop products outside the food chain. Moreover, the experts of all countries are in favour of monitoring systems based on modern biotechnology. On the other hand, they strongly oppose statements dealing with negative impacts on health (e. g. additional allergies) and the economy (e. g. reduction of employment). The most obvious differences between these two groups of countries emerge concerning statements which deal with the application of modern biotechnology in animal husbandry and animal breeding. The German, Dutch and Greek panels estimate the use of modern biotechnological approaches and genetic engineering in animal production rather critically, while the Spanish and Italian panel members regard these developments more positively.

Concerning the time of realization, more than 60 % of the statements are seen as realizable within the coming ten years. Similar to the category "personal attitude" two poles of opinion can be observed. On the one hand, the Spanish experts who believe in a rather short-term realization and on the other hand the German experts. Most of them predict a realization of the statements in the medium term. In addition, around 10 % of the German experts even regard the statements as not realizable at all. The response behaviour of the Dutch, Italian and Greek experts is placed between these two poles. However, with respect to the Netherlands, it is important to point out that a considerable number of statements is seen as not realizable at all in this country. These statements mainly refer to negative economic and environmental impacts of Agro-Food biotechnology.

For the future application of modern biotechnology, the differing framework conditions in the five countries have to be taken into account as well. The demand-related factors social and ethical acceptance of modern biotechnology as well as the conditions on the relevant markets got almost twice the weight in Germany and the Netherlands as in the three Mediterranean countries. In addition, government activities and measures are regarded as very important in the Central European countries. By contrast, in the Mediterranean area R&D infrastructure, technology transfer as well as the availability of qualified and skilled personnel, i. e. important "technology influenced" factors, are seen as rather important constraints. Moreover, the diffusion of background information on modern biotechnology is also regarded as more relevant in the Southern countries. In all countries the relatively low relevance of "ecology" and "international collaboration" as influential factors is rather surprising.

Developments in plant breeding and plant production are mainly assessed as most important for "knowledge creation in science and technology". This relates in particular to basic research activities in this field, the use of new monitoring and control techniques, and the genetic modification of complex regulated characteristics in plants. Moreover, a significant extension of scientific knowledge is expected from the finishing of basic research activities with relevance for food processing, the development of enzyme systems with higher environmental performance, as well as the investigation of long-term health impacts of genetically engineered products on consumers, farmers and employees.

The majority of experts of all involved countries points out developments describing positive impacts on modern biotechnology as important for the competitiveness of the national economy. They expect strong impacts on the economy if modern biotechnology contributes to the extension of the industrial innovativeness of small and medium-sized companies in the Agro-Food sector or to the reduction of production costs for agricultural bulk products. Additionally, high economic impacts are expected if new markets for agriculture and the food industry can be made accessible with the help of modern biotechnology. The latter relates e. g. to the production of pharmaceutical substances, optimized renewable resources or genetically engineered vaccines for farm animals.

Furthermore, scientific and technological developments mainly in food processing and plant production are seen as essential for the competitiveness of the national economy. In particular, this relates to an increase in productivity in food processing due to enzymes produced with GMOs and to the use of specific monitoring and control systems in food production. Besides, the application of genetic engineering in the field of enzyme modifications and in plant breeding are interesting from the economic point of view.

Statements classified as important in improving the environmental situation mainly belong to the domain "environment". This relates in particular to the application of modern biotechnology approaches in animal husbandry in order to reduce emission or organic waste problems. A significant positive environmental impact is seen as well if environment-friendly enzyme systems are developed for food processing. Clear positive impacts for the environment are also expected if genetically engineered microorganisms and plants producing biocides are used for biological pest control, if salt- or drought-resistant plant varieties are cultivated in arid areas and if new monitoring and control techniques for genetically engineered plants are used to improve the reliability of risk assessment in field trials.

Biotechnology approaches which offer economic advantages, as well as contributing towards improving the environmental situation, represent suitable opportunities to promote modern biotechnology in the included countries because on the one hand, it can be expected that such initiatives will not meet severe societal resistance due to anticipated environmental damage and, on the other hand, represent areas with high commercial interests. Those statements relate to the improvement of the environmental performance of conventional food-processing procedures, the use of genetically engineered microorganisms and plants producing biocides for biological pest control, the biotechnological reduction of emissions and waste from animal production, as well as the biotechnological transformation of organic agricultural waste into marketable products. In addition, the experts of all five countries estimate the production of renewable resources as well as the development of specific resistance mechanisms for plants, like genetically engineered field crops resistant to herbicides or plant varieties resistant to salinity and drought, as very important for the environment and for the respective national economy.

Likewise, the experts of all countries generally believe that scientific research has a significant impact on economic competitiveness. The developments identified as most important in this respect relate to application of biotechnology in the non-food sector, like the production of pharmaceuticals and high value proteins, for control techniques and for food production processes. Moreover, the production of enzymes in genetically engineered field crops, the improvement of resistant mechanisms and the use of genetic engineering in plant breeding are included as well.

Differences between the questioned expert groups

Since different expert groups were included in the survey, it was possible to analyse the response behaviour of these groups separately. All in all, a different response behaviour between the expert groups in the Central European and the Mediterranean countries can be registered. The German and Dutch expert groups show rather polarized answering patterns, while the expert groups of the Mediterranean countries mostly indicate a relatively uniform response behaviour, both concerning personal attitude and time of realization. The German panel is characterized by the most polarized response behaviour between the expert groups compared to the other countries. Concerning the assessment of the "personal attitude", the extreme poles are represented by the experts from industry and research institutions on the one hand and consumers and critics on the other hand. The response behaviour of farmers is placed between these two poles with clear tendencies towards the consumer/critics cluster. In general, industry and research experts tend to assess the stated developments more positively, whereas most of the experts from the farmer, consumer and critics side are more sceptical or reject single developments. The highest differences between the expert groups in Germany are found in statements dealing with the application of enzymes in the food industry which are produced or optimized with the help of genetic engineering.

Concerning the time of realization, industry/research and critics represent the two extreme poles among the expert groups in Germany. Areas in which the future realization of modern biotechnology is heavily disputed among the different groups, are the ecological effects of modern biotechnology on agricultural ecosystems, the integration of specific genetic engineering approaches in organic farming, and the economic impacts of modern biotechnology in the Agro-Food sector.

Future scientific and technical developments

The results of the Delphi survey confirm that the number of modern biotechnology-derived products and methods is likely to increase significantly within the next ten years, due to scientific-technical advances and the synergetic combination of different approaches, methods and technologies. In the field of enzyme production leading companies producing food and feed enzymes have made strategic decisions in recent years to use genetic engineering as a core technology, together with efficient and innovative screening procedures for new enzymes, rational and irrational protein design, and fermentation and downstream processing, thus re-engineering their R&D activities substantially. Investing in these core technologies and exploiting synergies between them makes it possible for the enzyme producers to drastically shorten the time to market for new enzymes and to make a much larger variety of different enzymes commercially available than before.

In plant breeding and plant production multinational "life science companies" invest heavily in "green biotechnology". Their strategic goal is to transfer the genomics approach from the pharmaceutical sector to the Agro-Food sector. Relevant scientific information stems e. g. from genome sequencing projects of plant genomes and from research elucidating plant-pathogen interaction and plant development and metabolism on the molecular level. However, the Delphi results indicate that the applied methods still have to be refined in order to allow the engineering of polygenic traits in crop plants.

At present, artificial insemination, embryo transfer, in vitro fertilization and cloning by embryo-splitting are established reproduction techniques which are practically used in animal breeding and animal husbandry. Genetic engineering of farm animals and cloning by nuclear transfer is mainly in the research stage and, if applied, is predominantly targeted to the pharmaceutical sector. In the coming decade it will most likely become possible to combine these different approaches, thus overcoming present technical difficulties and exploiting synergies between the individual technologies.

In the field of analytical tools and diagnostics based on modern biotechnology, research activities currently being carried out in elucidating the molecular and genetic basis of economically important characteristics of food, crop plants, livestock and food-relevant microorganisms will provide a multitude of information in the coming decade which can be used for new analytic and diagnostic tests in the Agro-Food sector. Moreover, technologies such as DNA chips, presently at the threshold of revolutionizing analytics and diagnostics in human health care, can also be transferred to the Agro-Food sector. This may result in the broad application of inexpensive, easy-to-use and automated assays in the Agro-Food sector.

Although scientific and technical advances are necessary prerequisites for innovations, other factors also play a significant role in the future introduction of products and methods based on modern biotechnology in the Agro-Food sector. Both for enzymes and for analytic and diagnostic methods based on modern biotechnology, the most important influential factor is a lack of industrial innovativeness on the side of the users of these enzymes and bio assays. In the European food industry small and medium-sized companies predominate with relatively low profit margins and only few R&D activities. In the case of enzymes and bioassays, technical competence in core technologies remains mainly confined to the supply industry or research institutions and is not broadly available in the food companies themselves. The required expertise can only be acquired via cooperation and participation in appropriate networks. Therefore, the further diffusion of assays based on modern biotechnology in the Agro-Food sector will depend significantly on whether personnel can be qualified, especially in the Mediterranean countries, whether technology transfer between "assay developers" and "assay users" can be improved, whether additional funding especially in Germany and the Netherlands can be made available, and whether the regulatory environment can be made more favourable for assays based on modern biotechnology in the food area.

For enzymes produced with the help of genetically engineered organisms, the future prospects are different: their use will most likely increase if the enzymes enable the food company to reduce costs significantly or to develop innovative products with high profit margins. The results of the Delphi survey indicate as well that the cost-saving potentials of enzymes, if employed in production-integrated environmental protection, are not yet widely known in the food industry. Therefore, there seems to be a need for education and for the dissemination of information on companies which have successfully integrated enzymes in their production process.

The use of enzymes produced with the help of genetically modified organisms is disapproved of by consumers and retailers, although to a lower extent than genetically engineered crops or animals. Therefore, food companies which technologically depend on "pro-genetic engineering" enzyme suppliers and economically depend on "contra-genetic engineering" retailers and consumers find themselves in a difficult situation. At present, many food processors have not yet defined their own position in this conflict. Therefore, it is necessary to act very sensibly and to readily provide information on how the applied enzymes have been produced.

There is considerable uncertainty about how long it will take to reach the goal of genetically engineering polygenic traits in plants – the time scale of realization ranges from a few years to two decades. This uncertainty might stem from significant deficits in the R&D infrastructure which are uniformly pointed out by experts from all countries. Although multinational "life science companies" transfer genomics into plant breeding and production, relevant R&D projects and cooperations are still at a relatively infant stage, especially in the publicly funded area. Against this background, improving the infrastructure and providing additional funding for broadening the knowledge of the molecular basis of economically important crop plant traits are required.

On the other hand, public resistance to genetically engineered crop plants and food derived from them is considerable, at least in the Central European countries, as can be seen e. g. from the opposition to field trials with transgenic plants. Moreover, the delay in putting into force clear and practicable labelling regulations for food produced from genetically engineered crops does not correspond to consumers' needs and does not at all contribute to transparency. It seems unlikely that scientific-technological development can take place, if the public's concerns are not seriously taken into account. This requires among others the improvement of the knowledge base for risk assessments, careful risk assessments of transgenic plants prior to their commercialization, sound monitoring programmes for rare adverse or long-term effects after their commercialization, and the integration of transgenic plants into broader concepts, such as e. g. sound pest management strategies.

A significant bottleneck in genetic engineering of livestock seems to be a lack of understanding of the early stages of embryonic development. Full benefits of cloning will only be achieved in combination with genetic engineering. If the commercial application of these techniques in farm animal breeding and reproduction is desired, emphasis should be given to an advancement of R&D infrastructure in this field.

However, ethics is a key issue in the debate on which scientific and technical developments in animal biotechnology are socially required and acceptable. Ethical aspects concern animal welfare, but it is even more feared that results obtained in animals will one day be transferred to humans. Therefore, the full and objective information of the public and an intensive discourse on the ethical aspects and the implications of the biotechnological methods for animals as well as for man is an indispensable requirement. If certain subfields of animal biotechnology turn out not to be tolerated in society, it should be checked whether legislation should be amended respectively. This is also of importance for commercial reasons: biotechnology firms need a clear legal framework in order to invest capital in this field and to establish their business activities.

Impacts on health of consumers and the environment

In the public debate on Agro-Food biotechnology in the EU, potential risks of this technology related to the health of consumers as well as to the environment, play a prominent role. In addition, the measures foreseen in official risk assessment activities as well as regulation procedures before marketing of gene food products mainly focus on these aspects.

Four main factors have been identified which are decisive for the assessment of the future health impacts of modern biotechnology. The development and marketing of food with specific positive health effects due to the use of modern biotechnology is expected within six to ten years, mainly depending on demand-pull factors. A big majority of the experts expect that adverse health effects due to the use of modern biotechnology will occur in future, even if there are differing estimations concerning the time of realization. One possibility to avoid adverse health effects is the development of new monitoring and test systems which might be based on biotechnology as well. The experts express great expectations in these systems and expect a short- to medium-term realization, mainly depending on funding possibilities and technology-push factors. In addition, enhanced information activities as well as knowledge creation in this field (e. g. epidemiological studies to investigate the long-term health effects of modern biotechnology) are seen as possible activities to prevent or minimize adverse health effects.

The results of the Delphi survey show that modern biotechnology is seen by the respondents as a factor of aggravation, but at the same time as a possibility to reduce allergies. There are strong convictions and expectations, especially for the contribution of genetic engineering to prevent allergies. But at the same time, there are clear concerns about harmful health effects of the same technology. In total, the experts express higher expectations in technical solutions (e. g. the development of new monitoring and test systems) than in organizational approaches (like enhanced information activities) in order to reduce or prevent such adverse health effects.

Almost the same picture emerges in the answer of the experts concerning the environmental impacts of modern biotechnology. On the one hand, they regard modern biotechnology approaches as one opportunity to reduce environmental burden, while on the other hand, they see the possibility of additional, possibly long-lasting environmental damage due to the use of these techniques.

The use of modern biotechnology to reduce organic waste in animal husbandry as well as its conversion into marketable products is strongly appreciated by the expert panels and regarded as highly beneficial, both for the environment and the national economy. This also relates to the development of specific enzyme systems to improve the environmental performance of conventional food-processing procedures. However, this rather optimistic experts' view is counterbalanced by the estimation that a complex set of technically and demand-/market-oriented activities are required to realize these developments in the medium to long term. In addition, the experts express rather high expectations in different scientific developments in plant production, of which they expect direct or indirect positive impacts on the environmental situation.

By contrast, all expert panels strongly reject negative environmental impacts of modern biotechnology like horizontal gene transfer, negative impacts on biodiversity or the loss of traditionally used varieties and organisms. Despite this negative personal attitude, a majority of the experts expects that these effects will occur in future, although there are differing estimations concerning the time of realization.

Future acceptance and regulation of modern biotechnology

The results of the Delphi survey show that the degree of acceptance of Agro-Food biotechnology varies considerably from one country to another. In line with the results of the three Eurobarometers on Biotechnology, in the Delphi survey the most "optimistic" countries are Spain and Italy, while Germany confirms its very low degree of acceptance and the Netherlands appears to have an intermediate position. The genetic modification of microorganisms appears slightly more accepted than genetic modification of plants. These areas of modern biotechnology are much better accepted than genetic modification of animals, which is generally considered very negatively (except when it is used for medical purposes). The results of the Delphi survey seem to confirm what has been found by previous population surveys in the EU. This finding is rather surprising, as it shows that the level of knowledge and the familiarity with the concerned issues (which are assumed to be higher in the Delphi survey) are not so decisive in shaping the main attitudes. Instead, cultural factors seem to be predominant in shaping the personal attitude towards modern biotechnology. In addition, the attachment of consumers to their national food traditions is seen as an important factor in the process of acceptance of food biotechnology.

The expert panels in each country express rather high expectations in future developments aiming at clear regulations of market approval and penetration of gene food products as well as enhanced information activities in this area. Such developments are expected to be realized in the short term. However, the existence of a direct link between information and public acceptance is doubted, at least by a considerable proportion of the respondents. From the answers of the experts, it can be concluded that the regulations already in force are not precise enough. An effort could be made to specify their extent more precisely and to speed up their implementation time schedule. Important issues in this context are the labelling procedures as well as relevant criteria for the market approval of gene food products. In this context, compulsory labelling is very much welcomed by consumers and other non-technical experts, but it is seen as a very problematic measure by researchers and entrepreneurs.

The majority of the experts appears concerned about the "freedom of choice" of the consumer, i. e. the consumer should be made aware of the available options through proper information and be allowed to make a rational choice on the basis of cost-benefit considerations. All developments involving a broad diffusion of gene food products receive a moderate consensus. Positive attitudes seem to be linked to the indication of clear and specific benefits for human health and the environment. This confirms that the existence of ethical purposes is a key factor in order to obtain the social acceptance of genetic modification. This expectation, however, is accompanied by a feeling of inevitability of the final adoption of genetic engineering techniques in food production and processing.

The most supported applications of Agro-Food biotechnology are those concerning monitoring and control activities and those involving some benefit for the environment. The perception of possible risks is clearly indicated by the very high support given to epidemiological research on possible long-term impacts on human health due to the application of modern biotechnology. Together with the high share of positive attitude of the statements in the domain "regulation", this suggests that the social acceptance of biotechnology is also linked to satisfying the public demand for control.

Economic impacts of modern biotechnology

An important argument for the promotion of modern biotechnology is the expectation, mainly expressed by political institutions, that the application of this technology will contribute to additional employment possibilities and the competitiveness of European industry. In accordance with the results of other studies, the results of this survey indicate that within the next five to ten years only limited job growth is to be expected in the specialist biotech service companies and in supplying companies of the food industry. However, it is also expected that a decrease in the traditional jobs in agriculture will take place, partly due to the widespread use of modern biotechnology, but also as a result of structural changes in this sector which are independent of the use of modern biotechnology.

Small and medium-sized companies (SMEs) can profit from biotechnical developments. The Delphi survey shows that in the medium and long term SMEs might be able to introduce a large variety of biotech products and processes. However, especially in Mediterranean countries, the realization of such scenarios is highly depending upon successful technology transfer from research institutions to industry. In addition, the expert panels expect that modern biotechnology will have a considerable effect on the production costs of agricultural bulk products like cereals and milk. However, the increase in productivity in agricultural production will only result in additional income for farmers if it is not overcompensated by decreasing price trends for agricultural products in the coming years.

In accordance with the results of other studies, the experts asked in the Delphi survey express some uncertainties and concerns about a strong market penetration of gene food products. This may result in delays of the market approval and diffusion of gene food products. Nevertheless, the experts expect considerable market shares and a kind of "habituation" of the consumers to this type of product in around ten years. In this context, food products which offer clear benefits to the consumers are best suited for taking the lead because they are the best opportunity to overcome existing acceptance barriers. One example in this direction are probiotic foods or other food products supporting health requirements.

The experts asked in the Delphi survey have very high expectations in the market opportunities of biotechnology-based products for non-food purposes which are expected to be realized in the medium to long term. This relates to renewable resources used in the chemical industry and for energy production, to pharmaceutical substances produced with the help of transgenic animals or plants, as well as to the use of biotechnology approaches to reduce agricultural waste. In general, the social and ethical acceptance of these products or approaches is by far higher than that of genetically engineered food products. However, innovations outside the food chain can only be speeded up in a "concerted action" of science, industry and politics, taking into account the interests of other groups (e. g. farmers) as well.

A clear regulation of the market approval procedure for gene food products is required for economic reasons as well. On the other hand, a considerable proportion of the experts sees little chances for clear regulation procedures due to practical reasons and other intentions of industry and politics. This expectation may lead to a kind of fatalism and result in additional frustration and resistance to genetic engineering approaches in the Agro-Food sector, especially by militant opposition groups. In order to prevent such a development, an implementation of the "Novel Food Regulation" in each member country is strongly recommended as fast as possible, thereby ensuring clear control and monitoring activities. In this context, there are some indications in the results of the Delphi survey that parts of industry may change their argumentation from "deregulation" in direction to a compromise with opposing groups in a sense that a certain regulatory framework including the labelling of gene food products would be accepted.

Biotechnology in fish farming and fishery is a very unfamiliar subject to the panels in the five countries, although fish is often eaten. Although the industrialized production of fish is growing and also biotechnological techniques are introduced for breeding and processing purposes, this has not been a subject of great public concern so far.

Recommendations for further activities

Based on the results of the Delphi survey the project teams formulate the following main recommendations in order to allow decision-makers in politics, industry and society to adapt to emerging problems in time or to react to future possibilities:

- The current regulations in the field of Agro-Food biotechnology focus on the potential risks of this technology related to the health of consumers, users and employees as well as to the environment. However, the results of this Delphi survey and other studies indicate that ethical and cultural factors are gaining increasing importance in shaping the acceptance of modern biotechnology. With respect to the future design of biotechnology regulation, these results point towards a possible shift in the underlying policies: a need for a strong focus on ethical concerns as driving forces for the design of regulatory frameworks might arise in addition to the present risk-driven approach.

- Further research activities are recommended, focusing on the development of practicable ways to include ethical aspects in regulation and decision-making concerning Agro-Food biotechnology.

- A clear regulation of the market approval procedure of gene food products is required for safety, social and economic reasons. Therefore, it is recommended that the passed "Novel Food Regulation" should be implemented in each EU member country as fast as possible, thereby ensuring clear and realistic control and monitoring activities.

- Fostering public participation in decision-making seems to be a way of increasing positive attitudes towards - and confidence in - government action and companies' performance. In consequence, it is recommended to initiate the development of activities for involving the public in the decision-making process concerning Agro-Food biotechnology. In addition to public information activities, the establishment of an independent institution may strengthen public confidence in the control and the guarantee of the quality and safety of food products.

- The estimations of the experts concerning possible adverse health and environmental effects of modern biotechnology are a clear message to politicians and the responsible authorities at EU and national level to take effective measures in order to minimize these effects in future. These activities are not necessarily limited to regulations and the legal framework conditions, but should incorporate development of appropriate scientific methods and techniques for monitoring and analysing health or environmental effects, and stimulation of relevant research programmes as well as public information activities.

- Public research policy should focus to a greater extent on the analysis especially of the long-term impacts of modern biotechnology in future. This relates to the development and/or the adaptation of existing instruments (e. g. the methodology of epidemiological surveys to the specific environment in the Agro-Food sector), as well as enhanced integration of these issues in publicly funded research programmes.

- New instruments should be developed in order to get a close and more realistic view concerning the employment effects of Agro-Food biotechnology. Sound scientific studies should be initiated by national governments and the EU Commission to analyse this aspect. Applying such new concepts would help to substantiate political arguments for stimulating modern biotechnology for employment reasons.

- Many applications of modern biotechnology can only be speeded up in a "concerted action" of science, industry and politics, taking into account the interests of other groups (e. g. farmers) as well. This relates in particular to the use of modern biotechnology to produce high-value components for non-food purposes, to reduce the environmental burden in animal production as well as the use of optimized enzymes in the food industry. In this context it should be checked if the network between the different actors is sufficient or if building up or optimizing such a network should be actively supported. In addition, the integration of the relevant actors should be made a fundamental principle in all activities carried out in these areas.

- Taking into account the existing technical constraints in animal biotechnology and in the genetic modification of polygenic traits in crop plants, emphasis should be given to improving R&D infrastructure and providing additional funding for broadening the knowledge base, if the use of modern biotechnology and potential commercial applications in these fields are desired by society.

8.   Literature

Abbott, A.; Roeper, B. (1998): Germany seeks 'non-modified' food label. Nature 39, 828

Abramowitz, S. (1996): Towards inexpensive DNA diagnostics. Trends in Biotechnology 14, 397-401

Agrafiotis, D.: Socio-cultural dimensions of health and illness. Athens: Litsas 1988 (in Greek)

Agrafiotis, D.: Tendencies and perspectives of the greek Sociology of Health (1997). The experience in the National School of Public Health. Research Monograph No. 5, Athens, Dept. of National School of Public Health 1996

Amfep (1995): Association of Manufacturers of Fermentation Enzyme Products, Bruxelles. Personal communication 18.10.1995

Amtsblatt EG (1997): Nr. L 43 vom 14. Februar 1997, 1

Anonymus (1998): Hawaiian researchers produce three cloned mice generations. Genetic Engineering News 18 (14), 1, 3, 36, 38, 52

Ashworth, D. et al. (1998): Nature 394, 329

Beachy, R. N. (1997): Mechanisms and applications of pathogen-derived resistance in transgenic plants. Current Opinion in Biotechnology 8, 215-220

Becher, G.; Schuppenhauer, M. R.: Kommerzielle Biotechnologie - Umsatz und Arbeitsplätze 1996-2000. Einschätzungen der deutschen Wirtschaft. Arbeitsbericht für das Bundesministerium für Bildung, Wissenschaft, Forschung und Technologie. Basel: Prognos AG 1996

Bergmeyer, H. U.: Grundlagen der enzymatischen Analyse. Weinheim: Verlag Chemie 1977

Bijman, W. J.; Enzing, C. M.: Agrarische ketens en biotechnologie (Agro-Food chains and biotechnology), Report written in the framework of the Technology Assessment Programme of the Dutch Ministry of Agriculture and Fishery. The Hague, Apeldoorn: LEI-DLO/TNO-STB 1995

BMBF (Bundesministerium für Bildung, Wissenschaft, Forschung und Technologie) (1997): Staatssekretärin Wülfing: Landwirtschaft setzt auf technologischen Fortschritt. Press release October 24th, 1997, Bonn

BMELF-Informationen (1998): Hinsken: Deutschland bei Biogas führend in Europa. BMELF-Informationen Nr. 26, 2

BMFT (Bundesministerium für Forschung und Technologie) (ed.): Deutscher Delphi-Bericht zur Entwicklung von Wissenschaft und Technik. Bonn: Bundesministerium für Forschung und Technologie 1993

BML (Bundesministerium für Ernährung, Landwirtschaft und Forsten): Bericht des Bundes und der Länder über Nachwachsende Rohstoffe 1995. Schriftenreihe des Bundesministeriums für Ernährung, Landwirtschaft und Forsten. Reihe A: Angewandte Wissenschaft. Münster: Landwirtschaftsverlag GmbH 1995

BML (Bundesministerium für Ernährung, Landwirtschaft und Forsten): Agrarbericht der Bundesregierung 1997. Bonn 1997a

BML (Bundesministerium für Ernährung, Landwirtschaft und Forsten): Die Grüne Gentechnik. Bonn 1997b

Bohnert, H. J.; Jensen, R. G. (1996): Strategies for engineering water-stress tolerance in plants. Trends in Biotechnology 8, 89-97

Brandt, P.: Transgene Pflanzen. Basel, Boston, Berlin: Birkhäuser-Verlag 1995

Briggs, S. P.; Koziel, M. (1998): Engineering new plant strains for commercial markets. Current Opinion in Biotechnology 9, 233-235

Brower, V. (1997): Conference offers a look at the implications of mammalian cloning for science and society. Genetic Engineering News 17 (14): 1, 6, 50

Centerick, D. (1997): Integrated chain management of food products. The IPTS/Report 20, 12/1997, 27-33

Charles, N.; Kerr, M.: Women, food and families. Manchester: University Press 1988

Commission of the European Communities (CEC): Growth, Competitiveness, Employment - The Challenges and Way Forward into the 21st Century. Brussels 1993

Cowan, D. (1996): Industrial enzyme technology. Trends in Biotechnology 14 (6), 177-178

Cozzi, E.; White, D. J. G. (1995): The generation of transgenic pigs as potential organ donors for humans. Nature Medicine 1 (9), 964-966

Crueger, W.; Crueger, A.: Biotechnologie - Lehrbuch der angewandten Mikrobiologie. 3. Auflage. München, Wien: R. Ouldenbourg Verlag 1989

Cuhls, K., Kuwahara, T.: Outlook for Japanese and German Future Technology: comparing technology forecast surveys. Heidelberg: Physica-Verlag 1994

Cuhls, K.: Technikvorausschau in Japan. Ein Rückblick auf 30 Jahre Delphi-Expertenbefragungen. Heidelberg: Physica-Verlag 1998

Cuhls, K.; Breiner, S.; Grupp, H.: Delphi-Bericht 1995 zur Entwicklung von Wissenschaft und Technik - Mini-Delphi. Karlsruhe: Fraunhofer Institute for Systems and Innovation Research 1995

Dalkey, N. C. (1969): Analyses from a group opinion study. Futures 2, No 12, 541-551

Dalkey, N. C.; Brown, B.; Cochran, S.: The Delphi method, III: Use of self ratings to improve group estimates. Santa Monica, Rand Corporation Paper RM-6115-PR 1969

Dalkey, N. C.; Helmer, O. (1963): An experimental application of the Delphi-method to the use of experts. Management Science, Journal of the Institute of Management Sciences 9, 458-467

Daniels, J. J. M. C.; Duijzer, G. (1988): Delphi: Methode of Mode? Amsterdam: Symposium verslag, SISWO-publicatie 327

Dederer, H.-G. (1998a): Novel Food im EG-Kennzeichnungsdickicht. Komplizierte Rechtssituation bei neuartigen und gentechnisch veränderten Lebensmitteln. ZFL 49, 5/98, 52-54

Dederer, H.-G. (1998b): Novel Food im EG-Kennzeichnungsdickicht. Trends zu einer Vereinheitlichung der Kennzeichnung neuartiger Lebensmittel und anderer gentechnischer Produkte. ZFL 49, 6/98, 46-50

Demicheli, M.; Laget, P. (1996): Produktion hochwertiger Biochemikalien mit landwirtschaftlichen Nutzpflanzen: Eine langfristige Alternative für die Landwirtschaft? The IPTS Report 4/96, 25-29

Dixon, B.: Enzymes make the world go round. p. 20. Bagsvaerd (DK) Novo Nordisk A/S 1994

Dixon, B. (1995): Public understanding doesn't always lead to public support. Bio/Technology 6/95, 532

dpa (1997): Press release of June 3rd, 1997

Engel, K.-H.; Schreiber, G. A.; Bögl, K. W. (edts.): Entwicklung von Methoden zum Nachweis mit Hilfe gentechnischer Verfahren hergestellter Lebensmittel. Ein Statusbericht. BgVV-Heft 01/1995. Berlin: Bundesinstitut für gesundheitlichen Verbraucherschutz und Veterinärmedizin 1995

Enzing, C. M.: Biotechnology companies in the Netherlands: Human resources in R&D and production. Report to the Dutch Ministry of Economic Affairs. Apeldoorn: TNO-STB 1991

Enzing, C. M. (1997): Biotechnology. Chapter 2 in Baseline report of EST to IPTS, Sevilla

Epping, B.; Horn, I.; Michaelis, H.; Miersch, M.; Nakott, J. (1997): Klonen - Was die Forscher wirklich können. Bild der Wissenschaft 6, 58-63

Ernst & Young: European Biotech 1996. Volatility and Value. Third annual report on the European Biotechnology Industry. The Hague, Stuttgart, Cambridge 1996

EuropaBio: Benchmarking the Competitiveness of Biotechnology in Europe. Business Decisions Limited and SPRU for EuropaBio 1997

European Commission: Eurobarometer 35.1. Brussels: European Commission 1991

European Commission: The Europeans and modern biotechnology. Eurobarometer 46.1. Directorate General XII Science, Research and Development: Biotechnology. Luxembourg: Office for official publications of the European Communities 1997

Evans, G.; Durant, J. (1995): The relationship between knowledge and attitudes in the public understanding of science. Public Understanding of Science 4, 57-74

Falch, E. A. (1991): Industrial enzymes - developments in production and application. Biotech. Adv. 9, 643-658

Fladung, M. (1998): Transgene Bäume - Perspektiven und Grenzen. Biologie in unserer Zeit 28 (4), 201-213

Fox, S. (1998): Mr. Jefferson is born. Genetic Engineering News April 1, 1998, 22, 43

Fraunhofer-Institut für Systemtechnik und Innovationsforschung (1998a): Delphi '98 Umfrage. Studie zur globalen Entwicklung von Wissenschaft und Technik. Zusammenfassung der Ergebnisse. Karlsruhe: ISI

Fraunhofer-Institut für Systemtechnik und Innovationsforschung (1998b): Delphi '98 Umfrage. Studie zur globalen Entwicklung von Wissenschaft und Technik. Methoden- und Datenband. Karlsruhe: ISI

Friends of the Earth (1997): Für eine gentechnikfreie Nahrungsmittelproduktion. Background paper of Friends of the Earth Germany (BUND)

Fuji, J.; Otsu, K.; Zorzato, F.; Deleon, S.; Khanna, V. K.; Weiler, J. E.; O'Brien, P. J.; McLennan, D. H. (1991): Identification of a mutation in porcine ryanodine receptor associated with malignant hyperthermia. Science 253, 448-451

Gale, J. S.: Theoretical population genetics. Boston: Unwin Ayman 1990

Geldermann, H. (1997): Biotechnologie in der Tierzucht. Begleittext zum Kompaktkurs vom 17.2. bis 28.2.1997. Universität Hohenheim, Institut für Tierhaltung und Tierzüchtung, Fachgebiet Tierzüchtung

Gelvin, S. B. (1998): The introduction and expression of transgenes in plants. Current Opinion in Biotechnology 9, 227-232

Georges, M.; Dietz, A. B.; Mishra, A.; Nielsen, D.; Sargeant, L. S.; Sorensen, A.; Steele, M. R.; Zhao, X.; Leipold, H.; Womack, J. E.; Lathrop, M. (1993): Microsatellite mapping of the gene causing Weaver disease in cattle will allow the study of an associated quantitative trait locus. Proc. Natl. Acad. Sci. USA 90, 1058-1062

Goddijn, O. J. M.; Pen, J. (1995): Plants as bioreactors. Trends in Biotechnology 13, 379-387

Godfrey, T.; West, S. (edts): Industrial enzymology. London: Macmillan Press Ltd. 1996

Gofton, L.; Kuznesof, S.; Ritson, C.; Hutchins, R.; Tregear, A.: Consumer acceptability of biotechnology in relation to food products, with special reference to farmed fish. Report to the European Commission, DGXII. University of Newcastle upon Tyne, Department of Agricultural Economics and Food Marketing, Centre for Rural Development 1996

Gotsch, N.; Bernegger, U.; Rieder, P. (1993): Impacts of future biological-technological progress on arable farming. European Review of Agricultural Economics 20 (1), 19-34

Gotsch, N.; Rieder, P. (1989): Future importance of biotechnology in arable farming. Trends in Biotechnology 7, 29-34

Gotsch, N.; Rieder, P. (1990): Forecasting future developments in crop protection. Crop protection 9, 83-89

Grove-White, R.; Macnaghten, P.; Mayer, S.; Wynne, B.: Uncertain World: Genetically modified organisms, food and public attitudes in Britain. Lancaster: The Centre for the Study of environmental Change 1997

Hampel, J.; Keck, G.; Peters, H. P.; Pfennig, U.; Renn, O.; Ruhrmann, G.; Schenk, M.; Schütz, H.; Sonje, D.; Stegat, B.; Urban, D.; Wiedemann, P. M.; Zwick, M. M.: Einstellungen zur Gentechnik. Tabellenband zum Biotech-Survey des Forschungsverbunds "Chancen und Risiken der Gentechnik aus der Sicht der Öffentlichkeit". Stuttgart: Akademie für Technikfolgenabschätzung in Baden-Württemberg. Nr 87/1997

Hamstra, A. M.: Biotechnology in foodstuffs. Towards a model of consumer acceptance. SWOKA research report No. 105, The Hague 1991

Hamstra, A. M.: Consumer Acceptance of Food Biotechnology: The Relation Between Product Evaluation and Acceptance. SWOKA research report No. 137, The Hague 1993

Hamstra, A. M.: Impacts of new biotechnology in food production on consumers. SWOKA research report No. 170, The Hague 1994

Heimig, D. (1997): Bakterielle Impulse. Markt für probiotische Produkte wächst weiter. Lebensmittelzeitung 48/97, 46-47

Heins, V. (1992): Gentechnik aus Verbraucherperspektive. Soziale Welt 43, No. 4, 383-399

Heldt-Hansen, H. P.(1998): Development of enzymes for food applications. In: Poutanen, K. (edt.): Biotechnology in the food chain. New tools and applications for future foods. VTT Symposium 177, 28-30 January, 1998. 45-55. Espoo: Technical Research Centre of Finland 1998

Henze, A.; Zeddies, J.; Geldermann, H.; Momm, H.: Auswirkungen biotechnischer Neuerungen in der Tierzucht. Reihe A: Angewandte Wissenschaft, Heft 443. Münster: Landwirtschaftsverlag GmbH 1995

Hicks, P. (1998): Transgenic protein expression. Genetic Engineering News, May 1, 1998, 10, 39

Hodgson, J. (1998): LION and Degussa apply genomics to fermentation. Nature Biotechnology 16, 715

Hoheisel, J. D. (1997): Oligomer-chip technology. Trends in Biotechnology 465-469

Hüsing, B.; Jaeckel, G.; Marscheider-Weidemann; F.: Potentiale und Entwicklungen im Bereich der Katalysatoren- und Enzymtechnik. Studie für das Büro für Technikfolgen-Abschätzung beim Deutschen Bundestag (TAB). Karlsruhe: Fraunhofer Institute for Systems and Innovation Research 1997

Hüsing, B.; Gießler, S.; Jaeckel, G.: Stand der Möglichkeiten von prozeßintegrierten biotechnischen Präventivtechniken zur Vermeidung bzw. Verringerung von Umweltbelastungen. Studie im Auftrag des Umweltbundesamtes Berlin. Karlsruhe: Fraunhofer Institute for Systems and Innovation Research 1998

Jaeckel, G., Menrad, K., Reiß, T., Strauß, E.: Die Zukunft des deutschen Gesundheitswesens aus der Sicht von Ärzten und Experten. Karlsruhe: Fraunhofer Institute for Systems and Innovation Research 1995

Kähler, W.-M.: Einführung in die statistische Datenanalyse. Grundlegende Verfahren und deren EDV-gestützter Einsatz. Braunschweig/Wiesbaden: Vieweg-Verlagsgesellschaft 1995

Karatzas, C. N.; Turner, J. D. (1997): Toward Altering Milk Composition by Genetic Manipulation: Current Status and Challenges. J. Dairy Sci. 80, 2225-2232

Kirschenmann, A.; Kirschenmann, F. (1998): Genetic engineering and organic food: The proposed USDA rule on organic agriculture. Biotechnology and Development Monitor 34, 3/98, 18-20

Knop, J.; Saloga, J. (1996): Discussion of methods for detection of food allergies. In: Deutsche Forschungsgemeinschaft (ed.): Food allergies and intolerances. Weinheim, New York: VCH Verlagsgesellschaft, 65-74

Knorr, C.; Schmoch, U.; Keen, P.; Agrafiotis, D.: Legal and institutional constraints and opportunities for the dissemination and exploitation of R & D activities. Final Report for the Commission of the EU, VALUE-Programme. Karlsruhe: Fraunhofer Institute for Systems and Innovation Research; Athens: Infogroup Consultancy Services S. A. 1996

Koschatzky, K., Maßfeller, S.: Gentechnik für Lebensmittel? Schriftenreihe Zukunft der Technik (Ed.: Grupp, H.). Köln: Verlag TÜV Rheinland 1994

Kristensen, N. H.; Nielsen, T. (1997): From alternative agriculture to the food industry: The need for changes in food policy. The IPTS/Report 20, 12/1997, 20-25

Kuchner, O.; Arnold, F. H. (1997): Directed evolution of enzyme catalysts. Trends in Biotechnology 15 (12), 523-530

Langer, L. (1998): MetaMorphix, Inc.: Program launched to increase muscle mass in livestock. Genetic Engineering News, May 15, 1998, 28, 33

Larkin, M. (1996): Immunoassay usage increases in ag and food safety sectors. Genetic engineering news 16 (6), 1, 12, 13

Lebensmittelzeitung (1996): Milchbranche mobilisiert ihre Kräfte. LZ 33/96, 18

Lebensmittelzeitung (1998a): Aus der Nische zum Massenmarkt. LZ 4/98, 62

Lebensmittelzeitung (1998b): Gesund gewachsen. Nielsen-Bilanz für Molkereiprodukte - Neben Licht auch viel Schatten. LZ 10/98, 60-62

Lindersson, M.; Lunden, A.; Andersson, L. (1995): Genotyping bovine milk proteins using allele discrimination by primer length and automated DNA sizing technology. Animal Genetics 26, 67-72

Löhr, W. (1998): Gentechnik mit Bio-Label. GID 6/98, 18-19

Loveridge, D.; Georghiou, L.; Nedeva, M.: United Kingdom Technology Forsight Programme: Delphi Survey. A report to the Office of Science and Technology. Manchester: Policy Research in Engineering, Science and Technology (PREST), University of Manchester 1995

Luong, J. H. T.; Bouvrette, P.; Male, K. B. (1997): Developments and applications of biosensors in food analysis. Trends in Biotechnology 15, 369-377

MacQuitty, J. (1997): The real implications of Dolly. Nature Biotechnology 15: 294

Madsen, C. (1996): Allergy to food - the significance of route of sensitization to risk assessment. In: Deutsche Forschungsgemeinschaft (ed.): Food allergies and intolerances. Weinheim, New York: VCH Verlagsgesellschaft, 75-80

Magpie genome sequencing project list: http://www-c.mcs.anl.gov/home/gaasterl/genomes.html

Marlier, E.: Biotechnology and genetic engineering: What Europeans think about it in 1993. Survey conducted in the context of Eurobarometer 39.1 on behalf of the Commission of the EU. INRA (Europe) 1993

Martens, T. (1997): Harmonization of safety criteria for minimally processed foods. Inventory Report 1996-7, European Commission, DGXII

Menrad, K. (1995): Öffentliche Akzeptanz der Gentechnik im Industriepflanzenanbau in Deutschland. In: Fachagentur Nachwachsende Rohstoffe e. V. (Ed.): Biotechnologie und Gentechnik in der Industriepflanzenzüchtung. Gülzow. 39-53

Menrad, K.; Koschatzky, K.; Massfeller, S.; Strauss, E.: Communicating genetic engineering in the Agro-Food sector to the public. Report to project No. BIO2-CT 94-0029 of the European Commission, DG XII. Karlsruhe: Fraunhofer Institute for Systems and Innovation Research 1996a

Menrad, K.; Koschatzky, K.; Maßfeller, S.; Strauß, E.: Communicating genetic engineering in the Agro-Food sector to the public. A guide for companies. Brochure to project No. BIO2-CT 94-0029 of the European Commission, DG XII. Karlsruhe: Fraunhofer Institute for Systems and Innovation Research 1996b

Merkt (1980): In: Comberg, G. (Hrsg). Tierzüchtungskunde. Eugen Ulmer Verlag, Stuttgart

Mikkelsen, T. R.; Andersen, B.; Jorgensen, R. B. (1996): The risk of crop transgene spread. Nature 380, 31

Millstone, E. (1995): Collaboration in research and development in food safety in the E.U. The IPTS/Report 18, 10/1997, 21-25

Milstein, C. (1980): Monoclonal antibodies. Scientific American 243 (4), 66-74

Moffat, A. S. (1996): Moving Forest Trees Into the Modern Genetics Area. Science 271, 760-761

Moufang, R. (1995): Bedeutung und Auswirkungen des Patentschutzes für biotechnologische Erfindungen auf Entwicklungsländer. In: Franzen, H.; Begemann, F.; Wolpers, K. H.; Urff, W. v. (1995): Auswirkungen biotechnologischer Innovationen auf die ökonomische und soziale Situation in den Entwicklungsländern. Bonn: ATSAF, 117-121

Mourgues, F.; Brisset, M.-N.; Chevreau, E. (1998): Strategies to improve plant resistance to bacterial diseases through genetic engineering. Trends in Biotechnology 16, 203-210

Mühlenberg, K. (1997a): Betreten verboten - keine Gefahr. Globus 4 - 5/97, 27-29

Mühlenberg, K. (1997b): Trommeln und Theater. Zahlreiche Freisetzungen und Proteste dagegen gab es im letzten Jahr. Ein Rück- und Ausblick. GID 118, 4/97, 10-12

Mühlenberg, K. (1998): Appetitmacher: Positivkennzeichnung. GID 124, 2/98, 32

Murphy, D. J. (1996): Engineering oil production in rapeseed and other oil crops. Trends in Biotechnology 14, 206-213

Neubert, S.: Neue Bio- und Gentechnologie in der Landwirtschaft. Reihe A. Angewandte Wissenschaft, Ht. 394. Münster: Landwirtschaftsverlag GmbH 1991

Niemann, H. (1997): Gentechnologische Herstellung von Nahrungsmitteln tierischer Herkunft. In: Lohner, M., Sinemus, K., Gassen, H. G. (Ed.): Transgene Tiere in Landwirtschaft und Medizin. Schriften der Pädagogischen Arbeitsstelle für Erwachsenenbildung in Baden-Württemberg Nr. 20. Villingen-Schwenningen: Neckar-Verlag 1997, 117-139

Niemann, H.; Meinecke, B.: Embryotransfer und assoziierte Biotechniken bei landwirtschaftlichen Nutztieren. Stuttgart: Enke Verlag 1993

Niestijl Jansen J. J.; Kardinaal, A.; Huijbers, G.; Vlieg-Boerstra, B. J.; Martens, B. P.; Ockhuizen, T. (1994): Prevalence of food allergy and intolerance in the adult Dutch population. Journal of Allergy and Clinical Immunology 93, 447-456

Novo Nordisk (1995): Umweltbericht 1994. Bagsvaerd, Novo Nordisk A/S

Novo Nordisk (1996): Liste über die derzeitigen 66 Enzymanwendungen. BioTimes 11(1), 14-15

OECD database of field trials: http://www.oecd.org/ehs/service.htm

OECD: Biotechnology, agriculture and food. Paris, Organisation for economic co-operation and development 1992

OECD: Safety considerations for biotechnology: Scale-up of crop plants. Paris, Organisation for economic co-operation and development 1993

Ortolani, C. E. A.; Pastorello et al.: Study of nutritional factors in food allergies and food intolerances. AIDS -93-8012-IT/DGXII, EVR 16893 EN, Brussels 1997

Otsu, K.; Khanna, v. K.; Archibald, L.; McLennan, D. H. (1991): Cosegregation of porcine malignant hyperthermia and a probable causal mutation in the skeletal muscle ryanodine receptor gene in backcross families. Genomics 11, 744

Peerenboom, E. (1998): Pharming cloning ban could spread. Nature Biotechnology 16, 321-322

Pen, J. et al. (1992): Bio/Technology 10, 292-296

Pen, J. et al. (1993): Bio/Technology 11, 811-814

Petzold, U. (1998): Sag niemals nie: Neues zum Klonen von Säugetieren. Biologie in unserer Zeit 28 (4), 194-200

Pezzatti, M.-G.; Anwander Phan-huy, S.; Rieder, P.; Lehmann, B. (1996): Ökonomische Auswirkungen eines Einsatzes von Nutzpflanzen mit gentechnisch erzeugten Resistenzen gegen Krankheiten und Schädlinge. In: Schulte, E.; Käppeli, O. (Eds.): Gentechnisch veränderte krankheits- und schädlingsresistente Nutzpflanzen - Eine Option für die Landwirtschaft? Basel: BATS 1996, 513-622

Pontenagel, I. (Ed.): Das Potential erneuerbarer Energien in der Europäischen Union. Ansätze zur Mobilisierung erneuerbarer Energien bis zum Jahr 2020. Berlin, Heidelberg, New York: Springer-Verlag 1995

Potera, C. (1996): Sex predetermined calves produced with selected semen. Genetic Engineering News 16 (2): 1, 11

Prins, M.; De Haasen, P.; Luyten, R.; Van Veller, M.; Van Grinsven, M. Q. J. M.; Goldbach, R. (1995): Broad resistance to tospoviruses in transgenic tobacco plants expressing three tospoviral nucleoprotein gene sequences. Mol. Plant Microbe Interact. 8, 85-91

Proops, S. (1997): A comparison between functional food markets in the EU, US and Japan. The IPTS/Report 12/1997, 34-37

Quarrie, S. A. (1996): New molecular tools to improve efficiency of breeding for increased drought resistance. Plant Growth Reg. 20, 167-178

Reiß, T.; Holland, D.; Menrad, K.: Wirkungsanalyse zum Programm "Förderung der Biotechnologie in der Wirtschaft". Karlsruhe: Fraunhofer Institute for Systems and Innovation Research 1995

Reiß, T.; Koschatzky, K.: Biotechnologie. Unternehmen, Innovationen, Förderinstrumente. Heidelberg: Physica-Verlag 1997

Rudolph, N. S. (1995): Advances continue in production of proteins in transgenic animal milk. Genetic Engineering News, October 15, 1995, 8-9

Rudolph, N. S. (1997a): Biotechnology firms investigate mammalian cloning for commercial transgenics programs. Genetic Engineering News 17 (9): 1, 14

Rudolph, N. S. (1997b): Technologies and economics for protein production in transgenic animal milk. Genetic Engineering News, October 15, 1997, 16, 36, 37

Sandermann, H.; Ohnesorge, K. F.: Nutzpflanzen mit künstlicher Herbizidresistenz: Verbessert sich die Rückstandssituation? Heft 6: Verfahren zur Technikfolgenabschätzung des Anbaus von Kulturpflanzen mit gentechnisch erzeugter Herbizidresistenz. Berlin: Wissenschaftszentrum Berlin für Sozialforschung 1994

Schäfer, T.; Ring, J. (1996): Epidemiology of adverse food reactions due to allergy or other forms of hypersensitivity. In: Deutsche Forschungsgemeinschaft (ed.): Food allergies and intolerances. Weinheim, New York: VCH Verlagsgesellschaft, 42-53

Schell von, T.; Mohr, H.: Biotechnologie - Gentechnik. Eine Chance für neue Industrien. Berlin, Heidelberg, New York: Springer-Verlag 1995

Schmidt, K. (Ed.): WHO Surveillance Programme for control of foodborne infections and intoxications in Europe. Sixth report: 1990-1992. Berlin: Federal Institute for Health Protection of Consumers and Veterinary Medicine (BgVV) 1995

Schmitt, A. (1997): Nutrition Policy as a means of health prevention. The IPTS Report 20, 12/1997, 38-44

Schnieke, A. E.; Kind, A. J.; Ritchie, W. A.; Scott, A. R.; Ritchie, M.; Wilmut, I.; Coleman, A.; Campbell, K. H. S. (1997): Human Factor IX Transgenic Sheep Produced by Transfer of Nuclei from Transfected Fetal Fibroblasts. Science 278, 2130-2133

Schuler, T. H.; Poppy, G. M.; Kerry, B. R.; Denholm, I. (1998): Insect-resistant transgenic plants. Trends in Biotechnology 16, 168-175

Self, C. H.; Cook, D. B. (1996): Advances in immunoassay technology. Current Opinion in Biotechnology 7, 60-65

Sgaramella, V.; Zinder, N. D. (1998): Dolly Confirmation. Science 279, 635-636

Shah, D. M. (1997): Genetic engineering for fungal and bacterial diseases. Current Opinion in Biotechnology 8, 208-214

Shuster, D. E.; Kehrli, M. E.; Ackermann, M. R.; Gilbert, R. O. (1992): Identification and prevalence of a genetic defect that causes leucocyte adhesion deficiency in Holstein cattle. Proc. Natl. Acad. Sci. USA 89, 9225-9229

Signer, E. N. et al. (1998): Nature 394, 329-330

Smirnoff, N. (1998): Plant resistance to environmental stress. Current Opinion in Biotechnology 9, 214-219

Sofres (ed.): Enquête sur les technologies d'avenir par la méthode Delphi. Montrouge 1994

Solter, D. (1998): Dolly is a clone - and no longer alone. Nature 394, 315-316

Spalla, C. : Le Biotecnologie in Italia e nel Mondo. ASSOBIOTEC 1996

Sparks, P. et al. (1994): Gene technology, food production and public opinion: A UK study. Agriculture and Human Values 6, No. 1, Winter, 19-28

Stokes, R. (1998): Cloning technology applications accelerate one year after Dolly. Genetic Engineering News 18 (2): 1, 10, 34

Strauss, E., Jaeckel, G.: Entwicklungspotentiale von Werkstoff- und Verarbeitungstechnologien in Nordrhein-Westfalen, Ergebnisse einer Delphi-Befragung. Karlsruhe: Fraunhofer Institute for Systems and Innovation Research 1996

Streck, W. R.; Pieper, B.: Die biotechnische Industrie in Deutschland: eine Branche im Aufbruch. ifo Studien zur Industriewirtschaft, 55. München: ifo Institut für Wirtschaftsforschung 1997

Tennant, P. F.; Gonsalves, C.; Ling, K. S.; Fitch, M.; Manshardt, R.; Slightom, J. L.; Gonsalves, D. (1994): Differential protection against papaya ringspot virus isolates in coat protein gene transgenic papaya and classically cross-protected papaya. Phytopathology 84, 1359-1366

Teuber, M: Exploitation of genetically modified microorganisms in the food industry. Proceeding: 2nd international conference on the release of genetically engineered microorganisms. New York: Plenum Press 1992

Thorndike, R. M.: Correlational procedures for research. New York 1978

Timmons, A. M.; Charters, Y. M.; Crawford, J. W.; Bum, D.; Scott, S. E.; Dubbels, S. J.; Wilson, N. J.; Robertson, A.; O'Brien, E. T., Squire, G. R.; Wilkinson, M. J. (1996): Risks from transgenic crops. Nature 380, 487

Uhlig, H.: Enzyme arbeiten für uns. Technische Enzyme und ihre Anwendung. München, Wien: Carl Hanser Verlag 1991

Van den Daele, W.; Pühler, A.; Sukopp, H.: Transgenic herbicide-resistant crops: A participatory technology assessment. Summary report. Berlin: Wissenschaftszentrum Berlin für Sozialforschung 1997

Videbaek, T. (1997): Personal communication. Department of food enzymes, Novo Nordisk, Bagsvaerd, Denmark

Wakayama, T.; Perry, A. C. F.; Zuccotti, M.; Johnson, K. R.; Yanagimachi, R. (1998): Nature 394, 369-374

Wall, R. J.; Kerr, D. E.; Bondioli, K. R. (1997): Transgenic Dairy Cattle: Genetic Engineering on a Large Scale. J. Dairy Sci. 80, 2213-2224

Weigel, D.; Nilsson. (1995): A developmental switch sufficient for flower initiation in diverse plants. Nature 377, 495-500

WHO, European Centre for Environment and Health: Concern for Europe's Tomorrow. Stuttgart: Wissenschaftliche Verlagsgesellschaft mbH 1995

Williamson, M. (1996): Can the risks from trangenic crop plants be estimated? Trends in Biotechnology 14, 449-450

Wilmut, I.; Schnieke, A. E.; McWhir, J.; Kind, A. J.; Campbell, K. H. S. (1997): Viable offspring from fetal and adult mammalian cells. Nature 385, 810-813

Wintzer, D.; Fürniß, B.; Klein-Vielhauer, S.; Leible, L.; Nieke, E.; Rosch, C.; Tangen, H.: Technikfolgenabschätzung zum Thema Nachwachsende Rohstoffe. Reihe A: Angewandte Wissenschaft. Münster: Landwirtschaftsverlag GmbH 1993

Wolfrum, R.; Stoll, P.-T.: Der Zugang zu genetischen Ressourcen nach dem Übereinkommen über die biologische Vielfalt und dem deutschen Recht. Forschungsbericht 101 06 073. Berlin, Umweltbundesamt: Erich Schmidt Verlag 1996

Wrotnowski, C. (1997): Current trends in immunoassay technology development and immunoanalysis. Genetic Engineering News, September 15, 1997, 14, 35

Wüthrich, B. (1996): Epidemiology of allergies and intolerances caused by foods and food additives: The problem of data validity. In: Deutsche Forschungsgemeinschaft (ed.): Food allergies and intolerances. Weinheim, New York: VCH Verlagsgesellschaft, 31-35

Yeaton Woo, R.; Digiulio, K.; Grawford, L. (1997): Early warning systems and technologies to prevent food-born diseases: The U.S. experience. The IPTS/Report 17, 9/1997, 12-17

Young, E.; Stoneham, M. D.; Petruckevitch, A.; Barton, J.; Rona, R. (1994): A population study of food intolerance. Lancet 343, 1127-1130

Zagon, J.; Schulze, M.; Broll, H.; Schauzu, M. (edts.): Methoden zum Nachweis der gentechnischen Veränderung in glyphosatresistenten Sojabohnen und zur Identifizierung anderer gentechnisch veränderter Pflanzen. BgVV-Heft 03/1998. Berlin: Bundesinstitut für gesundheitlichen Verbraucherschutz und Veterinärmedizin 1998

Zimmerman, L. et al. (1994): Consumer knowledge and concern about biotechnology and food safety. Food Technology 48, 11/19, 71-77

Zwick, A. (1997): Global climate change: Potential impact on human health. The IPTS/ Report 4/1997, 27-35

Annex

Questionnaire of the second round for Germany

Delphi-Survey

Future Impacts of Biotechnology on Agriculture, Food Production and Food Processing

SECOND ROUND

Fraunhofer Institute for Systems and
Innovation Research (FhG ISI)
Breslauer Str. 48 - 76139 Karlsruhe

March 1997

Delphi survey:

Future Impacts of Biotechnology on Agriculture, Food Production and Food Processing

1. Why do you get a second questionnaire?

As mentioned in the questionnaire of the first round of the Delphy survey we send you this questionnaire of the second round. It includes the aggregated and statistically analysed results of the German experts answering in the first round. We again thank you very much for your participation in this comparative international survey which follows the Delphy approach. Now we have reached the second round, in which you are asked to review your own estimations of the first round taking into account the opinion of your German colleagues. This feed-back process improves the finding of consensus as well as the identification of controversial judgements. You are free to keep your answers (and comments) of the first round or to change your judgements.

2. Why should you answer the questionnaire?

Although the answering of the questionnaire requires some time and efforts, it is worth for you, your company, your organisation or your institution to participate in the second round of the Delphy survey because:

- The overview of the estimations of the other German panel members represents a so far unprecedented expert-based view of the future. You can directly compare the results with your own expectations and judgements thereby receiving a first feed-back from the German experts.
- As a participant in the second round you will get a short report with the key results of this international survey. This report will be sent to you approximately in October 1997.
- By answering the questionnaire you contribute to an active shaping of the future application of modern biotechnology in the Agro-Food sector. Your individual answer will be part of a qualified, expert-based prognosis which will influence the public debate in this field significantly.

3. Modifications in the questionnaire

The estimations of the German experts who have participated in the first round are included in this questionnaire. For this purpose we present the results graphically and in percentage numbers for each statement. The answers of experts who estimated their own degree of knowledge as 'not familiar' were excluded from the statistical analysis. All results are given as relative frequencies except for the answers concerning the category 'importance of the statement'. In this case you will find diagrams showing the weighted tendency of the answers of the experts.

We slightly changed some statements and the description of some categories according to your comments in the first round. In addition, we tried to eliminate some misleading or unclear formulations. More general comments concerning the procedure or the structure of the questionnaire will be considered in the final analysis of the survey.

The distinction between 'modern biotechnology' and 'genetic engineering' in the statements was maintained. We again ask you to think about developments in the whole area of biotechnology, where the application of genetic engineering is one aspect in a very complex field. In contrast, statements focusing on genetic engineering aim at judgements about the decisive and specific influence of this technology on a certain future development.

4. General instructions to answer the questionnaire

We have mailed this questionnaire to all persons who have participated in the first round of the survey. The statements of this questionnaire should be judged by you *personally*. Only you can compare your personal judgements of the first round with the estimations of the German expert panel. You will find examples for the filling in of this questionnaire on the next page. Please, make your entries in the empty fields above the results of the first round.

Since we have included statements of different areas and subjects in the questionnaire, your individual level of knowledge may differ between the statements. Please, try to answer the statements even if you are not very familiar with a certain area or topic. You should estimate your *degree of knowledge* according to the definitions in the glossary. In case you are not able to give an estimation to a specific statement, please continue with the next one.

The *glossary* is now – for your convenience – available as an extra sheet which you may want to consult when answering the questionnaire.

We follow the same procedure for the protection of your personal data as described in the questionnaire of the first round. We restrict the personal data which we ask at the end of the questionnaire to the minimum necessary for the management and the scientific analysis of the survey. If you prefer an additional protection of your personal data, please send the return form for your name and address (but not the page with your personal information!) and the filled in questionnaire by separate mail. In case you wish the sending of a short report of the results of the survey please indicate your name and your address in the questionnaire.

5. Contact persons

If you have any questions about or remarks on the questionnaire please contact us. You can reach the project team of FhG ISI by mail, fax, E-mail or phone:

Fraunhofer Institute for Systems and Innovation Research
Delphi Agro-Food
Breslauer Str. 48
76139 Karlsruhe
Fax: 0721/6809-176
E-mail: me@isi.fhg.de
Dr. Klaus Menrad: Tel. 0721/6809-262 (project manager)
Dr. Bärbel Hüsing: Tel. 0721/6809-210
Dipl. Biol. Christa Knorr: Tel. 0721/6809-191

Delphi-Survey: Future Impacts of Biotechnology on
SECOND

Example:

Statement	Degree of knowledge Please mark with a cross				Personal attitude Please mark with a cross				Time of realisation Please mark with a cross						
	very familiar	average familiar	less familiar	not familiar	development / situation is positive	development / situation is indifferent	development / situation is negative	not able to express	in the following 5 years	in the following 6 - 10 years	in the following 11 - 15 years	in the following 16 - 20 years	not realised during the following 20 years	not feasible at all / no realisation	not able to judge
83 Technologies based on DNA-analysis and genetic engineering are widely used to choose the sex of livestock.	10	32	34	24	16	33	48	4	24	24	18	10	10	6	9
84 Due to modern biotechnology organs of animals are transplanted into humans as a matter of routine.	8	28	32	32	20	30	45	5	16	25	20	12	12	3	13

Please send back until:

18.4.1997
to
FhG - ISI
Delphi Agro-Food
Breslauer Str. 48
76139 Karlsruhe

Name, Adress
(required to send you
the summary of the survey)

Agriculture, Food Production and Food Processing

ROUND

Most important influencing factors in your country												Importance of the statement to			Comment
Please mark with a cross (max. 3 crosses)												Please mark each field with: + = very important, 0 = average or - = not important			
R & D infrastructure	personnel (education, skills)	technology transfer	industrial innovativeness	markets	funding	regulation / standards	policy	ecology	social / ethical acceptance	information	international collaboration	knowledge creation in science & technology	competitiveness of economy	protection of environment / sustainable development	
□	□	□ X	□	□ X	□	□	□	□ X	□ X	□	□	0 + -			*I think ethical concerns are relevant here.*
5	2	5	3	,9	7	9	10	8	38	5	0				
□	□	□	□	□	□	□	□	□ X	□ X	□	□	+ 0 0			*Transfer of other pathogens has to be considered.*
10	3	5	3	5	9	9	10	6	35	4	1				

Please consider that each field has to be marked with +, 0 or -.

- Explanation of the results of the first round
  - Relative frequencies of the answers to each statement are shown
  - Category "Importance of the Statement to": Diagram shows in the upper part the „very important" answers, in the lower part the "not important" answers. The arrow indicates the weighted tendency in the estimation.
- Please answer the questionnaire personally.
- Your present answer has not to correspond with your former estimation.
- In case you are not able to answer a statement please continue with the next one.

Please read the detailed information to this Delphi-survey on page 1 and 2 as well.

		Statement	Degree of knowledge				Personal attitude				Time of realisation						
			Please mark with a cross				Please mark with a cross				Please mark with a cross						
			very familiar	average familiar	less familiar	not familiar	development / situation is positive	development / situation is indifferent	development / situation is negative	not able to express	in the following 5 years	in the following 6-10 years	in the following 11-15 years	in the following 16-20 years	not realised during the following 20 years	not feasible at all / no realisation	not able to judge
Acceptance	1	A governmental institution which provides information on modern biotechnology in the Agro-Food sector to all interested persons and institutions is established in Germany.	17	46	28	8	74	20	4	1	68	20	3	0	1	3	4
	2	Most consumers in Germany have quickly got used to all kinds of food and beverages produced with the help of genetic engineering.	25	50	23	2	22	25	51	2	18	39	20	7	6	6	3
	3	The widespread diffusion of information on genetic engineering from different sources (e. g. public and private institutions, several interest groups and associations, media) has increased the acceptance of food products made with the help of genetic engineering in Germany.	25	50	22	3	33	26	39	2	28	41	15	4	4	4	3
	4	Research on genetic engineering of farm animals is not funded by public institutions in Germany due to ethical reasons (e. g. animal rights, animal welfare, preservation of Creation).	16	45	30	10	32	25	39	4	27	20	9	4	5	22	13
	5	An advisory committee with the objective to assess whether food products made with the help of genetic engineering meet the needs of consumers is jointly established by different interest groups (e.g. consumers, industry, retailers) in Germany.	21	43	30	6	57	26	16	1	52	24	6	1	1	10	6
	6	Food companies and retailers proactively create specific labels for food and beverages produced without the help of genetic engineering.	24	46	25	6	54	20	25	0	61	19	4	1	1	10	4
	7	A restaurant chain or catering service specialised in offering trendy meals with genetically engineered ingredients open branches in almost all large cities in Germany.	12	36	36	16	10	37	50	3	12	24	18	5	8	23	10
Regulation	8	The labelling of all food products made with the help of genetic engineering is compulsory and fully implemented in Germany.	26	49	20	5	63	19	18	0	40	25	6	1	2	22	3

Most important influential factors in Germany												Importance of the statement to			Comment
Please mark with a cross (max. 3 crosses)												Please mark each field with: + = very important, 0 = average or – = not important			
R & D infrastructure	personnel (education, skills)	technology transfer	industrial innovativeness	markets	funding	regulation / standards	policy	ecology	social / ethical acceptance	information	international collaboration	knowledge creation in science & technology	competitiveness of economy	protection of environment / sustainable development	
3	6	3	2	3	27	6	21	2	14	9	3	+			
1	0	1	2	9	0	7	13	8	35	24	1	+			
2	3	2	5	3	5	16	6	32	20	2		+			
4	1	2	4	10	6	6	23	5	26	9	4	+			
3	9	2	4	10	16	7	15	3	17	11	4	+			
2	2	1	6	19	8	16	16	2	7	11	10	+			
0	1	1	3	24	3	7	5	7	38	9	1	+			
2	2	1	4	18	4	16	24	2	7	7	14	+			

388

Statement	Degree of knowledge				Personal attitude				Time of realisation						
	very familiar	average familiar	less familiar	not familiar	development / situation is positive	development / situation is indifferent	development / situation is negative	not able to express	in the following 5 years	in the following 6-10 years	in the following 11-15 years	in the following 16-20 years	not realised during the following 20 years	not feasible at all / no realisation	not able to judge
9 All food and beverages which are produced with the help of genetic engineering are subject to an official case-by-case procedure for approval in Germany.	20	46	27	6	50	17	31	1	33	20	6	1	4	30	6
10 For the marketing approval of Agro-Food products made with the help of genetic engineering, concerns on their social impacts and potential ethical conflicts are generally taken into account.	21	46	28	6	54	20	24	2	30	20	7	2	6	28	7
11 The responsible authorities in Germany practically use specific technical equipment and standardized methods to identify food and beverages which were produced with the help of genetic engineering.	21	41	30	8	56	23	20	2	23	30	13	4	4	22	5
12 European and national authorities start initiatives which involve the public in the debate and decision-making on the application of modern biotechnology in the Agro-Food sector.	23	43	27	7	61	20	18	1	39	23	7	2	5	18	6
13 The uniform implementation of the EU biosafety directives in all EU countries has led to a higher attraction of the EU for companies based outside the EU.	13	36	34	17	37	29	26	8	25	29	15	4	4	11	11
14 Food made with the help of genetic engineering achieve a turnover share of 30 % or more of all food consumed in Germany.	17	46	30	8	17	39	40	3	11	24	25	16	13	5	5
15 The prices of food products in specific market segments of which the quality is significantly improved by the application of modern biotechnology are at least 30 % higher than that of corresponding conventional products.	15	42	32	11	9	33	52	6	9	18	9	3	7	41	13
16 The widespread use of modern biotechnology enables small and medium-sized companies of the Agro-Food sector in Germany to introduce a large variety of innovative processes and products.	18	44	32	6	48	24	25	3	8	28	26	9	8	17	6

Regulation: 9–13; Economy: 14–16

Most important influential factors in Germany												Importance of the statement to			Comment
Please mark with a cross (max. 3 crosses)												Please mark each field with: + = very important, 0 = average or - = not important			
R & D infrastructure	personnel (education, skills)	technology transfer	industrial innovativeness	markets	funding	regulation / standards	policy	ecology	social / ethical acceptance	information	international collaboration	knowledge creation in science & technology	competitiveness of economy	protection of environment / sustainable development	
3	8	2	3	11	16	15	19	1	6	4	11				
4	5	2	4	16	7	12	21	1	9	10	10				
10	15	6	3	5	26	8	15	1	2	4	8				
2	9	1	3	6	14	8	26	1	8	11	12				
4	3	3	5	10	5	13	24	4	13	5	12				
2	0	3	5	18	1	8	10	8	35	9	1				
1	0	2	5	41	6	4	6	2	22	6	4				
9	5	14	10	12	16	7	6	3	13	5	1				

	Statement	Degree of knowledge				Personal attitude				Time of realisation						
		very familiar	average familiar	less familiar	not familiar	development / situation is positive	development / situation is indifferent	development / situation is negative	not able to express	in the following 5 years	in the following 6-10 years	in the following 11-15 years	in the following 16-20 years	not realised during the following 20 years	not feasible at all / no realisation	not able to judge
17	The practical use of modern biotechnology in the food industry creates a small number of new jobs (e. g. in specialized service companies and the supplying industry).	10	45	37	9	37	41	17	4	25	44	14	4	1	4	7
18	The widespread use of modern biotechnology in agriculture reduces the production costs for agricultural bulk products (e. g. cereals, milk) by approximately 30 %.	22	45	26	7	53	23	19	4	9	31	24	7	8	15	7
19	Small farmers in Germany are not able to afford new genetically engineered plants and animals.	20	41	30	9	10	28	57	5	16	29	15	6	3	17	15
20	The widespread use of modern biotechnology in animal and plant production leads to an approximate 30 % decrease in traditional jobs in agriculture.	17	39	32	12	5	23	67	5	5	24	25	11	7	16	12
21	Due to the widespread use of modern biotechnology in EU fish breeding and aquaculture, the imports of fish and fish products from outside the EU are significantly reduced.	7	22	36	35	25	38	26	9	4	23	26	13	9	10	16
22	Due to the application of modern biotechnology, farmers produce renewable resources (e. g. biofuel, starch, fatty acids) which are used outside the food sector on 20 % or more of the arable land in Germany.	21	41	28	10	66	19	11	3	9	30	31	14	6	6	3
23	Pharmaceutical substances produced by genetically engineered animals and plants (e. g. proteins, enzymes, hormones, antibodies) achieve a turnover share of at least 5 % of the pharmaceutical market in Germany.	15	32	34	19	55	27	13	4	30	34	18	7	3	2	7
24	Due to the hesitant application of genetic engineering (compared to other EU member states) the Agro-Food industry cuts 10 % or more of the jobs in this sector in Germany.	13	42	34	11	6	19	71	4	23	31	12	6	2	14	12

Economy

Please mark with a cross

Most important influential factors in Germany												Importance of the statement to			Comment
R & D infrastructure	personnel (education, skills)	technology transfer	industrial innovativeness	markets	funding	regulation / standards	policy	ecology	social / ethical acceptance	information	international collaboration	knowledge creation in science & technology	competitiveness of economy	protection of environment / sustainable development	
5	5	8	11	17	12	7	10	3	18	3	2				
5	2	9	6	14	5	11	11	8	23	4	2				
3	3	8	4	19	23	6	10	4	14	6	2				
3	5	6	4	14	6	6	20	7	23	4	1				
4	1	6	7	20	6	6	10	10	20	2	8				
6	1	9	11	20	11	7	12	9	8	2	3				
10	1	11	12	9	5	14	9	4	19	4	2				
3	3	3	6	11	5	9	24	6	22	6	3				

Please mark with a cross (max. 3 crosses)

Please mark each field with: + = very important, 0 = average or - = not important

Statement	Degree of knowledge (Please mark with a cross)				Personal attitude (Please mark with a cross)				Time of realisation (Please mark with a cross)						
	very familiar	average familiar	less familiar	not familiar	development / situation is positive	development / situation is indifferent	development / situation is negative	not able to express	in the following 5 years	in the following 6-10 years	in the following 11-15 years	in the following 16-20 years	not realised during the following 20 years	not feasible at all / no realisation	not able to judge
25 Enzyme systems are specifically developed to improve the environmental performance of conventional food processing procedures. %	14	36	31	19	70	19	7	4	44	36	9	3	1	1	6
26 The widespread deliberate release of genetically engineered organisms results in a significant transfer and recombination of the introduced genes with unintended negative impacts (e.g. development of resistances to herbicides and pathogens in weeds). %	25	41	23	11	5	17	75	2	28	26	11	8	2	13	12
27 The public debate about environmental risks of the deliberate release and marketing of genetically engineered micro-organisms (e.g. uncontrollable spread, disturbance of natural balances) has led to a broad rejection of such activities in Germany. %	29	45	21	5	31	19	48	1	66	12	4	2	1	6	7
28 Genetically engineered micro-organisms and plants producing biocides are widely used for biological pest control. %	18	42	28	12	48	22	26	4	19	37	22	10	4	2	6
29 The widespread cultivation of genetically engineered field crops resistant to herbicides results in an approximate 50 % reduction of environmental pollution in plant production. %	24	43	24	9	64	19	14	4	13	26	19	8	6	19	6
30 Despite the widespread use of methods and techniques related to modern biotechnology, all in all, no negative effect on the maintainance of biodiversity in agriculturally influenced ecosystems can be discovered. %	22	44	25	9	47	23	25	4	12	21	15	11	4	22	14
31 Modern biotechnology is widely used to reduce emissions and waste from animal production (e. g. less manure due to enzymes as feed additives, improved anaerobic waste treatment, biofilter). %	13	37	31	19	58	17	21	3	17	30	25	9	6	5	8
32 Modern biotechnology significantly contributes to the transformation of 40 % or more of the organic agricultural waste into marketable products (e. g. energy, secondary raw materials). %	11	38	34	17	73	16	8	3	8	21	27	19	11	6	8

Environment

Most important influential factors in Germany	Importance of the statement to	Comment
Please mark with a cross (max. 3 crosses)	Please mark each field with: + = very important, 0 = average or - = not important	

R & D infrastructure	personnel (education, skills)	technology transfer	industrial innovativeness	markets	funding	regulation / standards	policy	ecology	social / ethical acceptance	information	international collaboration	knowledge creation in science & technology	competitiveness of economy	protection of environment / sustainable development	Comment
8	2	12	13	7	10	11	7	4	18	4	2				
5	2	4	3	2	1	13	14	19	26	10	2				
2	1	1	3	6	1	5	20	10	26	21	4				
7	1	7	8	7	6	11	9	14	24	8	1				
5	1	6	7	8	5	11	10	12	25	7	2				
6	1	5	4	6	3	7	11	19	23	11	4				
7	1	9	9	7	8	9	9	11	21	7	1				
11	1	11	13	10	17	6	8	6	10	5	1				

		Statement	Degree of knowledge				Personal attitude				Time of realisation						
			Please mark with a cross				Please mark with a cross				Please mark with a cross						
			very familiar	average familiar	less familiar	not familiar	development / situation is positive	development / situation is indifferent	development / situation is negative	not able to express	in the following 5 years	in the following 6-10 years	in the following 11-15 years	in the following 16-20 years	not realised during the following 20 years	not feasible at all / no realisation	not able to judge
Environment	33	Organic farmers in Germany are allowed to integrate specific genetic engineering approaches (e.g. use of genetically engineered biocides or animal vaccines) in their production process.	18	40	30	12	37	21	39	2	12	25	17	6	9	23	7
	34	The widespread use of genetically engineered organisms in food production significantly aggravates the loss of organisms traditionally used in Germany.	14	40	35	11	5	29	62	4	8	23	24	16	9	8	13
Health	35	Food companies inform the consumers in detail about the specific health relevant effects of food made with the help of genetic engineering.	17	45	30	7	74	15	10	1	40	25	8	3	3	15	5
	36	Allergy-sufferers are offered special food of which the allergenic potential has been decreased by genetic engineering.	13	38	32	16	72	16	10	2	20	35	21	10	3	5	5
	37	The increased use of modern biotechnology in food production and processing results in additional allergies in people actively involved in these processes.	8	37	34	21	3	14	75	8	23	25	11	3	2	16	20
	38	Food enriched with specific micro-organisms having positive health effects achieve approximately 25 % turnover share in their product group (e. g. yoghurt).	11	37	34	18	39	41	14	6	29	30	16	5	5	4	11
	39	A large variety of food and beverages of superior nutritional value (e. g. vitamin, protein and fibre content; fatty acid ratios) supporting dietary health requirements is produced with the help of genetic engineering.	15	41	31	13	53	23	22	2	16	35	23	10	4	6	6
	40	Unintended impacts on the health of consumers occur in some products or manufacturing processes due to the use of genetic engineering during food processing.	11	37	37	15	4	14	78	5	30	28	13	4	2	7	15

Most important influential factors in Germany												Importance of the statement to			Comment
Please mark with a cross (max. 3 crosses)												Please mark each field with: + = very important, 0 = average or - = not important			
R & D infrastructure	personnel (education, skills)	technology transfer	industrial innovativeness	markets	funding	regulation / standards	policy	ecology	social / ethical acceptance	information	international collaboration	knowledge creation in science & technology	competitiveness of economy	protection of environment / sustainable development	
1	2	2	2	10	2	11	9	11	39	10	1				
3	1	4	5	14	3	10	10	12	28	8	1				
2	6	2	6	17	7	11	11	2	13	17	5				
11	2	8	13	11	9	8	5	3	19	11	2				
9	6	6	8	3	2	16	8	6	16	17	3				
3	0	7	8	20	2	11	4	4	25	13	1				
6	1	7	8	14	4	9	5	4	29	11	1				
9	4	6	6	4	2	19	9	5	15	17	4				

	Statement	Degree of knowledge — Please mark with a cross				Personal attitude — Please mark with a cross				Time of realisation — Please mark with a cross						
		very familiar	average familiar	less familiar	not familiar	development / situation is positive	development / situation is indifferent	development / situation is negative	not able to express	in the following 5 years	in the following 6-10 years	in the following 11-15 years	in the following 16-20 years	not realised during the following 20 years	not feasible at all / no realisation	not able to judge
Health — 41	The long-term health impacts of the use of Agro-Food products made with the help of genetic engineering on consumers, farmers and employees are investigated (e. g. by epidemiological surveys).	9	40	38	13	79	14	5	2	20	32	16	13	6	6	6
Scientific / Technological Development — 42	The hygiene monitoring of food processing is significantly improved due to the widespread use of modern biotechnological analytical methods.	12	32	29	27	79	15	4	3	38	38	12	4	2	1	6
43	Techniques based on modern biotechnology are practically used for on-line control of quality parameters in food processing (e. g. content of micronutrients or harmful substances).	10	32	35	23	80	15	3	2	33	39	16	4	2	1	5
44	The genomes of most economically important bacteria used in food production are completely sequenced.	11	32	27	31	52	33	8	8	15	26	24	15	7	3	9
45	In Germany most of the beer is produced with genetically engineered yeast.	12	35	29	23	11	43	42	3	17	22	17	7	13	15	9
46	Genetically engineered micro-organisms and enzyme systems are widely used to improve the processing quality of food (e. g. prolonged shelflife of biological products, synchronization of the ripening process).	13	42	32	14	43	27	28	2	23	36	22	9	4	2	4
47	Modern biotechnology has failed to produce food and beverages satisfying traditional taste preferences of large consumer groups in Germany.	10	34	35	21	21	44	28	7	20	19	11	6	3	21	19
48	A large variety of genetically engineered micro-organisms which can be precisely controlled and regulated in their metabolic activities during food production processes has been developed.	9	32	35	25	47	33	16	4	17	27	23	14	7	4	8

Most important influential factors in Germany												Importance of the statement to			Comment
Please mark with a cross (max. 3 crosses)												Please mark each field with: + = very important, 0 = average or - = not important			
R & D infrastructure	personnel (education, skills)	technology transfer	industrial innovativeness	markets	funding	regulation / standards	policy	ecology	social / ethical acceptance	information	international collaboration	knowledge creation in science & technology	competitiveness of economy	protection of environment / sustainable development	
11	11	3	2	3	31	6	16	1	5	7	5				
11	9	13	11	4	23	9	5	2	5	5	3				
12	7	14	12	5	23	9	5	2	5	5	3				
22	9	8	4	2	34	2	4	1	6	4	6				
2	0	2	4	20	1	11	9	5	39	7	0				
6	1	8	7	14	4	11	8	5	31	7	1				
10	1	8	13	18	4	5	4	3	24	7	3				
15	2	13	12	7	11	8	5	4	18	4	2				

		Degree of knowledge (Please mark with a cross)				Personal attitude (Please mark with a cross)				Time of realisation (Please mark with a cross)						
		very familiar	average familiar	less familiar	not familiar	development / situation is positive	development / situation is indifferent	development / situation is negative	not able to express	in the following 5 years	in the following 6-10 years	in the following 11-15 years	in the following 16-20 years	not realised during the following 20 years	not feasible at all / no realisation	not able to judge
49	Enzymes optimized by protein engineering are practically used in specific sectors of the food industry (e. g. starch processing, bakeries, breweries, cheese / dairy production).	13	34	33	20	50	29	19	2	46	32	13	3	1	1	4
50	Approximately 90 % of the enzymes used in the food processing industry are produced by genetically engineered organisms.	12	33	31	24	29	36	30	4	20	25	19	14	9	5	8
51	Artificial polyfunctional enzymes are practically used for analytical applications in the Agro-Food sector (e.g. rapid quantitative analyses of metabolic activities or the composition of raw materials).	8	26	33	32	43	32	19	5	10	28	27	15	6	4	10
52	The production of industrial enzymes in genetically engineered field crops is developed for large-scale application.	11	32	31	26	42	32	23	3	13	30	25	12	8	4	8
53	For the development of hybrids, genetic engineering is practically used in breeding of most field crops.	18	37	28	18	38	31	28	3	23	27	24	10	5	4	6
54	Genetic engineering approaches are developed which allow the alteration of polygenic traits in most economically important plant species.	14	27	29	31	38	31	25	7	8	21	20	22	14	5	10
55	Genetically engineered plant varieties resistant to salinity and / or drought have significantly improved agricultural productivity in arid environments.	14	32	34	20	67	18	11	3	10	22	26	19	12	5	6
56	To prevent resistance against biocides in pathogens, new genetic engineering approaches for plant defence mechanisms have been developed (e. g. combination of different resistance genes, increase in pathogen tolerance).	15	30	31	24	58	22	16	4	22	25	23	13	8	1	7

Scientific / Technological Development

Most important influential factors in Germany												Importance of the statement to			Comment
Please mark with a cross (max. 3 crosses)												Please mark each field with: + = very important, 0 = average or - = not important			
R & D infrastructure	personnel (education, skills)	technology transfer	industrial innovativeness	markets	funding	regulation / standards	policy	ecology	social / ethical acceptance	information	international collaboration	knowledge creation in science & technology	competitiveness of economy	protection of environment / sustainable development	
7	1	9	10	11	5	11	6	5	28	6	1	+			
8	1	8	9	12	6	10	6	5	29	5	1	+			
17	2	14	13	4	15	8	5	5	13	3	2	+			
12	1	9	10	8	11	8	6	9	22	4	1	+			
9	2	8	6	8	7	9	8	9	27	5	2	+			
16	3	9	8	5	14	7	5	7	21	4	2	+			
12	2	12	6	7	20	4	6	7	11	4	9	+			
15	3	10	6	5	13	7	9	9	26	3	5	+			

		Degree of knowledge (Please mark with a cross)				Personal attitude (Please mark with a cross)				Time of realisation (Please mark with a cross)						
	Statement	very familiar	average familiar	less familiar	not familiar	development / situation is positive	development / situation is indifferent	development / situation is negative	not able to express	in the following 5 years	in the following 6-10 years	in the following 11-15 years	in the following 16-20 years	not realised during the following 20 years	not feasible at all / no realisation	not able to judge
57	New monitoring and control techniques for genetically engineered plants in open fields have improved the reliability of risk assessment significantly.	15	34	32	19	72	17	6	4	29	30	15	5	6	8	6
58	The molecular basis of virus-resistance mechanisms in economically important perennial plants like e. g. vine, olive and fruit trees is elucidated.	10	30	34	26	81	12	2	4	13	33	26	14	5	0	10
59	Plant cell cultures in large-scale bioreactors are widely used for the production of high value components (e. g. pharmaceuticals, fine chemicals, proteins).	9	30	32	29	59	30	8	4	21	32	22	11	3	3	9
60	The content of erucic acid in rapeseed reaches 65 % or more due to breeding systems based on genetic engineering.	14	27	25	34	48	26	16	9	30	32	12	3	1	6	15
61	After years of experimentation with genetic engineering, the majority of plant breeders strongly prefers the combination of marker-assisted breeding with traditional breeding methods, compared to the use of genetically modified plants.	14	25	27	34	48	33	10	10	31	30	16	7	1	2	12
62	Rapid test systems based on modern biotechnology are widely used for pathogen identification in plant production and animal husbandry.	14	29	30	27	85	11	2	2	35	36	17	4	1	0	6
63	Breeding procedures based on genetic engineering are practically used for those farm animals (cattle, pigs, poultry, sheep, goats) which are of major economic importance for food production in the EU.	12	34	33	21	24	30	44	2	16	26	22	11	13	4	7
64	The elucidation of the principles regulating the expression of polygenic traits in farm animals allows the alteration of such traits by genetic engineering with regard to time, extent, quality and localisation (e. g. formation of milk similar to human milk in cows).	7	26	30	36	22	32	39	7	6	17	20	16	20	9	12

Scientific / Technological Development

Most important influential factors in Germany												Importance of the statement to			Comment
Please mark with a cross (max. 3 crosses)												Please mark each field with: + = very important, 0 = average or - = not important			
R & D infrastructure	personnel (education, skills)	technology transfer	industrial innovativeness	markets	funding	regulation / standards	policy	ecology	social / ethical acceptance	information	international collaboration	knowledge creation in science & technology	competitiveness of economy	protection of environment / sustainable development	
14	6	6	6	3	20	9	10	5	13	6	2				
23	6	9	6	3	28	3	4	3	6	4	5				
12	2	13	14	9	15	8	7	4	13	3	1				
11	2	9	8	12	13	7	6	8	16	5	4				
6	4	12	7	8	13	6	5	8	19	9	3				
14	4	16	12	5	20	8	4	2	7	5	3				
6	2	4	2	9	6	8	12	8	37	4	2				
12	2	5	4	7	9	7	9	5	34	5	1				

		Statement	Degree of knowledge				Personal attitude				Time of realisation						
			very familiar	average familiar	less familiar	not familiar	development / situation is positive	development / situation is indifferent	development / situation is negative	not able to express	in the following 5 years	in the following 6-10 years	in the following 11-15 years	in the following 16-20 years	not realised during the following 20 years	not feasible at all / no realisation	not able to judge
Scientific / Technological Development	65	The use of cloned embryos of cattle, sheep and goats is a widespread technique for the reproduction of farm animals.	10	32	34	24	16	33	48	4	24	24	18	10	10	6	9
	66	The elucidation of the principles underlying the developmental processes of oocytes leads to the genetic engineering of early stages of embryo cells from farm animals.	8	28	32	32	20	30	45	5	16	25	20	12	12	3	13
	67	New cell-biological methods are developed for the production of polyploid fish.	4	14	27	55	13	36	37	14	13	30	19	9	5	2	24
	68	Genetic maps and molecular test systems are practically used for the identification of economically important traits and marker-assisted breeding of fish.	5	16	26	52	33	36	22	8	16	29	24	9	2	2	18
	69	Monogenic traits are specifically and stably altered in fish important for aquaculture by genetic engineering as a matter of routine (e. g. anti-freeze gene, growth hormone gene).	5	19	27	49	22	28	43	7	17	30	18	10	5	3	16
	70	Feed additives made with the help of genetic engineering are widely used to improve the productivity of farm animals and fish.	10	32	34	24	27	34	36	3	27	37	18	6	3	2	8
	71	A large variety of genetically engineered vaccines for farm animals is developed which can reduce the use of other pharmaceuticals in animal husbandry (e.g. antibiotics) significantly.	9	28	35	27	63	21	12	4	16	42	21	7	3	4	8

Most important influential factors in Germany												Importance of the statement to			Comment
Please mark with a cross (max. 3 crosses)												Please mark each field with: + = very important, 0 = average or - = not important			
R & D infrastructure	personnel (education, skills)	technology transfer	industrial innovativeness	markets	funding	regulation / standards	policy	ecology	social / ethical acceptance	information	international collaboration	knowledge creation in science & technology	competitiveness of economy	protection of environment / sustainable development	
5	2	5	3	9	7	9	10	8	38	5	0				
10	3	5	3	5	9	9	10	6	35	4	1				
10	2	6	4	11	10	8	6	10	30	3	1				
12	2	9	5	10	14	6	8	7	22	4	2				
8	1	6	4	10	7	9	7	10	33	4	1				
3	1	4	5	13	5	13	9	10	32	4	2				
11	2	9	9	7	13	13	7	5	17	5	2				

| Gender | Male | ☐ |
| | Female | ☐ |

Age	Up to 29	☐
	30 to 39	☐
	40 to 49	☐
	50 to 59	☐
	60 to 69	☐
	over 70	☐

Occupation at / Member of / Background in

(Multiple answers are possible)

Agricultural Industry Company	☐
Food Industry Company	☐
Retail Company	☐
Other Company	☐

University / Research Institution:

Research field Plants	☐
Research field Animals	☐
Research field Microorganisms	☐
Process Engineering	☐
Research field Ecology	☐
Research field Nutrition, Physiology/Health	■
Economics and Social Sciences (incl. Technology Assessment)	☐
Other Research Fields	☐

Politics, Administration	☐
Farmer Organisation / Farmer	☐
Consumer Organisation	☐
Hotels, Restaurants, Catering	☐
Environmental Organisation	☐
Education, Media	☐
Other Institution	☐

| **Are you practically working with modern biotechnological methods and techniques?** | Yes | ☐ |
| | No | ☐ |

Thank you for filling in the questionnaire.

Glossary

1. Statements

Modern biotechnology: Application of scientific and engineering principles to the processing of materials by biological agents to provide goods and services (techniques like genetic engineering, bioprocessing, monoclonal antibodies, protein engineering, tissue culture, protoplast fusion or immobilised enzymes are included).

Genetic engineering: Characterisation, isolation, new combination and multiplication of genetic material.

Elucidation: Scientific and theoretical identification of principles or phenomena.

Development: Attainment of a specific technological target or completion of a prototype.

Practical use (= practically used): First market introduction or use under practical conditions of an innovative product or service.

Widespread use (= widely used): Significant market penetration to a level that a product or service is commonly used.

Matter of routine: There is reliable experience with a product or service because it has been used for a certain period of time.

Food made with the help of genetic engineering: All food and beverages are included which have been in contact with gene technology methods during the production process.

2. Categories of the questionnaire

2.1 Degree of knowledge

For each statement you are asked to indicate your individual level of knowledge using the following criteria:

Very familiar: You actively work in this area or with this issue at present. (This is one of your regular fields of work).

Average familiar: You are not working in this area but you are very well informed about the arguments dealing with the issue in the statement.

Less familiar: You have read articles in the newspaper or in popular magazines about the area/issue covered in the statement.

Not familiar: You have insufficient knowledge about the area/issue. In this case continue with the next statement.

2.2 Personal attitude

How is your personal opinion on the development/situation mentioned in the statement?

Development/situation is positive: All in all I appreciate the development/situation.

Development/situation is indifferent: Altogether I have no negative or positive feeling about the development/situation.

Development/situation is negative: I reject the development/situation.

Not able to express: I am not able to express my personal opinion on the development/situation.

2.3 Time of realisation

Please mark the time period with a cross in which the mentioned development/situation will be realised most probably. If you think that the issue already has been realised, please cross the subcategory "in the following 5 years" and make a corresponding note in the "comment" category.

2.4 Most important influential factors in your country

How do you assess (in an as objective way as possible) the current conditions in your country to realise the development/situation mentioned in the statement. *Please cross the most important influential factors (up to 3) which are decisive in your country.* The possible influential factors are explained in the following:

R&D infrastructure: Quantity and quality of relevant scientific and technological institutions (R&D = research and development).

Personnel (education, skills): Availability of educated and skilled staff which is required to realise a certain topic.

Technology transfer: Organisation of know-how transfer among universities, research institutions and industry.

Industrial innovativeness: Ability of the national companies to make a commercial success or new/improved products, processes or services out of scientific/technological findings.

Markets: Conditions on the relevant markets (e. g. future perspectives, competition), entrance to these markets and chances of new/improved products, processes or services.

Funding: Availability of private and public research funds and investment capital (e. g. venture, risk capital).

Regulation/standards: The international and national standards, intellectual property rights and other laws and regulations which influence the topic mentioned in the statement.

Policy: Activities and measures of the national/federal government and other politicians.

Ecology: Environmental situation in your country.

Social/ethical acceptance: Social and cultural opinions, ethical concerns, attitudes arising from general public discussion or pressure groups.

Information: Access to and diffusion of scientific, technological and background information on modern biotechnology (e.g. articles, broadcasts in mass media, information material, databases).

International collaboration: Number and/or quality of cooperations with research institutions/companies in other countries.

2.5 Importance of the statement

The importance of the development/situation mentioned in the statement should be estimated according to the objective relevance for the following fields. *Please indicate in each field whether the development/situation described in the statement is very important, average or not important.*

Knowledge creation in science and technology: Research in the direction mentioned in the statement is an essential pre-condition of additional scientific and technological developments or it is an important contribution to scientific/technological progress by itself.

Competitiveness of economy: The issue can contribute to the competitiveness of the national economy (e. g. development of marketable products and/or creation of new jobs).

Protection of environment/sustainable development: The issue mentioned in the statement can contribute to improving the environmental situation in your country.

2.6 Comments

Comments of all kind (e.g. related to a certain statement, a specific category, the methodology) are appreciated. The comments will be included in the analysis of the results of this survey.

TECHNOLOGY, INNOVATION and POLICY

Series of the Fraunhofer Institute
for Systems and Innovation Research (ISI)
